Six Sigma QUALITY for Business & Manufacture

Six Sigma QUALITY for Business & Manufacture

By

M. Joseph Gordon, Jr.

Gordon & Associates

Palm Harbor, FL, USA

ELSEVIER

2002

Amsterdam - Boston - London - New York - Oxford - Paris
San Diego - San Francisco - Singapore - Sydney - Tokyo

ELSEVIER SCIENCE B.V.
Sara Burgerhartstraat 25
P.O. Box 211, 1000 AE Amsterdam, The Netherlands

Library of Congress Cataloging-in-Publication Data
Gordon, Joseph (M. Joseph)
Six Sigma quality for business and manufacture / by M.J. Gordon, Jr.
p. cm.
ISBN 0-444-51047-8 (alk. paper)
1. Quality control. 2. Process control. I. Title.

TS156 .G66 2002
658.5`62--dc21

2002075914

British Library Cataloguing in Publication Data
Gordon, M. Joseph
Six Sigma quality for business and manufacture
1.Quality control - Statistical methods 2.Production management - Statistical methods
I.Title
658.5`62

ISBN: 0-444-51047-8

The paper used in this publication meets the requirements of ANSI/NISO Z39.48-1992 (Permanence of Paper).
Printed in The Netherlands.

Preface

Six Sigma is business and industries newest recognized quality program. It was developed at Motorola by their engineers to assist them in reducing their business and manufacturing costs, improving profitability, preventing problems, producing products to meet customer requirements, and knowing when and how much to adjust a process for repeatable operation. This is done to ensure their customer base continues to grow through their own initiated cost reductions while reducing the "Risk" of both business and product errors while continually achieving manufacturing improvements.

All of these benefits are being achieved when a company implements a Six Sigma program that is built on a strong quality foundation of ISO9000 certification.

When Six Sigma programs were first developed they were initially targeted at high-end savings programs, typically able to achieve cost savings of $170,000.00 per program. The industry leaders, Motorola, AlliedSignal-now Honeywell after their merger in 1999, General Electric, Dupont and other major corporations were able to achieve these savings. Their success has filtered down to other companies both large and small. This required a lowering of the savings bar to include their lower valued programs that have resulted in substantial savings and product quality improvement and manufacturing savings for them and their customers.

Service organizations have also adopted the Six Sigma format to improve their customer service and business department quality from banks to accounting firms. These companies have identified areas in their structure they can apply the metrics to measure their operations effectiveness in customer service and support. Here-to fore a service company's success was based on their growth curve up or down, without a true measure or indication of what the customer was doing to create the graph of quality service.

Six sigma deals with more than measuring a system and establishing tighter control limits. Six sigma utilizes the full basket of quality methods and tools. Using trained personal knowledgeable in their application to the companies methods of conducting business, future plans for growth, reducing and preventing the "Risk of Errors" in all departments of their business, while achieving continued and increased profitability with a growing customer base.

Acknowledgments

I want to thank my wife, Joyce, for her love and support during my writing of this book. I also want to acknowledge the support received from Dr. Edward Immergut, of Hanser-Gardner Publishing Company and Harold Wolf, Quality Engineer for their comments and insight in the review of the material for this book. Also, for assisting and recommending to me additional quality information to assist companies for ensuring the quality of their manufactured products will always meet their customers requirements and specifications.

The prevention of all business, manufacturing, and service related problems are the key objectives of this text. When the information presented in the text is followed only positive manufacturing and quality operations should result.

March 2002

M. Joseph Gordon, Jr.
MSME

Contents

Chapter 1

How Companies Use Six Sigma to Improve Processes and Prevent Problems

The use of Six Sigma methods to reduce cost and continually improve product quality, business operations, and customer service and support begins with a defined and documented quality system. A company with ISO9000 or QS-9000 certification has the basic system built on their current prevention quality system that meets the ISO9000-1984 certification criteria. Their quality system should be able to assure the company's customers of obtaining repeatable product quality and service, order to order.

This quality level is not guaranteed as always being the highest, only that the quality system in place meets the requirements of the certification process. If the quality provided is minimal to their industry it will not change or improve, unless continued change and improvement are incorporated into the quality system during the pre certification process. ISO9000 is not a judge and jury to ensure the customer obtains the highest quality possible, only that the company quality system in place during certification, meets the 20-certification section requirements. At the start of the year 2000, ISO9000-2000, updated quality prevention and documenting version of ISO9000, will add 34 new and required statements to the existing ISO9000 standard. These new requirements will be discussed in greater detail in a follow on section.

A companies quality system is historically driven by their customer's quality requirements. This implies the quality obtained is only sufficient to meet their customer's product or service requirements, no more! The problem exists if no quality improvements are added such as metrics, charting for defects, yield, problem elimination, etc., the company's 'cost of quality' may be misdirected into correction, not prevention. As a result, cost of quality and Risk Management may waste personnel time and company assets in their business operations. Therefore, the Six Sigma quality program can show management where savings are possible to further reduce their cost of rework and scrap plus prevent problems from occurring.

This opens up a new unexplored and potential area for savings and expanding their customer base.

Programs That Can Be Initiated

1. Customer order entry problems
2. Purchasing order problems
3. Contract problems
4. Engineering for manufacture
5. Manufacture and processes
6. Materials meeting compliance
7. Maintenance for repeatable operations
8. Assembly reduction and savings
9. Decoration and coding
10. Packaging for protecting product
11. Shipping on time at reduced cost
12. Repair for service-ability
13. Service cost and requirements reduced
14. Warranty and liability considerations

The list continues for all and any operations performed by the company. Under each listed problem area there can be a future breakout of each individual problem with a count of occurrences, (Pareto charts), severity in time lost and money, plus solution analysis and if they were effective. The true solution to any problem is preventing it from occurring. Until a company makes a concentrated effort to identify, quantify, and document problem areas, real preventive solutions cannot be implemented. Problem areas must be tracked for ensuring the solution effectively corrected the problem. Also, monitoring the correction in "Real Time" to be sure a solution has not created another problem prior to or further down the line. By monitoring solutions, hopefully discussed, and approved by all involved departments and personnel, including finance, the solution and its monetary effects can be judged if it was successful. Any solution must be continually monitored in "Real Time" for ensuring it is still effective and variables in the process are within specification limits required to control the process and product.

QUALITY IMPLEMENTATION TOOLS

In all fields of business there are many identified and unidentified variables that affect the input and output of the operation. Fortunately there are simple and very effective quality tools to use to identify these variables. These tools essentially QFD (quality function deployment), FMEA (failure mode and effects analysis), check lists, and Ishikawa (fishbone) plus manufacturing process control plans and procedures are used to analyze, identify, quantify, and document all the business and process variables. Then after these variables are identified other personnel from different departments can be called on to discuss these variables, reliability for the process maintaining control, and how they are controlled and monitored in the new procedure for improvement. This is a form of a pre-audit as before a redesign is implemented to verify all questions and answers were obtained and discussed. A second pair of eyes and ears to offer suggestions from the perspective of their department in the program analysis.

This may involve employment of reverse engineering techniques to fully comprehend, determine, and access the variables and their effects on the system. In some cases, climatic and time of day effects may have to be considered on the variables control and stability if the process is sensitive to these items. The depth of involvement should extend to the point where the variable is known to be and if it can remain stable. Often during these discussions variables are listed that never had a tolerance established for them even though they can affect the process. If so, establish an acceptable tolerance for the variable. These and all variables should then be monitored to ensure they remain within tolerance or specifications for the operation. Personnel considerations should also be considered as management style, knowledge and caliber of training for personnel performing these operations.

SIX SIGMA PROGRAMS

Six Sigma must not be mystified as being so unique or special that only certain people can perform the tasks. A competent leader, trained in the knowledge and application of quality improvements can effect positive change within any organization. This has been proven by successful completion of Six Sigma improvement programs in companies both large and small.

The Six Sigma designate can be selected within the company or an outsider, already trained in Black Belt methodology; a person trained and certified in Six Sigma knowledge and program operations can be hired. This is management's decision to make, and results have shown a Black Belt from a similar industry is a very good fit as they know the industry problems already, possibly now occurring at this company. Black belts from outside their industry can bring in very good suggestions but can cause problems if they do not learn the business mentality and operating procedures of their new company.

SIX SIGMA RISK AND FAULT ABATEMENT

The basic definition of Six Sigma for each level of management can elicit a different response. Six Sigma is still being redefined in the business and manufacturing community. Some define it as cost savings, others as process improvements, and still others as defect reduction. All of these are part of the definition for Six Sigma.

Six Sigma is even more than these definitions! It can lower business risk while improving customer satisfaction and product and service reliability. It can provide a significant risk reduction for each group, department, team, and individual for a business and company.

Six Sigma is an organized method to strategically and tactically manage the total capability of a business enterprise in all its basic forms and operations. It has the capacity to deliver to both the supplier and customer a higher degree of business satisfaction not attainable before at a reasonable cost and input of effort to complete the assigned tasks. All levels of management and their employees easily and quickly understand six Sigma methodology and requirements, as there is a commonality of quality output at all levels of employees in the business.

This mutual goal is often defined as value entitlement for customer and supplier in all forms of the business relationship. This exemplifies that Six Sigma is more than defect reduction; it is a method, not an actual new one, of how companies can do business yielding the ideals for business success while optimizing the control of functions within an enterprise.

Six Sigma in its basic definition is having only 3.4 defects per million opportunities (DPMO) for producing a defect. By this definition Six Sigma is related to a single opportunity or single critical to quality item that does not meet the customers requirement or specification. This is a 20,000 times reduction in defects from three sigma performance.

Six Sigma therefore has different goals for employees and business unit management in an organization. Each department sets their goals and vision for attaining Six Sigma capability differently within their structure; business, operation, or process. It assists them in establishing their long and short-term goals and how they are going to achieve this vision using Six Sigma, strategy, tactics and tools. These methods are shown in Figure 1 across the entire organization of the company.

To be successful, each level of the organization must be keyed into the total company vision for achieving Six Sigma results within a set time period. The key is to lower and eliminate "RISK" levels in everything an organization does and delivers to their customer. Six Sigma's basic premise is all errors and defects represent risk. But, not all risk represents a defect in output, product, or service.

Six Sigma is now being recognized more as a business initiative for success, not just a new quality program. It is more aggressively aligned with the ideas of risk elimination and prevention than with the initial quality definition of defect reduction as illustrated in Figure 2.

DEFECT REDUCTION

Defect reduction focuses on risk reduction. Management must assess the potential results to attain the changes supplier and customer must achieve to yield positive results versus the input of cost for training and improvements in business, operations, and processes.

A supplier company focusing on the source of risk in their departments operations can reduce customer risk for the products and services they provide. The supplier benefits and gains by their reduction of operation cost and process risk with quality improvements. This is how each can have a win-win business strategy and performance.

As the principles of Six Sigma are applied to reduce risk exposure in operations, we increase our confidence of performance in all our business functions and units. It has been documented that poor information given to the business units often result in the wrong, incorrect, or inferior product or

Six Sigma Implementation

	Organization				
	Goals		Vision		
Function	Long-term	Short-term	Strategy	Tactics	Tools
Business	Benchmark as best in class within five years from baseline period	Attain entitlement performance within two years from baseline period	Utilize Six Sigma to achieve business goals	Development deployment and compensation plan	Metrics tracking and reporting system
Operations	Improvement at 78% per year for all Six Sigma metrics	Attain entitlement rate of improvement for key metrics	Acquire Six Sigma human resource capability	Define Six Sigma project selection criteria	Six Sigma project tracking and reporting system
Process	3.4 defects per million opportunities for the CTO's related to all processes	Attain entitlement rate of improvement for key metrics	Build Six Sigma human resource capability	Apply Six Sigma breakthrough strategy to all projects	Six Sigma breakthrough technologies and software

Figure 1. Six Sigma from an organizational perspective. (Adapted from reference [1])

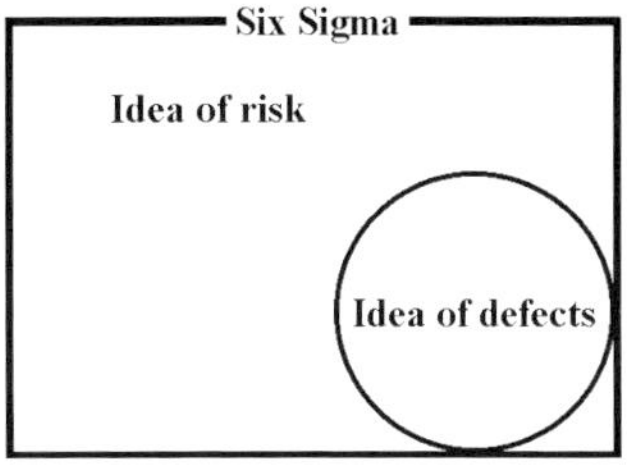

Figure 2. Relationship between risk and defects. (Adapted from reference [1])

part being purchased. This can occur because the engineering or manufacturing department did not correctly inform or specify the right material or product to be purchased for the use in the product or service. This does not imply that the business office, order entry, finance, purchasing, etc. did a bad job. When personnel are instructed to buy the least expensive, similar item, or product, they will never or seldom question the input from the specifying department. Personnel are trained to follow procedures and unless special conditions occur or instructions changed, will typically not question an order for a change in requirements.

As performance improves and productivity increases, product quality is enhanced including acceptable and shippable inventory levels stabilizing which results in on time delivery to their customers. This allows other key company functions to improve since less effort is spent in fixing a problem, since the problem has been prevented from occurring. Then as more company department operations improve, customer and supplier realize greater satisfaction, profits, and rewards in the exchange of value, the basic tenet of a successful business relationship.

It is interesting to note in Figure 1 that monetary savings for Six Sigma quality improvements are not projected out as long and short term goals. But, all major Six Sigma believers factor in the savings to be realized as risk decreases, defects reduced, and processes in business, manufacture, and services dramatically improve. The driving force is reducing cost and improving profitability. Remember this was, and still is, the basis for the Six Sigma programs at the major corporations that began this quality improvement program.

Six Sigma is actually an incentive to not only improve their business, operations, and process for their products, but to prevent problems from occurring in their business units, identified as risk or reoccurring problems.

The goal is for supplier and customer to concurrently achieve satisfaction. As the product and service risk of the supplier decreases, the customers risk decreases accordingly. This is achieved by the supplier using Six Sigma methods to achieve benefits in their operations and processes that lowers the risk to their customers.

WHERE CAN SIX SIGMA BREAKTHROUGH OCCUR

For significant breakthrough to occur, management must want to reduce customer and supplier risk, coupled with defect prevention. Where the quality dollars are spent is critical to success for any quality or Six Sigma program. Management must know and understand the tasks that lie ahead, commit to funding the reduction of risk, and be able to accurately and continually measure the success of the business unit's progress.

In prior quality programs, TQM, Kaizen, ISO9000 and others, companies have had difficulty in measuring their satisfaction and their customers. The reasons for this are no one physically mapped out a plan and strategy with goals to be achieved, monetary saving to be realized, and how they were to be measured from their operating units. Management must focus on the language of business, (opportunity, cost, time, risk, customer satisfaction, etc.) and allow their operating units to focus on the language of quality (errors, defects, prevention, analysis, solutions).

This raises the level of Six Sigma thinking to "Real Time" quality operations of personnel, machines, and systems. Management is committed to not only eliminate quality problems but also build a business on customer and personnel satisfaction in product, services, and job performance. As a result of reevaluating the areas where both business and operational break through can occur, these primary areas must be explored in greater detail to extract the most from the exchange. Executives must reevaluate their thinking in terms of solving problems and quality in terms of risk reduction with both speaking the same business language. Quality professionals are businessmen and women working to prevent problems and attain the long-term goal of customer satisfaction while reducing cost and risk.

To evaluate the risk versus defect analysis Figure 3 was developed. The evaluation of the organization is based on three methods for implementing Six Sigma change, managing, designing, and processing. Each organization in a company has a defined set of problems that can ultimately lead to

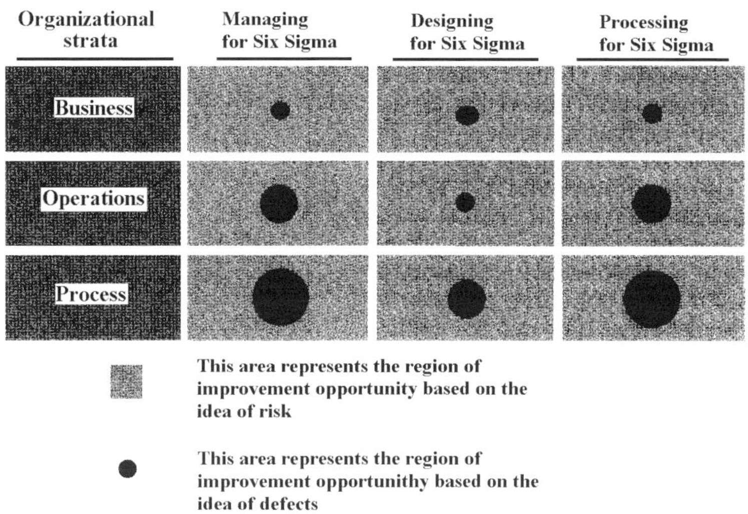

Figure 3. Vehicles for delivering Six Sigma Improvements. (Adapted from reference [3])

defects in products or services in the business, operations, and process unit areas. The key is to identify these potential defect areas and to develop methods of prevention. Once they are developed, they are then implemented as final solutions that will totally eliminate problems in their departments.

The tools of identifying these problem areas are available to the business and quality professionals within the company. They only need to be reviewed, taught, implemented, and finally evaluated for the effect on the departments operations. One problem some companies have is trying to solve problems before the actual cause is determined.

A company's management can often over react when these manufacturing and quality problems occur, especially if they are drastically affecting their financial bottom line. Upper management can exert an excessive amount of pressure on plant level management to speedily, without mention of time or cost to solve the problems. The problems are usually multiple in nature and no single cause can be assigned to the problem. In most cases the problems have been around for a substantial while and when the company is making money, can be pushed aside. But, when times get tight or new management is involved, the solution must be quickly found or people are replaced.

This replaceable level of management is given a goal for achieving a cost savings or cost of quality reduction. This goal usually affects each plant, department, or operating unit and it must be achieved within a designated, usually, very short time period.

The cost reduction/savings program is often implemented even though the operating unit may not have had the time or personnel available to establish a base line for evaluating the reason or root cause for their problems.

The biggest difficulty may be convincing management, especially in the business operation unit, to provide the time and resources to develop information to identify the root cause of the problems. They must be made aware that many problems are associated to personnel. The workers develop or are assigned procedures that result in defects during the performance of their job function. These can be eliminated after they are determined detrimental to the operation once the quality teams have spent the time to determine how to prevent them from occurring.

Using QFD (quality function deployment) with the customer and their business units, they identify needs and wants that can be addressed for improvement and used to evaluate their business unit for meeting the customer's satisfaction. The use of other quality methods, Ishikawa, fish bone, charting and FMEA's (failure mode and effects analysis) are not only for the manufacturing floor, but also for any business units operation that follows a set or specific flow of directives or instructions. Here check lists can prove invaluable to ensure no item is left unanswered from sales through proposals to contract completion, and agreement.

We have all experienced that sinking feeling that an important item was overlooked that sooner or later had an adverse effect on a program or product. Obtaining customer agreement on situations, even far in advance on how they will be resolved, is extremely important to both customer satisfaction and the suppliers bottom profit margin and success. I know of many situations where the supplier, due to not having a defined procedure for handling a customer's request, had absorbed the unanticipated cost of doing business to provide the desired change and keep the customer happy.

Too often finance is forgotten and not brought in early enough to assist in a major business decision or discussion. If a profit cannot be made, the business will ultimately fail. Resolution of disagreements should always be included in a contract so each party will know in advance how they will be handled and negotiated for the benefit of each party.

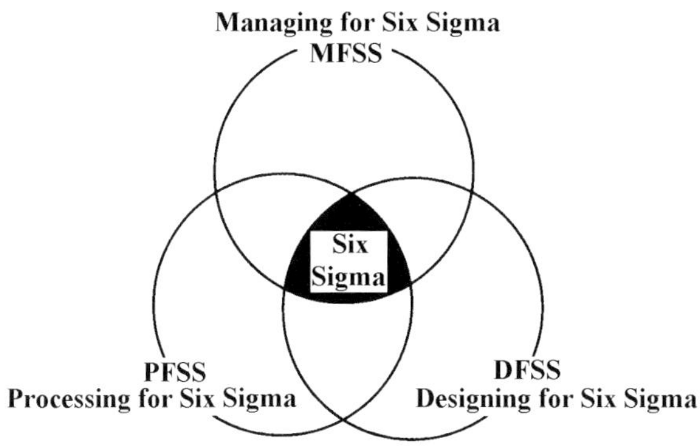

Figure 4. Primary vehicles for delivering breakthrough.
(Adapted from reference [1])

Therefore, the embodiment of bringing Six Sigma into the business in all units is necessary for a successful program. This is illustrated by the three overlapping value analysis methods as shown in Figure 4. These methods, managing, designing, and processing are used to reduce defects, improve output, and create a smarter and more efficient operation in the company.

Six Sigma quality methodologies are integrated into the business operations for product and process improvement. Six Sigma employs the following quality tools for implementing a program. There are four phases plus eight key tools used to perform a Six Sigma program as shown in Figure 5.

MEASURE – ANALYZE – IMPROVE – CONTROL

These terms, measure, analyze, improve, and control are called the (MAIC) phases. An important area to remember is that a well-designed and well-executed system maintains a business function or process under control at an acceptable or existing sigma performance level.

A baseline is established through the use of control charts that identify visually, how the system is performing and can predict when the process or system is becoming unstable and likely to go out of control. With the use of Six Sigma expanded quality metrics, it can show the operator when to leave the operation alone. But, when correction is needed how much adjustment or correction in a variable is needed to bring the process back into control.

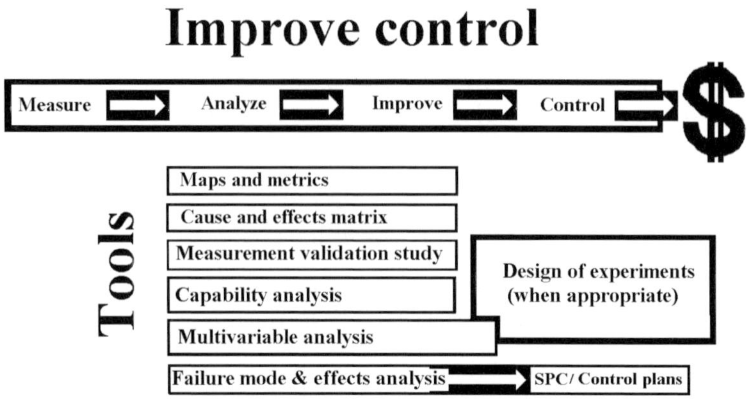

Figure 5. Four phases and eight key tools for Six Sigma. (Adapted from reference [2])

Then when a process correction is required and a problem identified, corrective actions can be initiated to stop the out-of-control trend and bring the process back into control. The difference here is using corrective action variable adjustment to improve or bring the process back to the target mean value for the process, versus fixing a problem that occurred as a result of not implementing preventative action. All processes are typically non-stationary and will shift during operation. Process is affected by the environment, material variance, mechanical faults, personal experience and knowledge. The process control system (either business or manufacturing) unless specifically programmed for analysis and system adjustment may not be capable of improving the operations, only monitoring and identifying trends during its operation. Improvement of a process or system can only be attained when the significant characteristics are; identified, understood how they may vary, optimized, and maintained at optimum performance levels consistent with the process.

This is where the trained six sigma black best can utilize their knowledge and training, working with the company teams personnel to improve a process and reduce cost.

Their tools include and are not limited to the following quality methods and tools as list in Table 1.

Imbedded in these tools are other time tested quality methods such as the fishbone diagram, vendor supplier audits, check lists, and company audits

Table 1. Quality Methods.

1. FMEA – failure mode and effects analysis
2. QFD – quality function deployment with customer and company
3. Maps and metrics
4. Measurement and validation studies
5. Capability analysis
6. Multi-variable analysis
7. Design of experiments
8. Statistical process control with control plans
9. Cause and effect analysis

for their existing quality system, and input from company personnel on methods and ways to improve a system. These are the primary tools and others can be used as required. But, whatever tool is used be sure it will supply useful data or information for the improvements of the process or program under analysis.

All of these tools can be used to collect and document the data necessary for business improvement. The controller's office input and support is required to develop the true or anticipated cost savings of all potential programs. The final selection of programs to be worked relies on the savings anticipated to be attained for the company.

PROGRAM ANALYSIS

A black belt trained team leader with a team or several teams selected for their knowledge, ability to affect change, and understanding of company operations in both business and manufacturing areas develops the Six Sigma programs. The selection and size of teams and their makeup of members is based initially on identifying Six Sigma type programs. These programs must have a significant and initial high rate of return to the company to provide continued management motivation for the Six Sigma program.

Initially the list of programs considered are characteristically the highly visible problem areas in the company that keep reoccurring or are serious enough to cost the company business by loosing customers or market share. It is the Japanese method to have an initial Kaizen program followed with possible daily Kaizen meetings to discuss these problem areas, assign a department team leader and have them, with assistance from knowledgeable personnel solve them within a form of critical path solution development. This form of programs must be coupled with identifying the root cause and eliminating the source of the problem, not just the occurrence on the manufacturing line. Preventive action, not corrective action is the key! As these programs are worked and improved, the lower value programs move up and will eventually be improved.

CHECK LISTS FOR SIX SIGMA ANALYSIS

A full set of check lists are in Appendix A for the topics listed in Table 2. These are just a few check lists that can be developed to ensure the operations are performed as required for your business and nothing is forgotten in the process.

Most check lists are only done once and modified if the requirements change. New employees can quickly comprehend the business or manufacturing requirements if a check sheet is used and followed in performing a task; in support of supplementing their jobs work instructions.

Tools and equipment needed can be scheduled and even reserved for performing the task without schedule delays. If specific items or tools are required they can be reserved and collected in advance.

The entire list of Check List should be entered in the company's computer for all business and manufacturing operations. They can be used to identify who did the work, when, amount of time spent, and results to inform other departments when their functions are completed. This information is entered on the master-manufacturing schedule so all personnel will know when their services are required to have the operation proceed on schedule.

A master equipment and machine scheduling program can show what assets are required for a job, if in current use, when they will be available,

Table 2. Six Sigma Check Lists for Business and Manufacture.

Sales and Contracts	Design and Development
Material	Manufacturing
Assembly	Price Estimating
Decorating	Packaging and Shipping
Program Scheduling	Purchasing
Quality	Product Development
Product Design	Warranty Problems

and if due to business demands, additional equipment is required to meet the schedule.

The methods used to select the initial Six Sigma program can use a checklist developed for the business function or process operation. They are used to evaluate and make sure the necessary questions are asked to determine the importance of the program selected and if it has the potential for improvement. There is also problem solving check list that can assist in developing information about a problem either in plant or with a customer. The checklists developed are general in nature to cover all sections of the company business areas and departments. Once this information is developed and documented, finer tuned checklists can be used for selection the initial programs to be improved. Examples of the initial program evaluation check list and follow up design and manufacturing check lists are shown in Appendix A.

Some of these check lists were originally developed for injection molding of plastic products and others, more generic to their title, can be used as is or modified to fit your company's business structure, goals, and manufacturing requirements.

SELECTING THE SIX SIGMA PROGRAM

Initially programs with high visibility and large cost savings potential must be implemented. Also, factored into the analysis should be the probability of success, time line (how long to obtain savings), and assets required, and anticipated cost and capable personnel required to complete the program.

These are the first items for improvement upper management will consider and answers must be specific, well planned, and documented in your proposal for potential six sigma programs. A discussion of the time line projected to achicve Six Sigma in a company is discussed separately along with "Risk" reduction, the key to success for customer and supplier quality improvements.

The Six Sigma black belt teams are formed to gather data are typically selected from the staff and personnel of the company. The six sigma teams coordinator or leader must be trained in the use and application of six sigma methods and tools. Hiring or sending an employee with the background and educational knowledge to become a black belt is advised. This person, if staff, must be knowledgeable in the business and manufacturing capability of the company. This person is typically selected from the quality assurance staff but may come from another department. This can be the QA manager or ISO9000 coordinator if applicable. The black belt team leader can also come from any department as long as leadership and team management coupled with good personnel interaction personality qualities are characteristics present in the individual.

Should no staff personnel be selected and an outside hire is undertaken, try to find a certified black belt who has experience in or close to your industry. This will assist in an easier transition of knowledge and business practices to your organization. Black belts coming from large companies to smaller or different type of industries will have been schooled in a different business philosophy and method of doing business that can affect positively or adversely their new companies methods of operation. To circumvent these problems from occurring, the black belt coordinator must adopt initially the operation methods of the company and QC department. Implementing different or totally new methods too soon into the company operations can cause serious unintentional problems if they implement changes too quickly and without understanding the business, manufacturing, and quality culture currently in place. Over time when their success is apparent the positive changes can be implemented after personnel and management understand it is the best intention for making the changes in the system. Current personnel can react defensively to change even when it can assist them in their operations. Always try to reinforce change as a positive influence on their prior management of operations under their control. Bring them into a positive offensive position to implement positive savings in their departments.

The personnel selected for the Six Sigma teams must be trainable in Six Sigma principles. They need to understand the requirements for analysis, accuracy, and quality methodology then can be taught and then implemented. There also needs to be a desire to improve the company's business base and quality of operations. These personnel can be trained and later certified as first green belts, then when fully trained, as a team leader black belt and quality team member. This is done by an outside training program or by implementing an in-company black belt training program; whichever proves most cost effective and productive for the company.

The black belt team leader's time is dedicated 100% to the duties of their Six Sigma team responsibilities. They should not have any collateral duties assigned, as being a Six Sigma team member is devoted full time to the programs. The lead coordinator black belt will typically only perform the six sigma program operations and projects. The team members may not always be on a Six Sigma team program and when not needed can return to their original jobs. Therefore, the black belt selected must be trained, a self-starter and able to operate independently when teamwork is not required and during the program direct their efforts as required. A separate job profile, performance description, with anticipated results should be written to select this person. A typical job profile and expectations desired is shown in Figure 6. This profile should be followed even when the quality manager is selected as the black belt team leader.

An example of this was a black belt hired from an aerospace company to be the QA manager and Six Sigma coordinator along with responsibility for maintaining the company ISO9001 and QS-9000 certification program. The company was in the fire suppression system area for commercial and industrial buildings, valves and high-pressure water distribution systems.

The problem created was a missed shipment due to his telling an incoming inspector to reject any incoming material not meeting specifications as was the standard practice and procedure at the aerospace company. Valve housings were received from their supplier with a key way slot 0.07″ to deep on several housings in their inspection AQL (acceptable quality level) sample. The inspector rejected the housings and the entire lot was sent back to the supplier without consulting anyone. This is what he was told to do, as it was standard policy the black belt had followed at the aerospace company. This caused the company to again miss a critical shipment to this customer. This was caused by the new QC manager instructing the incoming inspector to reject and then return any lot of parts not meeting specifications.

This action was concluded without calling for a MRB (material review board) to review the seriousness of the problem and what disposition the company should make to meet their commitment to their customer.

In this case, the key way slot depth could have been adjusted by making, a replacement keyway 0.07″ higher to compensate for the incorrect and deeper keyway slot. The major problem would also have been adverted if an MRB review of rejected incoming material or other situation occurring in the manufacturing operation had been specified in their operation procedures. Then manufacturing, engineering, and quality assurance could review the problem and if possible implement a corrective action, after notifying and gaining the customers approval.

These are some of the desired attributes for the following position:

Team builder and leader
Proactive, self-starting and decisive in making decisions
Open to suggestions from others
Mentor to others
Respected in the company
Knowledgeable in quality
A Preventor of problems
Good common sense
Trainable
Good communicator and listener
Good problem solving skills
Able to ask for assistance when necessary
Process control oriented
Investigative and structured in operations
Goal Achiever
Likes people and wants others to excel
Gains respect early from peers

The job performance for a Six Sigma Black Belt:

The Six Sigma Black Belt (BB) will be the leader of the companies quality improvement team program. The BB will serve full time in this capacity reporting to the CEO/President of the company and collaterally to the "Company Champion". The BB will select a team of personnel who will assist the BB in solving and preventing manufacturing and quality programs and problems as directed by the CEO/Champion. The BB will teach team members in the Six Sigma methods and work with them on programs providing direction and guidance in all areas of the programs. The responsibility of all programs is the BB and they alone will make decisions on how and where to proceed in all programs for improvement and prevention of problems. The BB will ensure the team remains focused and on schedule for the successful completion of all programs.

Anticipated Six Sigma Program Results:

The BB is charged with obtaining the best results from all programs and ensures the team members have the assets and support for them to successfully complete their program objectives in a satisfactory amount of time and cost to the company.

Figure 6. Six Sigma Black Belt job and performance profile.

This is the only and correct way to handle this type of situation. The customer must always be the final decision point in accepting or rejecting the product or solution. The aerospace method to deal with this problem was to reject and return any part not meeting specification. His current company did not require this as a modification could be made with engineering and customer approval. This would have avoided the return of the parts resulting in a missed shipment to their customer. Therefore, a leader must know or quickly learn the product requirements of his new industry and avoid old company procedures until the new companies methods of controls are learned. MRB's are always a wise quality method to employee as long as the staff is not abusing them, setting specifications not attainable by suppliers, and with suppliers held to a higher standard than required by engineering and manufacturing specifications and customer requirements.

ESTABLISH QUALITY IMPROVEMENTS

It is assumed a company implementing a Six Sigma program is currently an ISO9000 certified facility. If not, it should be in the process of becoming certified. The use of Six Sigma procedures expands and compliments the ISO program for continuing to improve the company quality system.

Six Sigma fits the continuing quality and business requirement plus process control for manufacturing operation. The revised ISO9000-2000 standard can build on the accomplishments achieved with a Six Sigma program. Personnel training both in work and quality operations are necessary to be documented and become an ongoing history of the company improvement program.

Six Sigma builds on and combines the new ISO9000-2000 requirements of analysis. Stronger emphasis on the use of data for analysis of effectiveness of the quality system, process, and customer satisfaction can be achieved by:

Continual Improvement

Expand requirements for the use of improvement activities such as:

1. Improvement of the quality system
2. Reducing non-conformities; corrective action

3. Action to avoid potential non-conformities, preventative action

Customer satisfaction can be expanded with emphasis to address:

1. Achieving customer confidence
2. Understanding and complying with customer requirements, QFD
3. Customer satisfaction to be documented and to be part of the measure of the quality systems effectiveness and management reviews.

Measurements: More specific requirements on the frequency of measurement to evaluate the system for prevention and solution of problems.

Achieve Results by Establishing the Company Culture

Quality culture begins and must be maintained by upper management. How they perceive the quality process, achievements, and savings drive the programs. Success must be achieved to further achievements tied to profitability, reduction of business and product risk, and leading to improved customer satisfaction.

Six sigma projects must be selected and presented to management with specific goals, cost reduction numbers, time schedules for accomplishment, and both employee and customer satisfaction expectations.

Ownership of the Six Sigma program begins at the top executive management team and must be shared by the department managers and their quality teams and all workers in the company. Each employee must be trained in how his or her actions and suggestions can be a part of the programs success. The majority of General Electric's plants and operation, business and manufacturing alike want all of their employees to have a minimum, green belt trained status. Management must attend their specifically designed for management, Six Sigma training seminars, to understand their role in the Six Sigma programs. Time must also be made available to attend, when invited; black belt designated planning meetings when their input is required. Management must delegate authority to the Six Sigma coordinator and the team members to achieve their goals of improvement with the necessary assets to meet the programs objectives.

Management often interfaces with their peers to find out what results their quality and six sigma programs have achieved at their respective companies. They can also share data, experiences, and training methods when good results are obtained and savings realized. The major Six Sigma user corporations are now doing this. They train and lend support to their sub suppliers to train and assist them in achieving six sigma success, originally on their products, but when completed they are recommended to take the program throughout the rest of their company for their other customers.

Achieving Customer Satisfaction

Customer satisfaction must be achieved, maintained, and improved by any company to remain in business. Measurement of customer satisfaction has in the past been measured negatively. This has historically and still to date been determined by the number and type or seriousness of the customers complaints. To produce a successful quality response system the department must factor in and compare the ratio of problems that have been successfully solved by personnel in the company and stay prevented to the number or old problems that reoccur. This can take a negative and turn it around into a positive with the customer and aid in training new employees on problem prevention and solving.

The solution is to compare a problem situation with a satisfactory response and solution to the customer's problem tempered with a sound corrective action that solved and will prevent the problem from reoccurring and improve the company's quality score.

Many companies do not know what their customer's satisfaction requirements are and as such find it often difficult to meet. Unless a two way conversation is initiated with specific question asked with honest responses, does a supplier really begin to know their customers needs, requirements, and wants. In fact, this is often the best time for understanding customer needs and how to evaluate the actual requirements they have for their products. Often, requirements are carried over from prior programs and said to apply to the new program or product.

QUALITY FUNCTION DEPLOYMENT

Using the QFD (Quality Function Deployment) method of communication will give the supplier the opportunity to discuss with the customer their requirements to verify if actually required, and if not, can they be reduced or eliminated. This could save both time and money to bring a product to market while opening a discourse for suppliers to offer other options combining of multiple parts in one to reduce cost that would make an even better and more salable product.

The use of check lists, surveys and QFD techniques can ascertain the customers needs and wants from all business, product and quality viewpoints. Too often no communication from the customer is assessed as "good news" when in reality the customer may be so "turned off" they are seeking a second and better source for their products. Never ignore and assume a customer is satisfied if you have not determined their requirements and are evaluating them in your company for meeting and exceeding their wants and requirements.

The QFD analysis can also be used at the suppliers company to evaluate their capability in terms of ability to meet your own product and service requirements. Likewise, QFD is a cross-functional evaluation form to determine what the customer wants and then evaluate if you, the supplier, can meet their requirements with the systems now in place. Or if your evaluation is unsatisfactory, what you must improve to continue being their supplier for the long term.

SIX SIGMA PROGRAM IMPLEMENTATION GUIDELINES

Step 1. Customer and supplier focus

Supplier and customer focus is necessary for knowing and understanding their individual needs and requirements. In analysis, it will be discovered that each are nearly identical in their business relationship to meet their customers needs and requirements and make a profit. During initial negotiations with the customer and supplier both sides of the partnership learn what product requirements and/or service is desired, with price, delivery, quality, service and support hopefully discussed and defined in the contract. Depending on the product, service, or work required, only sales is typically involved at the start. As the program proceeds, other company representatives may become involved to handle questions specific to their areas of responsibility and expertise.

The customer contract and sales, design, product and program development, scheduling, and manufacturing check list can assist in answering these questions. Visits to each companies facilities and specific business and product discussions can ensure a better understanding of each companies needs, requirements, ability to supply, and product specifications required to be attained for the product or service. In some situations it may prove helpful to send in early the check lists you will need answered, so they can prepare answers by the time you arrive to discuss these items with your customer and/or suppliers.

Customers are also becoming more selective in their suppliers and their "real" capabilities of meeting their product and service requirements. Customers have steadily been shrinking their supplier base with suppliers selected that were pre-qualified for meeting the customer's requirements for price, delivery, product quality and support. Most companies also monitor their suppliers but their rating system can vary greatly in actually judging a suppliers ability to meet their goals. Each customer will have their own set of criteria for judging the ability of a supplier to remain in their supplier base. This must be determined so the supplier knows ahead of time what is actually required to stay their supplier. An example of two typical (self and supplier) system evaluation check lists are shown in Appendix C, Supplier Evaluation, pages 1–16 and Self Evaluation, pages 1–3.

Rating systems should be specific, qualified for requirements, and quantified as to what occurs if products do not meet their standards and the action plan required for their solution or prevention of any problem. This should include an understandable scoring system plus any outstanding or new problems noted in the report to inform their suppliers of these true company ratings in "Real Time". In many cases a customer's problem is seriously lacking in problem detail, such as lot number, when made, when shipped, type of failure, why it failed, failed parts available for analysis, etc. All of this information is required to satisfactorily diagnosis, solve, and prevent future problems.

Step 2. Data driven

Programs without hard data to back up the information is virtually useless in today's business operations. The methods used to generate the data should also be reported. Within this report the number of samples for each point, time line and how gathered and reported (by hand or instrumentation) will give the audience a feel for the confidence to be attributed to the reliability of the information. Questionable data, sloppy procedure, not gathered at set internals, unreliable inspectors, poorly maintained inspection equipment, poorly trained inspectors, etc. is always useless in attempting to solve a problem. Data must always be carefully gathered, cataloged, identified, and documented to be useable in problem solutions and prevention.

Data is collected manually or automatically using process control tools (calibrated measuring and counting equipment) that rely on a trained operator, process or quality engineer setting up the system to accurately collect and report the data necessary for a program.

Then how the data is used to obtain the results is also necessary to be reviewed to ensure the methods and results are in agreement. How calculations were made and were they checked and verified for Six Sigma accuracy?

Six Sigma is a data driven quality improvement system for business and manufacturing areas. The current, three sigma method, most often discussed and used for control of processes and charting of production data assumes a fit within the bell curve of 99.73%. This is the area under the curve between the two (upper and lower) control limits as shown in Figure 7, on the left hand curve.

The three sigma limits allows an acceptable defect (if defects can ever be called acceptable) rate per million parts of 67,000 pieces that is very substantial. These could include parts for rework or scrap depending on the method and material of manufacture. In some cases the cost of rework, repair, is often so high and labor intensive that it is not economical to even consider this as an option. Any way you count it, these are not acceptable product for the company to ship. This also assumes only a 1.5 sigma shift, left or right of the bell curve target median to generate this substantial loss.

The Six Sigma method of control is a much tighter bell curve. The rate of defects drops to only 3.4 defects per million parts, a 20,000 times defect reduction. This also, includes a ± 1.5 sigma shift from the bell curve median as shown in Figure 7, referring to the tighter Six Sigma curve on the right hand side of the figure.

Notice with Six Sigma the bell curve is much tighter in comparison to the three-sigma curve with spread limits out ± 3 sigma where as the Six Sigma curve is only allowed to extend ± 1.5 sigma.

Six sigma results are very dramatic with savings potential, even for small programs, well disciplined in quality and financially wise to employ. Table 3 shows the defects per million for 1.5 to 6 sigma limits.

Step 3. Management involvement in the Six Sigma program

Selling quality improvement to upper management can be presented in several ways to obtain their commitment for excellence. This program begins by the presenter doing their homework and thinking through the total

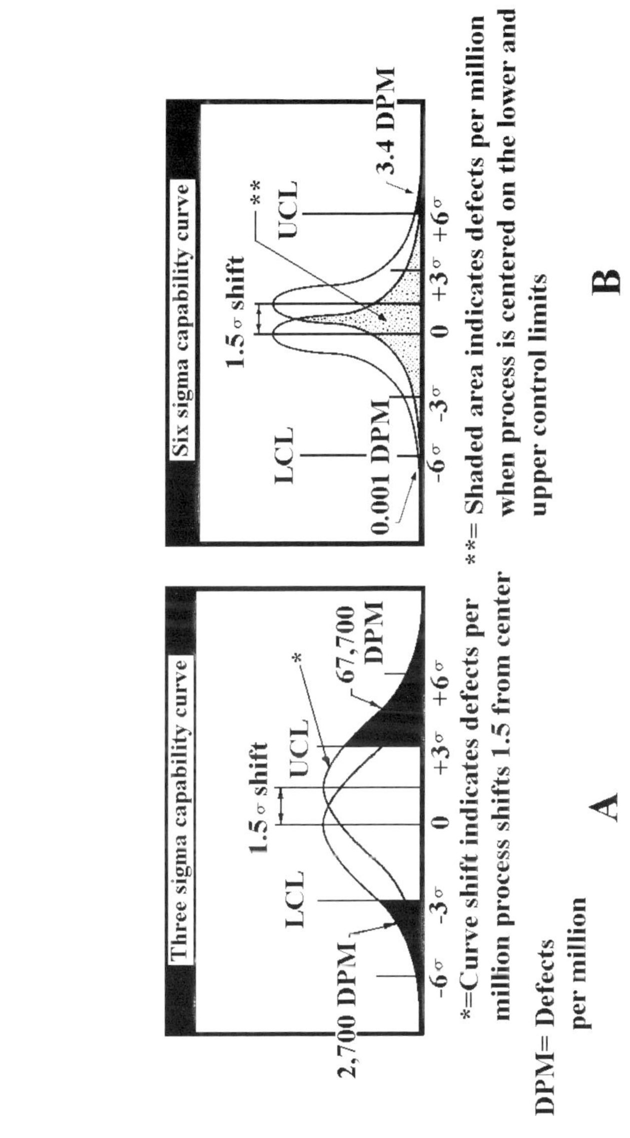

Figure 7. Sigma capability.
(Adapted from reference [2])

Table 3. Significance of Sigma on Results.

Sigma numbers	Defects per million
1.5 σ	500,000
2.0 σ	308,300
2.5 σ	158,650
3.0 σ	67,000
3.5 σ	22,700
4.0 σ	6,220
4.5 σ	1,350
5.0 σ	233
5.5 σ	32
6.0 σ	3.4

Six sigma is more than twice as good as three sigma, it is close to 20,000 times better.
(Adapted from reference [2]).

effect for their investment in a Six Sigma quality program. Quality improvements must result in a payback to the company for the funds, personnel, and time involved. The old saying, "Quality must be designed in, not manufactured in" is always true as long as the correct procedures are implemented and used during this process. It is assumed that data has been accurately gathered that can provide insight into viable Six Sigma programs. If not, then the data must be gathered before proceeding further in the program.

Outline, and present in detail, the quality improvements desired, their cost in output dollars, personnel time and selection of the personnel team to manage and run the program. Also estimate the desired time and savings and payback to the company in reduced scrap and rework costs plus on time deliveries and customer satisfaction resulting in increased orders. Remember the black belt and team in a Six Sigma program are dedicated "full time" (100%) to the improvement program.

All programs need a start and end date as close to operation time lines as deemed possible within the company operation departments and systems. Define the anticipated results, improved quality and profitability. Profitability can be broken down into three areas, how much money the program will cost and then save. Plus how much additional business it can generate with current and anticipated new customers. Document these potential new customers and why you can now make them your customer. What did your actions accomplish that your plan will now achieve? All good ideas must be coupled with a precise definition of dollars as it helps strengthen the reasons to initiate and implement the project.

The black belt and the company champion, to be selected later, for obtaining buy-in and any necessary assistance begins by presenting ideas that management can understand and share in their development supported with strong program documentation. Discuss your ideas with management in one-on-one meetings and why they are good for the health and growth of the company. This method develops management (department) support and allows the individual managers to comment and add their own ideas and recommendations into the potential quality and manufacturing improvement programs. It should also identify any of their possible objections, and help you understand in a "friendly" environment any objections or serious doubts they may have that do not contaminate others in management.

Objections can then be addressed and supportive information developed to support any convictions over their objections when the management team meets later to discuss all options for Six Sigma programs. Objections will be addressed in future meetings, building on the strength of other suggestions to make them less of an obstacle to the Six Sigma program.

Coupled in the Six Sigma program presentation, explain how it will eliminate costs, scrap and rework, for the company. With a good quality program, focus on how your customers are first satisfied. Next how employee retentions will be enhanced for several reasons; less rework required and scrap generated, signs that the companies profitability is being enhanced by their good work and productivity, which can result in pay and benefits, bonuses, and rewards directly to them.

Use case histories if available or talk with your suppliers or similar companies who have benefited from like programs. Also, check into what your competition is doing or has done to improve their quality and profitability for your customers or future customers.

Once all of this information has been gathered and discussed individually, call a staff meeting to present your and their views in a documented, clear, and direct presentation which can develop into an agreement for implementation of business and quality improvements. Recommend and insist on training programs for any and all staff that may have negative reservations and include in the attendees those favorable for continuing an upgradeable quality program for the company. Always be positive but discussion of negative items is good as long as solutions to any objections have been developed by you and are considered and discussed. It is extremely important to have all management agree on the programs for improvement, as they have to provide the assets and training for the success of the programs.

Step 4. Involvement of company personnel
Personnel selected for quality programs can come from all departments within the company. Ownership for the success of the Six Sigma program is the key and important for all personnel to have in the program. Training is mandatory for personnel to know not only what they are doing, but why, to achieve the anticipated and desired results.

TRAINING

All personnel are trainable and should be selected for their knowledge and experience they can add to the program. Personnel will be utilized during their normal work time and must be selected so that they can devote full time to the program. Alternates must be selected to perform their daily tasks without affecting the normal departments work output while the team members perform their Six Sigma quality team functions. If critical tasks are required, by a team member, then alternate team members must be selected to ensure the team is always staffed to perform their quality function.

In large companies, the six sigma team members total time is devoted entirely to gathering, monitoring, and documenting the Six Sigma information to improve the business or manufacturing operation. This allotment of their time for small companies could vary with company size and responsibility of the person. For maximum results full time is always recommended for the Six Sigma team members.

Six Sigma team members should be told that continuous improvement of all company business and manufacturing operations are their goal. All operations can usually be improved, and often a person outside the group or department can question why an operation is currently performed as it is. Process flow diagrams with control plans can map out the manufacturing or business functions. This diagram will show all operations and/or workstations in the process as shown in Figure 8.

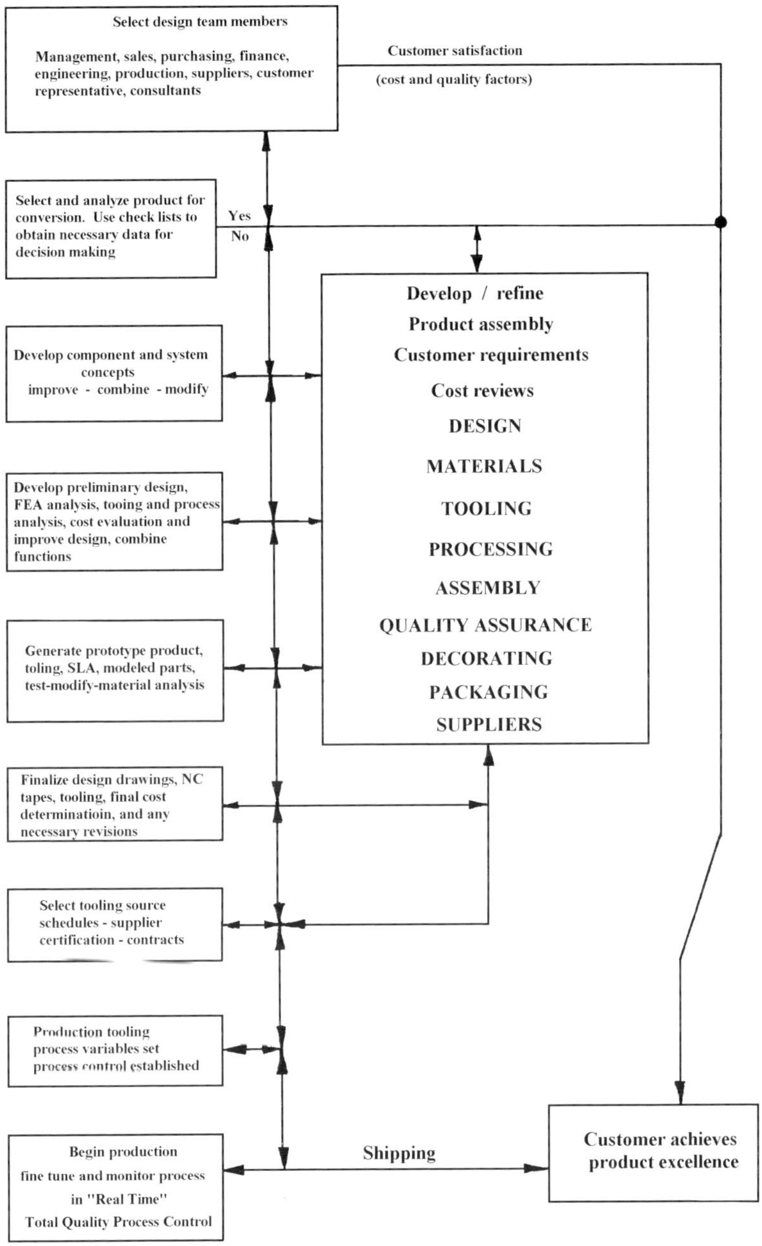

Figure 8. Product development path.

Inquisitive minds are necessary along with understanding why operations are carried out in the order they are performed. Hopefully, with continuous improvement changes in the process, may make earlier or later operations fall within the scope of their capability and will also be improved. But, remember change should only be applied if improvement is possible. Work habits and instructions not requiring change need not be affected if continuous improvement of an operation is not immediately possible or should prove counter productive to prior or later operations.

Each program scheduled for continuous improvement must be evaluated for its effects on other operations or departments. Also, discussions with personnel in how they perform their job will lead to discussions on how it can be improved and quality increased. Initially continuous programs may be ranked in importance by potential money savings, reduced time or personnel required for an operation, product scrap loss reduced, improvement of other departmental operations or just better acceptance of the product by your customer. These will have to be evaluated by the program review team before selection for a Six Sigma project.

COMMUNICATION

Keeping management apprised of the program progress is very important. Meeting time and cost schedules during analysis, evaluation, testing, consideration of data collected, and then recommendation and evaluation of solutions to the group is very important. How the programs communication is documented, reported, and frequency can be a contributing factor of program success. Good documentation of all results is the key to success.

Results must be reported, either good or bad. All programs once begun may not initially move toward success. If new equipment, test, or measuring devices are needed, then the program may have to be "put on hold" until assets arrive. This information must be communicated to the group, departments, and management.

A critical factor analysis and involvement of your company "champion" may have to be made by management to expedite the flow through the company to keep the program on schedule and within cost projection

Communication within the Six Sigma quality team is very important. The team leader is typically the reporter of program progress. The team members gather the data, perform the tasks, and report their findings to the team leader for evaluation by the team and management. A secretary, note taker, should be chosen and attend all meetings to ensure an accurate reporting of information and results is documented for the team.

The data initially collected for a program or system analysis is very important no matter how good or bad it says the system is performing. This data will become the base line of the program and communication of results will give the team and upper management an indication of where they are and where they can move to improve the operation.

Meetings of the team members occur daily and progress reports are issued on a time schedule or when important results are learned or changes required in the system for improvement. Meetings with management on scheduled internals must have a time and date and agenda, preferably with the subject matter and any supporting data attached, for all to consider prior to the meeting. The importance of communication must never be diminished as all information to personnel and plant performance is necessary to keep the program alive and well in the minds of staff and employees.

REFERENCES

1. Harrold, D. "Optimize Existing Processes to Achieve Six Sigma Capability." Control Engineering March 1999: 87–103.
2. Harrold, D. "Designing for Six Sigma Capability." Control Engineering January 1999: 62–66.
3. Harry, M. J. "Abatement of Business Risk is Key to Six Sigma." Quality Progress July 2000: 72–76.
4. Schonberger. "Work Improvement Programs."

Chapter 2

Six Sigma Implementation Process

Six Sigma is the newest quality initiative program to focus concurrently on all the prior quality methods and techniques. Some businesses found some of these existing quality methodologies and tools are not always totally capable of achieving the quality of success desired. The single major fault of these management decisions were the problems they were trying to solve did not have a single solution.

Each quality method they implemented never completely yielded the results desired. As a result of these earlier quality solution fixes not always working and yielding the desired results, management has not always totally supported their in-house quality programs as was necessary for success. Therefore, to solve the entire company or corporation quality problem they need a brand new approach.

Six Sigma provides this answer to their quality problem by initiating a program for analyzing their entire business, manufacturing, and quality operation to implement a permanent solution.

The matrix soup of prior basic quality methods, programs, and techniques; such as SPC, quality circles, Kaizen's, Total Quality Control, TQM, FMEA's, etc. were successful in their specific use but not entirely melded together to form a total quality system. A comprehensive list of the current quality systems is shown in Table 1. Many of these tools were often misunderstood or not able to be implemented by a company for various reasons. Some of these reasons were no management support, non-department support, poor facillator, no assets, or time made available. Six Sigma combines all elements in the matrix of quality methodology and technique, plus new ones, under one management supported banner that provides a complete framework for turn around and implementing a balanced and profitable quality system.

Quality must be redefined as, "A state in which value entitlement is realized for the customer and provider in every aspect of the business

Table 1. Quality Improvement Methods.

Program Name	Worker Involvement	Specialist Oriented	Group	Individual	Procedure	Work Methods	Quality	Prod Design	Morale Enhancement	Motovation
Qual Circles	X		X		X	X	X		X	X
Zero Defect	X			X			X			
Employee Suggestion	X			X		X	X	X	X	X
Work Simplify	X		X		X	X				X
Quality of Work Life	X		X			X	X	X	X	X
Scanion Plan	X		X			X				X
VE/VA		X	X		X			X		
IE Work Study		X		X	X	X				
QA/QC		X		X	X		X			
Organizational Developmnt	X	X	X	X	X				X	X
Fish Bone	X		X	X	X	X	X	X	X	X
SPC	X			X	X		X			X
DOE	X		X	X		X	X	X	X	X
CP/CpK		X		X	X		X		X	X
FMEA		X		X	X	X	X	X	X	
PAP	X		X		X	X	X	X		X
PPAP	X		X		X	X	X	X		X
QFD		X	X	X	X	X		X	X	
GMP	X		X		X	X	X		X	X
Kaizen	X	X	X		X		X		X	X
ISO9000	X	X		X	X	X	X		X	X
QS-9000	X	X		X	X	X	X		X	X
CEA	X		X		X	X	X		X	X
8-D	X		X		X	X	X		X	X

(Adapted from reference [2]).

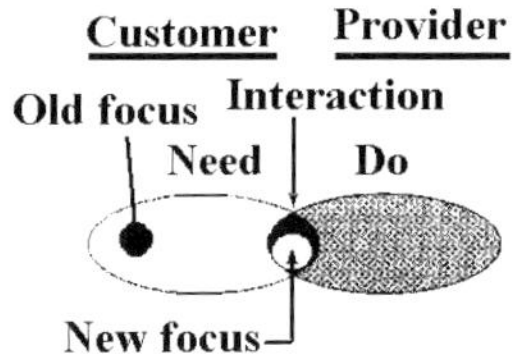

Figure 1. The need–do interaction.
(Adapted from reference [1])

relationship". This definition now considers both customer and supplier, as the supplier incurs the cost and awaits the customer's approval with continued business. Should this not happen, management can blame their increased costs and an unappreciative customer of failure for their profitability to not increase. Companies are in business to make money and this is often overlooked, when and where quality improvement dollars are spent, while trying to improve quality for the customer. A two state relationship is necessary, meeting the customer's quality requirements and value of manufacture for supplier to provide products.

The use of Six Sigma must involve an interaction between customer and supplier of "Required and Provided" interaction. Both have to contribute to the plan – Satisfaction – of each entity as shown in Figure 1. Quality has focused on the customer satisfaction requirement with the supplier, typically forced to meet this or lose the business.

With Six Sigma the new definition of quality looks at both sides for their individual contributory input within a larger scope of quality improvement. The new state of quality looks at both the real and perceived satisfaction of each. It expands the application of performance metrics and standards to the real and perceived regions of the business relationship for satisfaction to be attained.

Under the current quality systems of most companies, their goal is to meet the customer's quality requirements that were measured by post-delivery audits of products or services received. These reports gave a score for defects, delivery, service, and often price competitiveness. This report, usually lacking in specific detail, did not stress the degree of the problem, only the number of occurrences without any real problem explanation, and was only the surface of the business relationship.

Quality departments exchange specific occurrence problem data on a regular basis with remarks to occurring problems as the customers require this corrective action response for their paper work and usually accepted these written responses. Too many companies wrote responses that anticipated the elimination, but not always the ultimate prevention of the problem. This was often the case as the same problem kept reoccurring month to month. The customer then, if their tracking system was good, wanted a real and permanent fix of the reoccurring problems. They often required SPC data that the problem was in control and this was the proof if the data was generated during the manufacture of their product. When the company was good they fixed the problem correctly the first time and never again had to answer these questions. But, many were not and satisfactory answers were difficult or impossible to obtain unless major work or reorganization of the suppliers quality and manufacturing systems was performed to improve the process and reduce defects and other problems.

There were often never any pre-delivery customer satisfaction requirements with audits performed to verify quality was capable of being attained or the consideration of a suppliers real or expected satisfaction factors to meet their customers expectations.

In the majority of cases, a company's classical cost of quality was not considered or tracked in any detail. The supplier only worked to ship on time products to meet the customer's satisfaction. This could involve changes in manufacture to meet very rigid customer specifications while using the in-place quality system to meet their normal output. This was not financially satisfactory for the supplier with only a few customers, the top 20%, achieving good results.

There existed a discontinuity of information within a suppliers operating departments and between the customer's quality requirements. This caused confusion with plant personnel if the customer order was not identified and closely monitored during manufacture. Too often if not identified early, many problems occurred, defects, rework, and remanufacturing was required to meet the specific customer specifications.

Six Sigma focuses on aligning the satisfaction factors of supplier and customer for front end understanding of Real and Perceived requirements plus reduced the risk of errors to increase satisfaction for both companies as shown in Figure 2.

A method to evaluate the business Entitlement factors between customer and supplier was therefore required and is shown in Figure 3. The main reason successful companies have used this evaluation process is to ensure

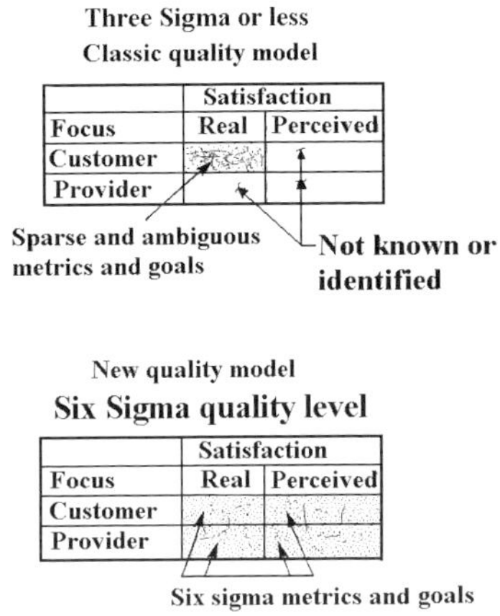

Figure 2. Expanded focus.
(Adapted from reference [1])

Number	State	Focus	Value	Sigma: Target (T)	Sigma: Entitlement (E)	Sigma: Performance (P)	Performance gap: P-T	Performance gap: P-E
1	Real	Customer	Utility	6.0	5.3	5.0	-1.0	-0.3
2	Real	Customer	Availability	6.0	4.9	3.7	-2.3	-1.2
3	Real	Customer	Worth	6.0	5.2	4.6	-1.4	-0.6
4	Real	Provider	Utility	6.0	5.0	5.5	-0.5	0.5
5	Real	Provider	Availability	6.0	4.6	4.3	-1.7	-0.3
6	Real	Provider	Worth	6.0	5.7	4.7	-1.3	1.0
7	Perceived	Customer	Utility	6.0	5.0	4.2	-1.8	-0.8
8	Perceived	Customer	Availability	6.0	4.6	5.8	-0.2	-1.2
9	Perceived	Customer	Worth	6.0	4.9	3.7	-2.3	-1.2
10	Perceived	Provider	Utility	6.0	4.7	3.1	-2.9	-1.6
11	Perceived	Provider	Availability	6.0	4.3	4.7	-1.3	0.4
12	Perceived	Provider	Worth	6.0	5.4	5.1	-0.9	-0.3
			Quality =	6.0	4.7	3.7	-2.3	-1.0

Figure 3. Business entitlement matrix.
(Adapted from reference [1])

the world-class quality characteristics they will employ are the most critical to the value of customer and supplier.

Selecting processes to be within Six Sigma control, if not justified by each, is good business but can incur high expense for the reward of increased profitability. During the evaluation process the company must determine the right level of expectation required for their operations. The desire to yield 3.4 DMPO (defects per million output) of less than, equal to, or greater than must be determined, and this is how to evaluate the customer requirement and company manufacturing and quality capability.

The graph of customer entitlement versus the target for six sigma versus the company capability for maintaining dimensional control, is represented in Figure 4. The value built into the product must meet customer entitlement requirements to keep the customer satisfied. But, initially short term, one year or less, for meeting supplier goals of Six Sigma are not initially realistic.

Customer entitlement targets must have a rational benefit and meet mutually agreeable measurable dimensions that are attainable and agreed to by each party. This is shown in the Figure 5, for the customer and provider of the service or product.

As the gap between customer and provider closes it is apparent that the Six Sigma quality system is more than defect reduction. It is also a management system. Six Sigma also is concerned with recognizing the real and perceived needs and requirements of the customer for the provider to furnish. Six Sigma analyzes the gap between current performance, the

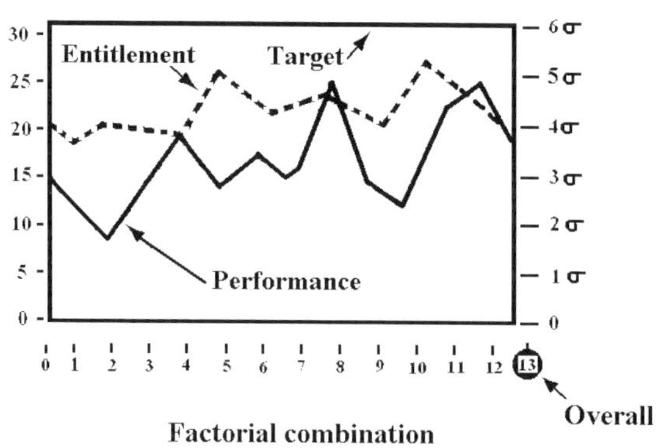

Figure 4. Performance gaps.
(Adapted from reference [1])

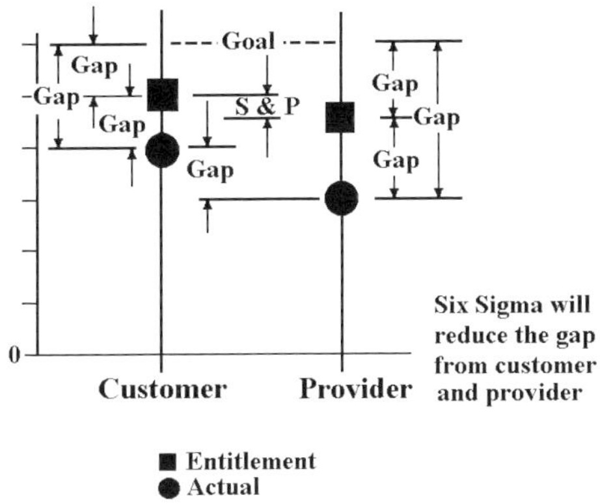

Figure 5. Linking goals, entitlement and performance. (Adapted from reference [1])

entitlement the customer believes is due them, in the short term, and what is the long term goal to maintain the business and keep the customer as a loyal customer. This is illustrated in Figure 5, as the analysis progresses on a daily basis for the systems, operations, and process levels of the business relationship.

The use of the Six Sigma business and manufacturing methods is used as a means for closing performance gaps and controlling key factors that contributed to the reason the gaps existed. Gaps can occur any where in the business relationship of the provider and customer as perceived entitlement of the customer and the real entitlement of the provider. This can also be seen in the provider's actual performance and the provider's real entitlement. The important factor is that business wide improvements can only be made by closing all performance gaps that are rational and germane to the actual business relationship. This is one of the main reasons for using the QFD method for analyzing the business relationship both at the beginning and during the life of a business arrangement with the customer and provider to be discussed shortly.

There is also a loss factor to consider when the performance gap (related to customer entitlement) is either plus or minus. A negative gap for entitlement of 5.2 versus supplier performance of 4.6 clearly indicates an improvement is necessary. But, if reversed, the positive number states the business is doing better than anticipated. This benefit could translate to the customer that no actual direct benefits are to be realized. It could constitute a loss on the supplier's balance sheet in pure operating costs of doing business except less scrap and product loss is realized.

The supplier must try to operate a quality program to raise a negative Six Sigma identified controlling processes while not initially concentrating on process areas already meeting or exceeding customer entitlement requirements. Deploy assets in a controlled and knowing manner to raise any low performance values to meet entitlement values. Employing the Business Entitlement Matrix will identify the areas of exceeding or not meeting the customer's requirements before any Six Sigma programs are considered for implementation. The ultimate supplier goal is to over time achieve an absolute performance level of Six Sigma, reaching this goal at the rate of Six Sigma.

RATE OF SIX SIGMA IMPLEMENTATION

Six Sigma is not achieved in a crash program. Moving to Six Sigma involves many factors, time, personnel, and funds. Moving toward Six Sigma from a four-sigma quality level requires a quality improvement at a rate of 78% annually to be achieved in 5 years. This requires improving monthly at 12% those areas needing improvement. This is based on a standard logarithmic learning curve and represents a length of time to gradually move a program into higher compliance. The time line can be reduced if required by increasing the assets into a program to meet the requirements, but is it really necessary. This is the question only upper management can answer based on reducing Risk of Error in their business with their customers.

THE KEY IS MANAGING FOR SIX SIGMA

Management is the key support for Six Sigma by providing the leadership and foundation for breakthrough and success. Management must understand the key operations, support and initiative required while providing the assets for achievement for both short and long-term goals they establish with support of their black belts.

Six Sigma requires the participants to be involved in the creation, installation, initialization, and utilization of the deployment plan for their company. Also, required is a reporting system and implementation process that supports the design and processing operations of changing within the

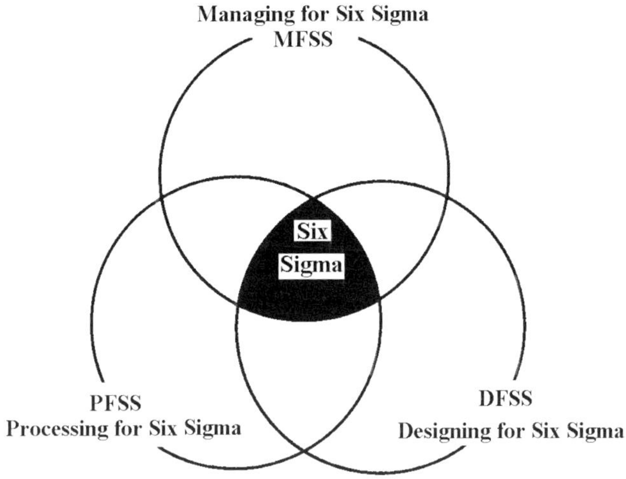

Figure 6. Primary vehicles for delivering breakthrough.
(Adapted from reference [1])

company. This is shown in Figure 6, as overlapping functions of design, process, and managing for Six Sigma operations.

How fast the goal of achieving "best of class" in business performance is determined by setting a time goal for improving the operational capability of the organization at an annualized compounded rate of about 78% improvement over a five year time period as shown in Figure 7.

This is an estimated company Six Sigma learning rate when starting at a four-sigma quality base line. The achievement rate will vary with each company based on the complexity of their starting business base. A conservative estimate established from prior results and studies for a company to achieve Six Sigma unit goal performance in the five year time period. The time line estimate for the implementation of the Six Sigma program mechanics and training is realistic for all participants in the quality improvement program.

Good management decisions will ensure the overlapping and unifying component of the Six Sigma program will bring together the business, design, and processing operations to run concurrently with maximum confidence of success. This will aid them to achieve value entitlement for both the customer and supplier in every business aspect of their relationship.

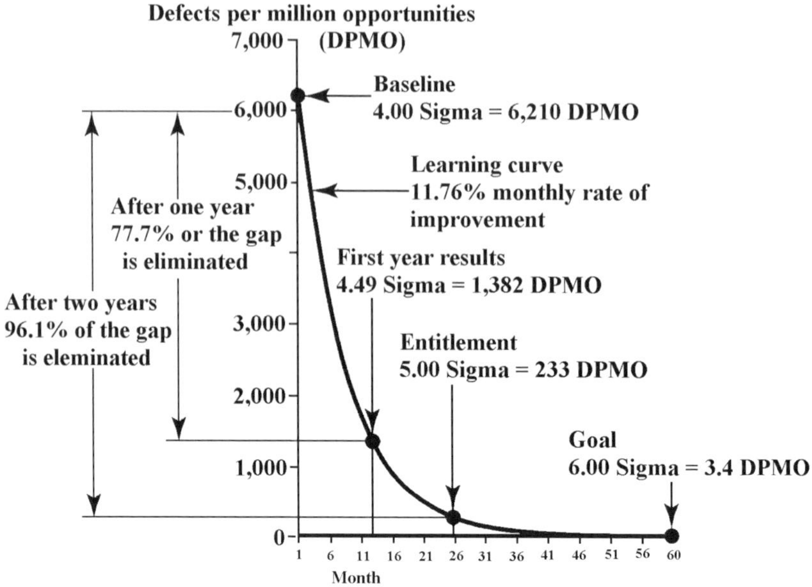

Figure 7. Six Sigma learning curve.
(Adapted from reference [2])

DESIGNING FOR SIX SIGMA

Once management support is validated, the design and development of your Six Sigma system can begin by analyzing all of the companies business units. This means focusing on reduction of any and all various forms of risk involved in the operation, function, and design of a product, process, or system. This is the main focus of your personnel regardless of its business base; commercial, industrial, or service operation.

Designing a process, product, or system must be dedicated to reducing the amount of risk exposure a product or service may inherently possess for its functioning, performance, and physical attributes. First, you want to reduce customer and supplier liability that will immensely improve customer satisfaction.

Second, the suppliers want their risk of doing business and manufacture variability kept within controllable and tolerable managerial levels, for provider satisfaction.

A competent product design and manufacturing system will maximize confidence that will ensure supplier and customer the end use product will perform as advertised. This will ensure the breakthrough strategy goal of having a design risk limited to no more than 3.4 risk exposures per million opportunities for a critical value risk item associated with processibity of the product or functionality of the design.

PROCESSING FOR SIX SIGMA

Once the design process or operation (here used similarly) is established, the process must be evaluated for capability and repeatability to reduce further the risk factors in business, industrial or commercial operations. Processing involves all work functions and as identified in ISO9000 and QS-9000 there must be work instructions for all operations within an organizational unit. They must also be enforced and reinforced daily for repeatability and risk reduction of their manufacturing system.

This will increase the confidence of the system being repeated ably capable of risk reduction and able to remain in control in "Real Time" to achieve the long-term goal of Six Sigma operations.

Manufacturing a product in a company versus a business operation has the highest visibility within an organization. When the process is tuned and controlled for each value risk point critical to the operation it should have no more than a 3.4 risk exposures per million operations. All risks should be identified during the design stage for manufacture analyzed using process control plans and then a FMEA. In companies developing a Six Sigma program this is where the initial action begins. But, remember there is more to just producing a product. The product can only be as good as the design, tooling, material, and control of the manufacturing operation. Having an in control process cannot make the product better if the design is not capable. Also, quality cannot be measured into a product if manufacturing is not in control. All operations must work together to achieve success.

Quality and business operations must be very strong joint partners. One cannot succeed without the other. Quality must work with all departments to ensure the amount of risk is minimized. Management must work with all departments to ensure risk is reduced, customer satisfaction increased, and all company personnel from CEO on down; understand Six Sigma must prevail in all dimensions of their company.

Six Sigma is the language of business now and into the future to control risk, time and cost while working on elimination of defects and errors in all areas of a company. Quality professionals must focus on business risk reduction with upper management realizing quality is their tool and path to achieving customer satisfaction and increased business growth and profitability.

RATES FOR THE EMPLOYMENT OF SIX SIGMA METHODOLOGIES

Once Six Sigma methodology is planted in a company it begins to grow. How fast it can produce fruit is a function of the items shown in Table 2.

Six Sigma initially stated a program must have a potential saving to the company within a predetermined time period of $170,000. Training was expensive and being a new methodology, slow at first to gain acceptance and support within the management structure. But, once the major companies, Motorola, General Electric, Honeywell, etc. started realizing substantial

Table 2. Six Sigma Functions Affecting Company Growth.

1. Size of the company
2. Management commitment and understanding of the Six Sigma methodology
3. Existing baseline level of quality within the company
4. Knowledge and training of personnel to perform the work
5. Management expectations for completion (time line)
6. Selection of initial Six Sigma program for the company
7. Selection of the initial Six Sigma team members
8. Capability of Black Belt to implement program(s)
9. Black Belt Champion program selection and support
10. Preprogram planning and strategy development for success
11. Data and information development
12. Analysis of data and ranking it
13. Developing preventative solutions
14. Implementing preventative solutions
15. Monitoring and evaluating the preventative solution
16. Establishing procedures and work instructions for the solution
17. Training of personnel
18. Audition and reinforcement of preventable solutions
19. Determining monetary and quality performance objectives
20. Recognition of the improvement teams success

(Adapted from reference [2]).

savings in process improvements with these results, the word was spread and other major companies joined in the improvement and saving program.

Companies, large and small, started programs mostly as suppliers to the larger corporations, who often offered training for their engineers in Six Sigma. Benefits were realized and risk decreased for supplier and customer. Finally enough programs were completed to draw together enough information to project a realistic time line for Six Sigma improvement programs.

Management wanted results as soon as possible to prove the Six Sigma methodology really worked. This was not always possible based on selection of programs, reorganization of plant and process structure, equipment repair, newly ordered, and then installation and startup training of equipment and personnel. In fact Motorola's initial programs ran over anticipated budget and time estimates. But, with renewed effort and often a new direction, the results were realized, and with time so were the profit savings and quality improvements. This new methodology had to have a learning curve associated with the program and management did take the time and spent the assets for the return promised by the program.

Nothing really proceeds as planned, but with perseverance and a good leader, your black belt, results can be obtained to prove the process really works and saves money while increasing output and quality.

A very important question must also be answered early in the beginning of a Six Sigma program. This is, "is an actual Six Sigma level of quality necessary for the company to meet their customers level of satisfaction?"

The goal of Six Sigma as a management tool is to have their companies operations reach a level of quality entitlement at a rate of Six Sigma progress.

ESTABLISHING THE BASE LINE FOR SIX SIGMA IMPROVEMENT

The base line the company establishes is the starting point for Six Sigma and follows the plan of objectives listed in Table 2, item 3 'Existing baseline level of quality within the company'. A company with a four-sigma quality level has already expended a lot of hard work to reach a defect risk level of 6,210 per million opportunities per operation. The elimination of a defect begins with having every manufacturing operation capable of producing a defect free product that is passed on to the next operation or station. The same response in the business or accounting section of the company is also required to eliminate and prevent business problems.

Determining a company's actual Sigma rating can be difficult. Preplanning is required for error and defect analysis. Once they are identified, their cost to the operation is determined by cost accounting. This can be difficult, as each manager wants to shift the responsibility of defect and errors from their operation to another. Six Sigma is used to establish root cause and identify where preventative action can eliminate the problem at its source.

In most operations defects or errors are identified and sorted out after, or during, each operation. This aids accounting in establishing the actual cost of quality for defects on the company's bottom line. Then a tally of all production in the WIP (work in process) is used to determine the true cost of quality and Sigma level. Always count defects that are removed or corrected (reworked) before being passed on to the next operation since there are individual costs associated with each operation.

Therefore, the true level of Sigma quality, if based on final acceptance count may be to high. All material entering the manufacturing cycle must be counted as parts with defects were cut out. Analyze and be thorough in determining exactly how manufacturing handles their defect and rework process. Are product defects analyzed using Parato charting or do they even record a problem existed or a business operation cause the problem? Control charting is only helpful when the information on the charts is actually used in monitoring the process and assisting in developing the solution to an existing problem.

Deming visited a company once and when shown control charts on the wall, commented that they must not be of any use as they were perfectly clean and look like they were never touched. Defects during the normal routine of doing business or manufacture must be counted in each departments operation. Defects must be identified where they occur so they can be prevented in the future!

DETERMINE YOUR BASELINE QUALITY LEVEL FOR SIX SIGMA ANALYSIS

After the quality base line is established, improvements can be determined, implemented, measured, and monitored for success of your Six Sigma program methods. The nine learning curve guidelines are presented in Table 3, for focusing attention on the rate of improvement anticipated within an established time period for the change to be implemented.

Using Figure 7, as a time line and function guide for a Six Sigma learning curve, several key assumptions are made:

1. Most corporations have already established a CTQ (critical to quality) risk factor of four sigma or 6,210 defects per million opportunities (DPMO).
2. In the short term, with a major effort, an optimum quality level of five-sigma is attainable without changing technology or major capital expenditures for a 233 DPMO quality level.
3. The quality goal is established at Six Sigma, the highest quality level attainable with today's experience and knowledge for manufacture.
4. In accordance with the company's five-year plan, Six Sigma is attained with minimum expenditure of assets to achieve the long-term strategic business and technology plan.

Table 3. Establishing the Six Sigma Learning Curve.

1. Establish the baseline performance level as defects per million opportunities as a sigma level
2. Establish the final goal for Six Sigma performance
3. Determine the time period, number of months to achieve the goal (convention is 60 months)
4. Calculate the monthly rate for improvement needed to reach the goal in the allotted time
5. Determine through historical data and experience how fast prior change and improvements were implemented
6. Estimate the time expected to be spent for achieving the program improvement with assets on hand
7. Establish the first year target program with input from all departments involved, first to last
8. Organize your improvement team and deploy resources in accordance with the target program
9. Repeat the cycle starting with the new performance baseline. Audited to confirm confidence that the improvement is established and performing as required on a daily basis.

(Adapted from reference [2]).

5. Major gains are achieved in the first three to six months, the usual management threshold for achieving significant monetary and process improvements.

These early gains are typical to a "Kaizen" approach to quickly rearrange an area for optimal and speedy results with minimum expenditure of assets. The Kaizen is performed in a time period of a week. The only problem associated with this short implementation period is not all programs and major projects may realize long lasting and positive results. In fact, Motorola and others did not always experience the anticipated return in all of their initial programs. In each company there is a learning curve that upper management must recognize and not always demand a positive return in the short term. In any system, there are many separate and independent operations or variables that are affected by operations performed before the product reaches the final acceptance and inspection station. A change in a variable at any point in a system can later affect a workstation down stream.

Therefore, to achieve positive results management must begin a Six Sigma program at the correct point in their operations to obtain the short-term results required for support and continuance of the program in the company. This means controlling all variables and processes in a system from their starting point that may begin at the raw material supplier's plant.

6. The strategic aim of the Black Belt is to achieve 70 to 80% of the goal within the first year. This is attainable with a tightening up of resources, personnel training, and changing or writing improved procedures and work instructions for an existing operation. This includes working with your suppliers to ensure the products supplied to your operations are in specification and delivered on time.

Depending on the companies beginning base line quality level, the time line and results graph, Figure 7, shows within the first two years, the majority of change and improvement (readily visible) has occurred. As the easiest to fix quality items are brought to a higher quality level, it becomes more difficult to effect a change and then control the change of the system. In a non-changing environment Six Sigma would be obtained easier and faster. This is never true as suppliers change, personnel leave, or promoted, equipment wears, process and product demands are revised, etc. Six Sigma process control is a moving target in a companies business and manufacturing operations and only attained with a specific goal and set of objectives determined before the program was implemented for quality improvement.

As the curve in Figure 7 illustrates, 80% of the quality improvements occur in 20% of the time period specified to reach the goal of Six Sigma. To really impress management, 50% of the improvements actually can occur within a six month time period for a five-year plan dependent on the company quality base line. Always plot and verify results as decision points are reached. Continue the improvements on your established time line to keep up enthusiasm and progress for the reduction of operation risk.

With all planned operations in control after one year, 78% of the goal is reached. Two years of effort brings it to 96%, an 18% improvement for a defect reduction from 6,210 to 233 DPMO. The next three years, if true Six Sigma is the company goal, will require renewed effort, tighter process control, and more accurate analysis of the process to reach 3.4 DPMO.

As processes are improved, better analytical methods and monitoring tools for controlling the improved process are required. This involves more training of personnel to know when to adjust the process and when to leave it alone. If the process does require an adjustment, they need to know just how much to adjust it and exactly what variable is controlling for process improvement.

DOE (design of experiments) can assist in this analysis as can other quality methodology as fishbone (Ishakawa) and CAE (cause and affect) analysis. Then using control charts, a process or operation can be judged, and a decision reached, if and when a process change is necessary. Then using the Box Jenkins method the major controlling variable of the process is adjusted to achieve improvement. All of these methods will be discussed in greater detail as to what they are, how used and when to apply their use to a program. In conjunction with these quality methods they may require more accurate monitoring equipment plus an updated set of variable effect questions to be answered. Also required is more in-depth and specific process analysis training to go that last short distance for process improvement.

As the move from five to Six Sigma is started, the design of the process to reduce the risk of change that produces a defect is addressed in the system, from beginning to end. This is the primary reason the last 4% is the hardest to achieve.

How fast this achievement of five to Six Sigma is reached depends on how quickly the quality gap is to be closed. In this example, to close the gap

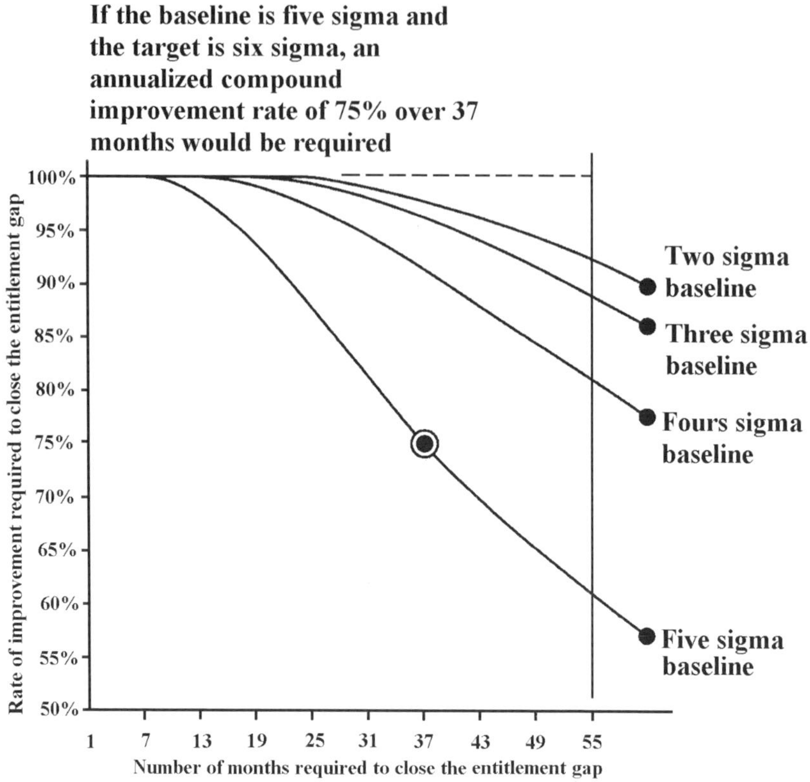

Figure 8. Improving Six Sigma learning curves.
(Adapted from reference [2])

in 36 months, an annualized rate of improvement of 75% is required as shown in Figure 8.

MONTHLY IMPROVEMENT RATE SIGNIFICANCE FOR CHANGE

To obtain an annualized rate of improvement of 75% the monthly rate of quality improvement must be 11% as shown in Figure 9. Depending on your rate of closing the quality performance gap and time established for it, Figure 10 and Table 4, give you a realistic base line for the measurement of improvement required to reach the goal of Six Sigma.

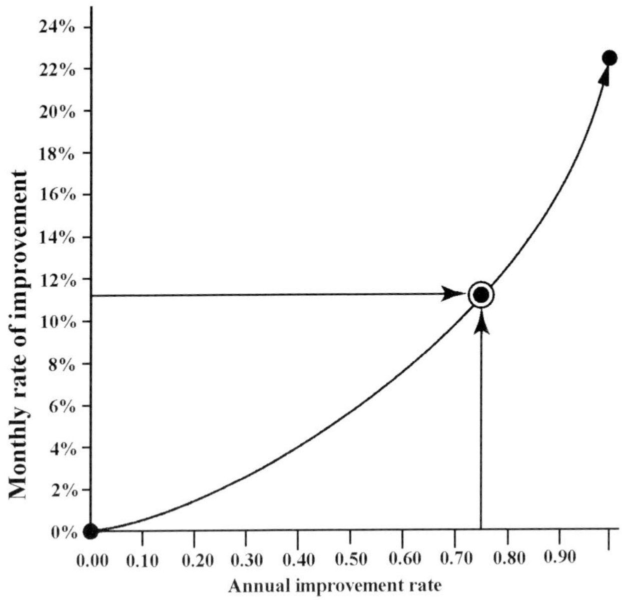

Figure 9. Monthly rate of improvement.
(Adapted from reference [2])

Many companies using QFD analysis realize not all of their customers require Six Sigma quality for their products. But if their manufacturing operation can maintain Six Sigma quality limits everyone is ahead on the quality curve. As the company progresses from one sigma quality level to the next, their operational improvement is enhanced and the risk to customers for producing a defect is dramatically reduced. There is an additional cost factor to consider since more stringent quality evaluation methods must be employed to see additional progress in operation and product quality improvements.

The additional cost is due to more accurate sensors and controls to monitor the specific variables in the process. During the improvement period, the operational control of the process must be improved at even higher rates. This could require equipment more capable of performing at the current improvement level. The product and process engineers will determine how tight the manufacturing process must be held to achieve an even greater increase in quality improvement and risk reduction.

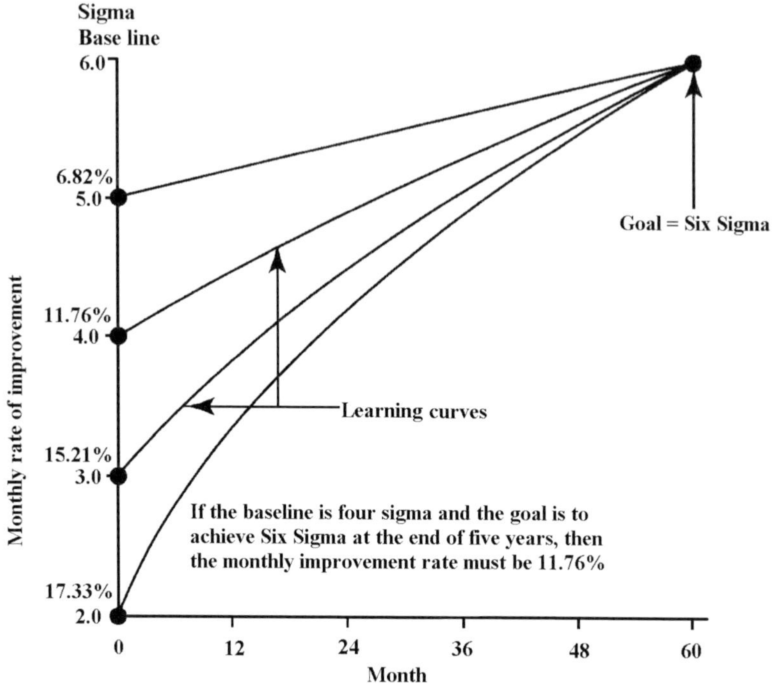

Figure 10. Closing the performance gap at a Six Sigma pace. (Adapted from reference [2])

Manufacturing equipment and tools must be analyzed on a periodic schedule to ensure they are capable of producing products to the specification. Preventative maintenance must be performed as specified and operators trained to ensure the system is controlled as required to achieve and remain at the higher quality levels. Supplier quality must always be improved to reduce the risk of defects to your customer as your own manufacturing processes improve. Time and resources must be spent to ensure they improve as quality requirements also improve. Once you achieve a known quality Sigma level, it will require more intense effort to remain at this level. All supporting input process variables within the organization must attain the same quality level as the main process.

Management must know how their support equipment, processes, and operations influence the company's rate of achieving Six Sigma quality. A projected achievement time line within a company's five-year business plan

Table 4. Six Sigma Pace Rates of Improvement.

Annualized rates of improvement

Goal	Baseline 2.0	2.5	3.0	3.5	4.0	4.5	5.0	5.5	6.0
2.0									
2.5	0.1246								
3.0	0.2636	0.1588							
3.5	0.4063	0.3219	0.1938						
4.0	0.5421	0.4770	0.3782	0.2287					
4.5	0.6626	0.6145	0.5418	0.4316	0.2630				
5.0	0.7626	0.7288	0.6776	0.6001	0.4815	0.2965			
5.5	0.8407	0.8180	0.7836	0.7316	0.6520	0.5278	0.3288		
6.0	0.8980	0.8835	0.8615	0.8282	0.7773	0.6978	0.5705	0.3601	

Duration of improvement effort = five years

Monthly rates of improvement

Goal	Baseline 2.0	2.5	3.0	3.5	4.0	4.5	5.0	5.5	6.0
2.0									
2.5	0.0110								
3.0	0.0252	0.0143							
3.5	0.0425	0.0319	0.0178						
4.0	0.0630	0.0526	0.0388	0.0214					
4.5	0.0866	0.0764	0.0630	0.0460	0.0251				
5.0	0.1129	0.1030	0.0900	0.0735	0.0533	0.0289			
5.5	0.1419	0.1324	0.1198	0.1038	0.0842	0.0606	0.0327		
6.0	0.1733	0.1640	0.1519	0.1365	0.1176	0.0949	0.0680	0.0365	

Duration of improvement effort = 60 months

(Adapted from reference [2]).

and forecast is a wise and reasonable goal to set for achieving Six Sigma control.

The charts showing the effort required to achieve a company wide Six Sigma quality level from a four sigma base line is illustrated in Figure 11. The rate of monthly improvement is aggressive for the five-year plan allowing for any unknown variables, equipment changes, and modifications of processes to achieve Six Sigma.

The rate of 12% seems low until you factor in what changes are required and how soon you or your suppliers can design, implement, debug the processes and then control the system to meet the Six Sigma requirements of 3.4 DPMO.

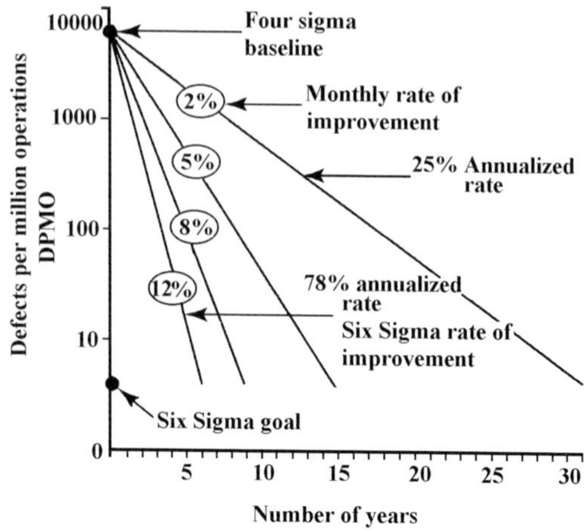

Figure 11. Years for improvement from four to Six Sigma. (Adapted from reference [2])

Also, remember if your Six Sigma goal stretches out farther than planned, as it may due to unforeseen delays, the company is still "performing within Six Sigma improvement operations".

As the company moves at the rate determined capable within the organization, the amount of "Risk" for them and their suppliers is decreased as improvements come on line. This will heighten expectations and drive improvement at an even faster improvement rate.

The addition of more Six Sigma improvement teams on the company organization can drive improvements using Ishikawa and FMEA charting to analyze the company operating systems even before a Six Sigma program is selected. The initial and accurate information gathered for selecting a Six Sigma program can save time, support faster rates of improvement, and focus the company resources of time, personnel, and money in the correct improvement direction.

This is also the time to focus on problems identified by your customer's quality personnel and internal quality audits used to analyze where the problems originated. Then using quality techniques, track the suspected problem variables to their source of origin, either supplier based or in-plant operations and fix them.

Any critique of a company's operating department requires an honest and current input and concerns of on-going problems areas yet unsolved, and with each concern a short method of solving the problem is supplied as positive input from the department. During this, and for any critique, no opposition or rebuttal from the department personnel is permitted. Only the problem should be stated and not who is causing the problem. The intent is to solve and prevent the problem from reoccurring in the future. Therefore, state only the problem and potential solutions.

The details can be worked out later using such techniques as the Ford developed 8-D problem solving method. The department manager and supervisors should not have to feel or be attacked during the problem solving analysis. Since the methods used were developed in the past and was how things were done. Now in analysis, they can always be improved. This input must be stated and reinforced at the beginning of the critique and maintained throughout the critique session by management. Personal feelings should not be allowed to influence the critique in any way!

The critique can focus on the amount of quality risk each department faces in performing their operations. The final focus will be how to reduce risk and where to focus the personnel support required for positive change. Management need to know the personnel, time and assets required to reach the Six Sigma level of operation and company quality performance.

The last point is realization that the Business Entitlement Matrix is not just a guide to meeting one customer's requirements, as it has been the traditional supplier focus. But, is a review of both customer and supplier performance focus areas. There must be a balance of all dimensions of quality, optimized over time to achieve value entitlement on a business wide scale.

How a customer perceives the suppliers efforts to meet their expectations is predicated on the entitlement standards as they apply to the supplier. When the business is based on entitlement and entitlement is achieved, both customer and supplier are satisfied. This then is the true goal of Six Sigma process and business control.

Most companies begin their Six Sigma program by establishing an absolute and strict program allowing no more than 3.4 DPMO. When achieved they often do not know if their programs actually added value to their customer's products. Management can be misled by using this as a base line goal Six Sigma management tool with the statistical (3.4 DPMO) as a goal for absolute quality using the existing quality tools for achievement. The company should know their internal cost and improvement and often fail to ask their customer if the improvements were ever noticed by the customer and if so how appreciated within their manufacturing system. The companies cost may not always justify the end results obtained.

Therefore, a management and company break through with Six Sigma must be directed from the CEO level including all business practices, not just a product, service, or quality perspective. Six Sigma's greatest results have come from creating economic worth for the customer base and corporation. The Six Sigma management system is focused on recognizing the perceived and real needs of the customer and provider. It is concerned with defining entitlement that is the level of expectation for each of the quality factors.

Six Sigma also analyzes the gaps between current performance, what it should be to meet customer requirements (entitlement) and the long-term goal (Six Sigma target). This analysis progresses on a regular basis at the operations, process, and systems level of the company. This analysis can then focus the means to close performance gaps and control the key factors contributing to these analytical estimated gaps in the companies business operations.

All gaps between performance and entitlement do not deal with achieving a product 3.4 DPMO. If current manufacturing processes meet the customer's requirements they do not now need work. The analysis may show a business area that when improved, could raise other departments' quality level and eliminate problems that could assist in raising engineering, design and even manufacturing to Six Sigma levels as shown in Figure 5.

Gaps in the analysis are the difference between what customers can really expect and what suppliers can really provide based on inherent capabilities. This is one reason why companies should draw on their in-house manufacturing, and tooling personnel and any outside product and material suppliers, before finalizing new products to ensure all are in agreement and the product is capable for manufacture by the supplier. This agreement should extend from pre-design to final agreement including sales, contracts, purchasing, etc. so all parties are in agreement on expectation and service/product to be provided within a scheduled period.

If gaps must be closed, moving directly to Six Sigma may not be required. To close the gap a company with a 3-sigma process capability may have to go to a 4.8 sigma level to achieve entitlement that is still short of the Six Sigma goal.

The 4.8 sigma level may be sufficient to capture the business and produce product within cost. After award of a contract the company can continue toward Six Sigma capability for long-term market leadership and growth. It is being proved by six sigma quality companies that they are the lowest price bidders on business with the highest quality output.

QUALITY FUNCTION DEPLOYMENT

Establishing Guidelines and Systems for Business Relations with QFD Evaluating both your capability, and customer expectations and requirements, using QFD (quality function deployment), should be the first step in a Six Sigma program. Six Sigma performance is always the long-term goal for a company quality system. It may not be possible or beneficial to operate at this level immediately since most companies operate at three or four sigma initially, and to reach Six Sigma requires extensive personnel time and effort. Providing customers with Six Sigma dimensioned parts when only a four sigma product is required, can be initially very time consuming and a possible misuse of company assets. But, if all operations of the company can be brought to Six Sigma operational status, a considerable savings can be realized in overall operations and manufacturing capability, less wasted assets as time and energy and products.

To develop the information for the Business Entitlement Matrix an understanding of the customers wants and needs is required. Coupled with this understanding the supplier must know their capabilities to meet these needs and wants of the customer.

Supplying a product or service, definition beyond the stated contract requirements needs to be analyzed and this initial information can be generated using a function or operation check list specific to your industry and customer base. Examples of these fourteen business, design, and manufacturing check lists are in the Appendix for reference. Knowing the customers business, product, and service requirements compared to their importance and your capability to meet them are necessary.

QFD (quality function deployment) is the tool for listening to the voice of the customer and your internal departments capability to meet the needs with minimum risk. Then deploy supplier quality into the design of the product, process, production, and service so that all functions of the organization are all customer driven.

Each customer will have their specific methods of conducting their business with their suppliers. How requests for, and replies to, a RFQ (request for quote) are handled, information gathered so all parties know what is required by the customer. The RFQ should be specific to what is required for all services desired. It must be specific for the quoting companies' business and manufacturing departments to document the services and/or products they will provide to the customer.

It is important to remember just because a company is ISO9000 certified, does not mean they can meet the business and manufacturing requirements of all their customers. Certification to ISO9000 only means the companies quality system meets the ISO9000 requirements as interpreted by them and their auditor. The suppliers manufacturing may not always meet the customers specifications in all areas, is always a suppliers consideration. All companies are in a state of flux, meaning, seeking ways to improve their business and profitability base, while retaining current customers and adding new ones. Therefore, their methods of communicating information from and to their customer, and in-house personnel may not always be sufficient. Accuracy to meet requirements is necessary to eliminate customer and production problems. Plan ahead of an operation for the personnel and procedure to use to communicate the needs of the customer and your in-house requirements.

ESTABLISHING QFD OPERATIONS

Marketing and sales begin the gathering of information using appropriate check lists and schedule meetings with department and managers of each company to ensure understanding and compliance to requirements is met in all phases of the business relationship.

Six Sigma methodology is used to assist in this undertaking. QFD has the steps in place to complete the tasks of gathering this information leading toward customer understanding and satisfaction. Concern by the customer for the supplier to meet their needs must be satisfied prior to the release of a purchase order. This takes understanding and asking the correct questions and supplying the same to the customer will generate the capability to answer the questions in the Business Entitlement Matrix.

Identifying the customer quality criteria involves being able to establish a relationship grid called the house of Quality Matrix, see Figure 12. A relationship grid is employed to ask and then record the import information and requirements the customer and provider will determine necessary to produce a lasting and prosperous business relationship. The matrix is used to quantify the importance of customer perceived product and service quality items. The matrix can be used by all manufacturing and service related companies by adjusting the relative importance factors. The importance factors are derived from discussions with the customer on this perceived critical business and product issues and items.

Assume analysis, for example, of the business section of the quality program matrix from sales, excluding production at this time, to shipping, including order entry, finance, billing, etc, documentation in the business correlation matrix.

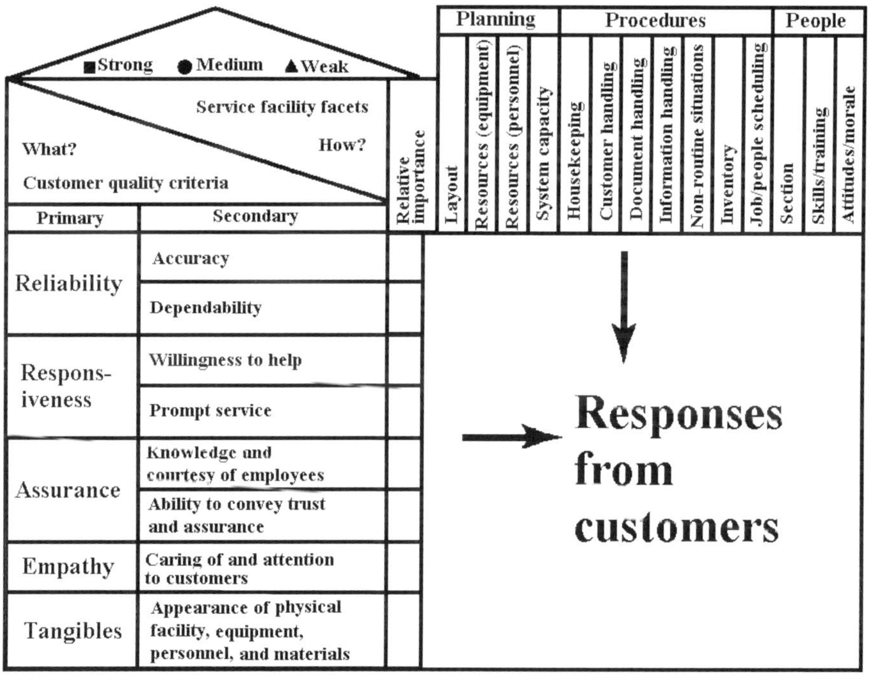

Figure 12. Quality function deployment relationship grid.
(Adapted from reference [4])

Knowing before production what the customer wants and expects, you can then analyze your system as to its ability to meet their needs. QFD converts the "Voice of the Customer" into product and service definitions to your personnel. From this base, design, production, and delivery decisions can be developed, analyzed, and brought into compliance to fulfill the needs of the customer. QFD discussions are used to gather the necessary data for implementing early in a contract/program to aid your manufacturing planning for product and process design, process control, testing, and training of personnel to ensure that all stages, business to manufacture, are meeting the customer and you the suppliers requirements. QFD analysis should also be conducted with current customers to find areas meeting and possibly needing improvement that could further enhance the business relationship between each other.

To often these discussions only occur when a problem has reoccurred to frequently and the supplier is in danger of loosing the business. QFD is also used as a concurrent and proactive tool for analyzing and meeting the customer requirements. Business and manufacturing operations do not have to exceed customer needs, just meet them based on the company ISO9000 procedures already in-place. Business operations such as, in this example, order entry, must meet all your customers needs, whereas, products may have varying levels of customer requirements based on the product specifications. For example the diameter of a medical syringe must be very exact versus a computer cover in some dimensions. But, the aesthetic requirements for each could be different requiring a higher degree of color match for the cover versus a clear virgin resin for the syringe. Requirements vary with the product and customer final specifications.

The QFD process uses a series of interrelated matrices to analyze customer needs in company process and response steps. The data in the matrix asks customer "What" concerns versus supplier "how" responses and inserts a 'grade' of how the "What" is met and from this, possible ways to improve the system must be developed or agreed to be provided as shown in Figure 12, the relationship grid.

STEPS IN THE QFD IMPLEMENTATION PROCESS

To analyze the importance of quality for "order entry" what the customer perceives begins with:

Step 1. Identify the Customer Quality Criteria

The type of questions used in the QFD analysis are based on the type of business involved and the product or service provided. Each industry will develop their own set of questions that are specific for their business relationship with their customers. They will also develop the information they need to repeatedly produce products or services to successfully meet the customer's requirements and specifications. In this example I used a service related analysis framework as shown in Figure 12. The customer quality criteria is then listed as the primary and secondary items. The primary analysis items specific for this analysis are accuracy, reliability, responsiveness, assurance, empathy, and any company entitlement tangibles they may receive. These groups were further broken into secondary items, as deemed necessary; for each of the primary items, Accuracy, Dependability, etc. for each primary requirement.

Step 2. Determine Service Operations

Determine the service operations as existed within the company under analysis provided the following areas of interest; planning, procedures, and personnel. Each of these three operations have their own specific set of related question areas such as; layout, resources, (equipment), etc.

In the initial QFD analysis it is always better to have too much information that can later be better defined as required or an asset to the customer for doing business. This information request data is placed on the right hand side of Figure 12.

Step 3. Establish a Numerical Score for Needs

Establish a numerical score range for each customer-quality criteria. Use 5 for most desired to 1 for the least. This numerical ranking will be used in all sections requiring evaluation. The scoring can be better defined if the scoring numbers are identified as to the criteria selected for their use in scoring. This is left up to the company to establish the level of scoring and the attributes relating to the scoring as shown in Table 5.

Table 5. Example of QFD Customer Scoring.

Score	Critical response need
5	Required to sell and hold as customer
4	Major benefit to keep customer
3	Good to have to maintain customer
2	Could offer if needed to get business
1	Not recognized as a benefit to sell

Step 4. Rank Customer

The customer's quality critical criteria were ranked from reliability (5) to tangibles (1). This ranking implies that the customer requires a higher importance, or lower Risk potential for reliability from their suppliers. How this is achieved by their supplier using what ever physical resources to improve their reliability, using computers with real time data, updated inventory control data, etc., tangibles, is not their concern and therefore has a lower ranking as they have no control over what assets the supplier uses to achieve higher order entry reliability.

Step 5. Document Incidents

Be sure all customer critical and non-critical incidents and any occurring problems are documented. Typically the quality department does this initially and follows up with the offending department to ensure the problem is corrected and prevented for all future business relationships. They also, reply to the customers quality deficiency report and copy all departments in the company to communicate the corrective action implemented in the company.

Hopefully these items have been documented but in most cases they have not for most companies. They just know there were problems but no one wrote them down as important and how they could be solved when identified as a quality order entry problem. Often these were management-type decisions or responses given to quality assurance for responding back to the customer. These were often buried in the operations systems or procedures not developed. But, if developed, not monitored for follow up, ensuring they were implemented and were effective corrective actions, not always preventative.

Critical incidents are complaints from their customer regarding an error in services or products provided to their purchase orders. The wrong product was shipped, late delivery occurred, the product was packaged wrong, or the pricing for the material and quantity was incorrect requiring them to pay a higher price for material, until it was found in auditing their accounting documentation. How the supplier responds to these problems is critical and how often they occur, plus what is being done to eliminate the problem. The customer does not want to continually check their supplier's invoices for accuracy. The critical operational items are established for all customer service quality assurance problems.

Order entry is performed in many varied ways. The more up to date systems use computer generated order entry information for the customer service representatives often split screens, for aiding the CSR (customer service representative) for entering required information for generation of the order, checking inventory, scheduling the plant, selecting delivery dates, packaging, and how product is marked, priced, and handled prior to and how it is shipped to arrive within the customers window of delivery.

The preferred method for order entry uses a form, or sequential order entry computer screens to input all the required information for the customers order. When an order entry item is omitted, it can cause confusion, a problem for the plant, and even result in a lost customer. Therefore, the CSR must be persistent in obtaining accurately all the data. Confirming order information can be simplified when the computer is used as it verifies data inputted for correct identification and if inventory is available for shipment when requested. Therefore, any problems perceived by the system can be addressed while on line and solutions developed for the customer to meet their needs in "Real Time".

Ensure when problems occur, they are documented, how solved, and frequency, so lasting preventative solutions are created and customer satisfaction is maintained.

Step 6. Competitive Bench Marking

This information is gathered from your customer on how their other suppliers are thought of and they comply with the customer's requirements.

Once the program is in place or by searching through back problem areas, a benchmark for the severity and quantity of problems per customer, order entry representative, and what could have been done to prevent each occurrence is established.

This data uses 1 to 5 ranking values and is recorded on the right of the 'Customer Quality Criteria', Figure 12, under the "Relative Importance" column. The scoring for this column for the Primary and Secondary, 'Customer Quality Criteria' is explained in Step 8. Important information can be learned at this stage to evaluate your progress in the customers business as a supplier. It opens up areas you can discuss if necessary to reinforce your good business policy and how to improve it.

Step 7. Documenting the Relationship Matrix

Ranking the Customer Quality Criteria versus the company facility assets now begins. Use the three ranking symbols strong, medium, or weak on the graph in only the matrix areas where they apply. Other areas are left blank. If unsure rank the area but draw a circle around it until satisfied it is of value in the evaluation. Good exchange of information is developed during this phase of the analysis.

Step 8. Ranking the Total Weighted Score

The ranking of the company analyzed used the weighted score for each service asset analyzed and discussed for the process. This involves ranking each item of importance with a quantified value of intended importance for each item discussed in the evaluation matrix. Their importance is derived from ranking each quality data facet as follows.

S = equals the importance ranking score from the following:

S = (relative importance) *

The score strength is based on the following items:

* [strength of relationship, critical incidents, competitive benchmarking]

In this example, accuracy was the customer's critical incidents service criterion. It is significant if only one of the customer-quality criteria is judged as major importance during the evaluation. When this occurs, the cell values in the other cell, for other criteria are set at zero. This identifies accuracy as being the most important customer-quality criterion.

As a result, the quality aspects for the customer can be simplified as requiring only those service factors that have a direct relationship with accuracy for customer service order entry. The house of quality contains three relationships in this example: customer service personnel, methods used for order entry, and information handling. Therefore, all three relationships were evaluated as being very important and all other aspects are equal and have the same rank of importance.

The body of the matrix is ranked using the three rating markers, strong, medium, and weak to quantify the respective quality qualifiers. Rank only those areas that pertain specifically to the customer's criteria and the area being evaluated.

For example; Accuracy: resources (personnel) document and information handling were specific to the customer in evaluating your order entry system/personnel. If a problem is noted that improves equipment, not a direct customer concern, resources (equipment), but known to affect order entry accuracy, it should be noted as an internal problem and addressed to increase the personnel order entry accuracy score with the customer. The other items in the matrix are then completed following the same guidelines.

After completing the main matrix, rank by averaging the scores for your company responses from the customer. Whenever possible, request your customer to rank your competitors in a like matrix. This is more positive response than using just their numerical response, as they must think out how they are treated by the other supplier. This questionnaire gets them to consider the problems they have with their other suppliers, your competitors, and each of you are being ranked using the same questions in about the same time period. If completed accurately you should find out any areas you excel in or need improvement to raise your ranking with your customer. As you satisfy and meet the customer's requirements you will be a better supplier to them and as a result increase your business share.

Step 9. Completing the Roof of the House of Quality

The correlation matrix Figure 13, or often identified as the roof of the QFD grid, is now completed assessing the relationship between planning,

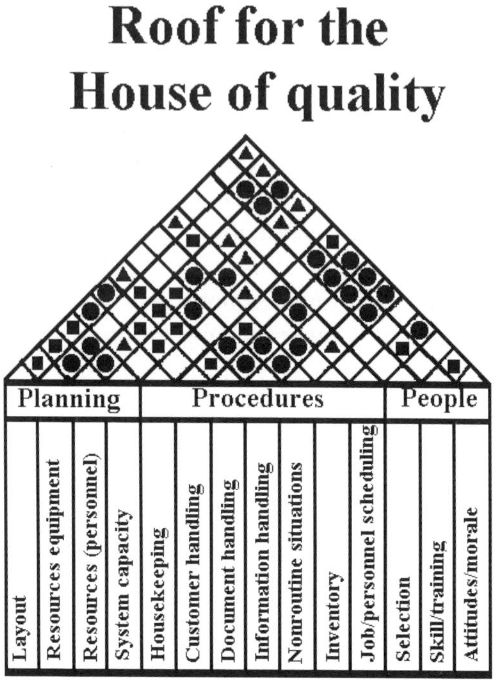

Figure 13. Correlation matrix.
(Adapted from reference [4])

procedures, and personnel. In this analysis a strong relationship exists between resources, personnel, and order entry. A review of the completed house of quality reveals these reinforced with information handling to be the key customer perceptions of order entry service quality. When redesigning and improving the current order entry system emphasis should be on these areas. But, do not ignore the other areas, as these should be evaluated for any impact of the key customer and normal order entry systems. If this is your first evaluation of your operation, note improvement areas and include them in your Six Sigma business plan for improvement in the near future.

QFD ANALYSIS CONCLUSIONS

The completed customer accuracy analysis, or the house of quality, is shown in Figure 14, in completion of the analysis, indicates a recommendation to incorporate a better order entry system. Orders for quick shipment are not getting the attention required to meet the customer needs. The customer could be persuaded to forecast better if they operate on a JIT (just-in-time) mode that puts greater emphasis on their suppliers to meet their needs. The customer must address actual forecasting product requirements.

The suppliers company may also devise a method, blanket customer purchase order with firm scheduled delivery dates, to ensure sufficient stock is always on hand. A critical order can be highlighted for their suppliers to meet their needs. This will ensure that management attention can be brought to the situation to ensure the company is capable of on time supply to meet

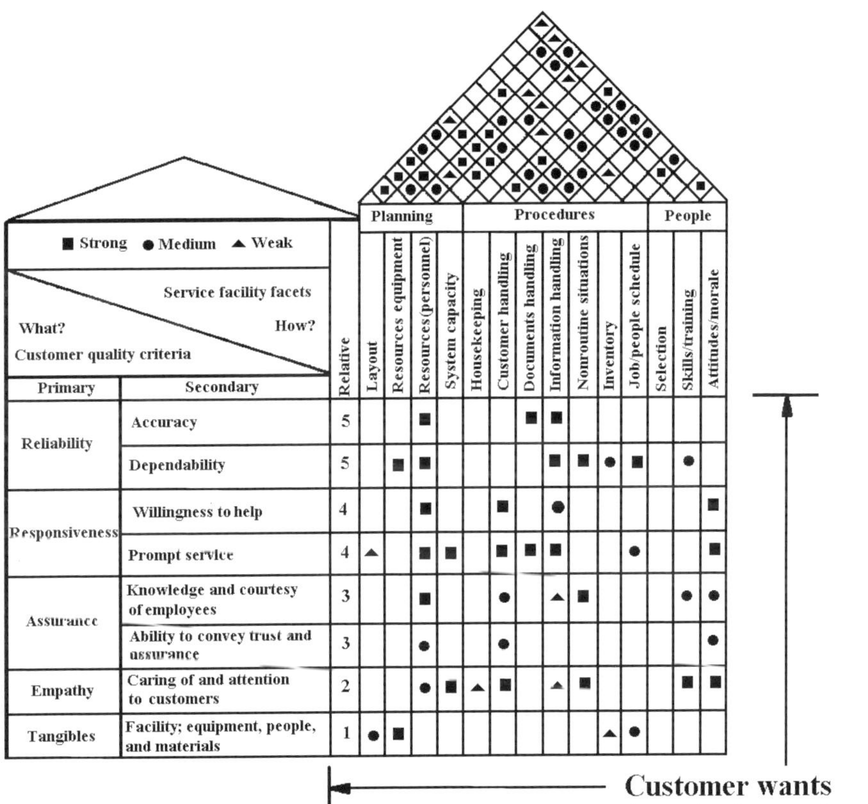

Figure 14. Customer house of quality matrix.
(Adapted from reference [4])

the customer's demands. If unable to perform, then an agreement must be developed with the customer so they are aware of the situation and can develop an alternative solution to the problem with their supplier. Always keep the door open for discussions as the customer may have a way to assist you to make a delivery, if both companies discuss the problem before the situation becomes critical.

In these situations a method can be developed to ensure inventory is available, possibly financed by the customer to be held as ready inventory by the supplier. Then as inventory is drawn down to a specified level, new product is manufactured to replenish inventory levels. These actions depending on lead-time and cost factors by flagging these special requirements. These methods depend on improving communication systems and procedures to meet the requirement within the content of each company wanting to meet their customer's requests.

Any improvements depend on the type of order entry system used, if minor programming can eliminate delays and extra CSR time to get information and speed up these orders to manufacture so schedules can be adjusted. Often a short-term solution may have to be developed to meet the customer's special request, described as reserve inventory, JIC (just in case).

Another QFD key concern was personnel. If the CSR's have specific accounts and they are absent, a backup CSR must be assigned who is trained in knowing and understanding any special customer requirements for the specific customer. Also, special training or specific instructions should be documented in the order entry system to be readily available for the fill in CSR to accurately assist the customer.

The layout of the customer service area, arrangement of files, so all systems are the same including order entry procedures, will assist the CSR's and plant personnel in ensuring information for customers is available, easily located, and accessible to those needing the information to complete an order.

Also, their work area, conducive to a quiet work area free of outside noise and inter cubical sounds, so as to accurately communicate with the customer and other CSR's when necessary. The work environment, free of outside interference is critical for accuracy and speedy order entry.

QFD is focused on "listening" to your customer for implementing quality into your operations for business, product, process, and production. Service to answer questions is equally important to keep the customer apprised of their order status. Also, solving problems and answering the customers requests for assistance.

Equally important is ensuring your internal operations, internal departments, (purchasing, scheduling, manufacturing, etc. and other operation departments) have accurate and timely information to perform their assigned tasks. Therefore, listening to your external and internal customers for implementing "Real Time" quality into your operations.

SELECTING SIX SIGMA BLACK BELT AND TEAM CANDIDATES

Selecting a Six Sigma (black belt) candidate to manage your quality improvement program is a demanding personnel decision. Who the candidate will be and how you begin the program of continued improvement involves many management decisions. But, first where did the term "Black Belt" originate? The name Black Belt envisions a karate, "martial arts" type of person. Mikel J. Harry, president of the Six Sigma Academy, Scottsdale, Arizona, coined this term. He envisioned a Six Sigma leader, Black Belt, who is humble, yet possessing a sense of humor with a very precise command of simple and effective quality tools. They are tasked with accepting major challenges, making important decisions daily, and developing creative solutions through rigorous application of qualitative quality methods.

This definition has stuck and created other Green, etc. color belt skilled quality personnel who develop their knowledge and skills while learning and working up to Master Black Belt status.

There are numerous training companies, like Harry's who can train candidates for a fee using a defined course and objectives curriculum. Some major companies, General Electric, will train their supplier's black belt candidates at a reduced fee. This is done to ensure General Electric will be able to reap the rewards of their efforts when they return to their parent companies. This means, they are a supplier of products to a General Electric company and require high quality within Six Sigma requirements that can lead to a lower product price after implementing cost savings at their supplier plants. A just reward for the services provided that the supplier can, in time take across the company line, to further reduce their costs and increase customer loyalty and profitability with their other current and future customers.

These programs have resulted in a lower product cost, improved quality and delivery plus product output resulting in reduced cost to General Electric for products delivered to their plants. A win-win situation for each company is created with the supplier for increased profitability, lower total production and scrap costs, and an increased business base.

SIX SIGMA BLACK BELT SELECTION

The selection of your Black Belt team leader is vital to the success of your Six Sigma programs. The leader must be self-starting, creative, and diligent in seeking out the critical variables for any problem or process undertaken. They must be well trained and knowledgeable in what quality method to use or combination of, to obtain the data for success. Being able to inspire, lead, motivate, and train other members of the team in improving a process is vital. Couple this with being very good at developing questions and listening to their team's answers. The Black Belt must also know where to go for answers in the company and also outside the company where to seek assistance and support for the success of the Six Sigma program. The ability to later analyze the data collected, sort out unnecessary information, and then arrive, with the teams support, a method to improve or solve a very complex problem or improve a manufacturing process.

The environment the black belt was trained in, plus where they worked, and with whom, will prejudice how they lead their own team. Large companies will be able to provide a larger base of trained personnel in varied disciplines often not available in smaller companies. This does not imply smaller companies have less capable quality personnel but not always the reserves to dedicate a full time, person or team, to a Six Sigma program, as necessary.

Also, methods of operation or procedures of the company must be known and the team instructions very explicit as to what instructions personnel use to perform their duties. This includes procedures personnel are told to follow when certain conditions occur during the performance of their duties. It is important the black belt use, initially, only the current companies methods of operation and not try to implement new methods they have used at their last company. This can cause confusion and lead to mistakes if personnel are not familiar or instructed in the new procedures and instructions of the black belt. Once, the team and black belt are working well together, and training is provided, this method can be employed. It is important to have the team operation together on initially familiar methods so cohesion of the team is possible.

Ensure the black belt and the team, composed of members from both management, supervision, and operators, know how each department reacts to conditions and situations when a problem requires a decision to be made by the team. The black belt must never micro-manage the team. They must ensure all ideas, good or poor, be solicited, evaluated, and analyzed for their input on the problem or process. Never hinder free thinking, always ask "what if" this were tried for an effect on the problem until a workable solution is developed.

The candidates selected within the company for Six Sigma black belt training are taught the basic tools of achieving quality through quantitative methods. These methods employ the tried and true quality techniques used by outstanding quality departments for years. Six Sigma is based on process measurements to establish a base line and all improvements are gauged from this initial measurement. If you cannot measure it, you cannot control or improve the process as a requirement for quality improvement in any company.

The real power of Six Sigma is achieved by searching for new ways of thinking and doing operations and then applying the existing quality tools with a focus on delivering bottom line cost, and process improvements. The lead black belt candidate selected needs to posses the following characteristics as listed in Table 6.

These are a few of the basic personal attributes a candidate should possess. The person does not always have to have an analytical, engineering, or quality background. Some of the more successful candidates have been floor technicians who understand the manufacturing process and have the network to investigate many possible factors to improve a process or operation. They can then, from their team, draw on the technical expertise of others to collect the data to prove the team's theory of the solution being developed and investigated.

The example of a typical job profile for the Six Sigma Black Belt is shown in Figure 15.

An important asset a black belt must possess is in the selection of team members. Compatibility among personalities is important so the team can function amiably without dictatorial status or position hindrances. Members of the team leave their titles behind when working on a Six Sigma program.

Table 6. Six Sigma Black Belt Desired Attributes.

1. Proactive and self-starting
2. Team builder with project management skills
3. Centered on completing a program
4. Open to suggestions
5. A leader
6. A networker
7. Good problem solving skills
8. Trainable
9. Good person ability
10. Able to seek assistance when required
11. Methodical
12. Structured
13. A decision maker
14. Investigative nature
15. Process control driven
16. Goal achiever
17. Not a micro-manager

An open and unregulated exchange of information must exist without team members feeling overpowered by a very strong personality or higher-ranking member in the company structure. All are equal team participants in solving a problem and their input is valuable to all members of the team.

The black belt is the unquestioned leader seeking support from team members. Members can seek outside assistance as long as the black belt approves and within the normal operation structure of the team and company.

The black belt leader after selection is sent to an accredited black belt training facility, sponsored by the company, or a major supplier, such as General Electric, if the supplier makes their products. General Electric and other major corporations have their own black belt training program and has allowed and welcomed their suppliers to send candidates to their training school with the intent of realizing cost and quality savings when their programs are worked on at the supplier company.

Typical black belt training involves a four-month training and knowledge application and learning period. Initially a one-week intense training session is taught with a three-week break in between for the black belt in training

These are some of the desired attributes for the following position

Team builder and leader
Proactive, self-starting and decisive in making decisions
Open to suggestions from others
Mentor to others
Respected in the company
Knowledgeable in quality
A Preventor of problems
Good common sense
Trainable
Good communicator and listener
Good problem solving skills
Able to ask for assistance when necessary
Process control oriented
Investigative and structured in operations
Goal Achiever
Likes people and wants others to excel
Gains respect early from peers

The job performance for a Six Sigma Black Belt:

The Six Sigma Black Belt (BB) will be the leader of the companies quality improvement team program. The BB will serve full time in this capacity reporting to the CEO/President of the company and collaterally to the "Company Champion". The BB will select a team of personnel who will assist the BB in solving and preventing manufacturing and quality programs and problems as directed by the CEO/Champion. The BB will teach team members in the Six Sigma methods and work with them on programs providing direction and guidance in all areas of the programs. The responsibility of all programs is the BB and they alone will make decisions on how and where to proceed in all programs for improvement and prevention of problems. The BB will ensure the team remains focused and on schedule for the successful completion of all programs.

Anticipated Six Sigma Program Results:

The BB is charged with obtaining the best results from all programs and ensures the team members have the assets and support for them to successfully complete their program objectives in a satisfactory amount of time and cost to the company.

Figure 15. Six Sigma Black Belt job and performance profile.

to return to their company and begin to train their team and apply the knowledge they have acquired on specific programs. The black belt then returns for three additional, one week, intense training sessions to learn more Six Sigma knowledge and then return to share their new knowledge with their team members working on a Six Sigma program at their company.

The training program focuses monthly on one of four phases of the Six Sigma problem solving methodology termed: MAIC (measure, analyze, improve, and control). Other Six Sigma trainers have expanded the MAIC training curriculum to include; recognize, define, measure, analyze, improve, control, standardize, and integrate into their training programs. These two beginning and ending identifying areas of training can assist the team in selecting and then controlling the process while investigating how it impacts on both up and down stream operations.

Many long lasting problems are not created at the point of rejection and must therefore be taken back in the process to identify (recognize) where the effects actually occurred to define the solution of a problem. The Ishikawa, fishbone, technique is an excellent tool for creating a visual map of the process with all the interactive variables, where they occur in the process, are shown visually and identified at the point of interaction on the process. An example of the use of a fishbone diagram is shown in Figure 16, for a manufacturing operation analysis for injection molding.

During the time between black belt's training sessions and when the black belt returns to their company, the trainee team members work on the selected Six Sigma program. They setup procedures to begin gathering data to solve an actual business or process problem for improvement at their company. The results expected from the team, besides demonstrating knowledge of information taught and then applied, is a monetary bottom line savings that is expected to yield positive quality results from the actions and performance of the black belt's Six Sigma team.

Assisting the new black belt may be an in training Green Belt team member. The green belt trainee is selected by the Black Belt as their alternate while they are in training to lead the team. They may assist and train more than one team when several Six Sigma teams are implemented at the same or later time period at the company.

The Green belts are members of the Six Sigma quality improvement team who have been trained by the black belt in an abbreviated 40-hour training program. They work with the black belt to increase their knowledge for

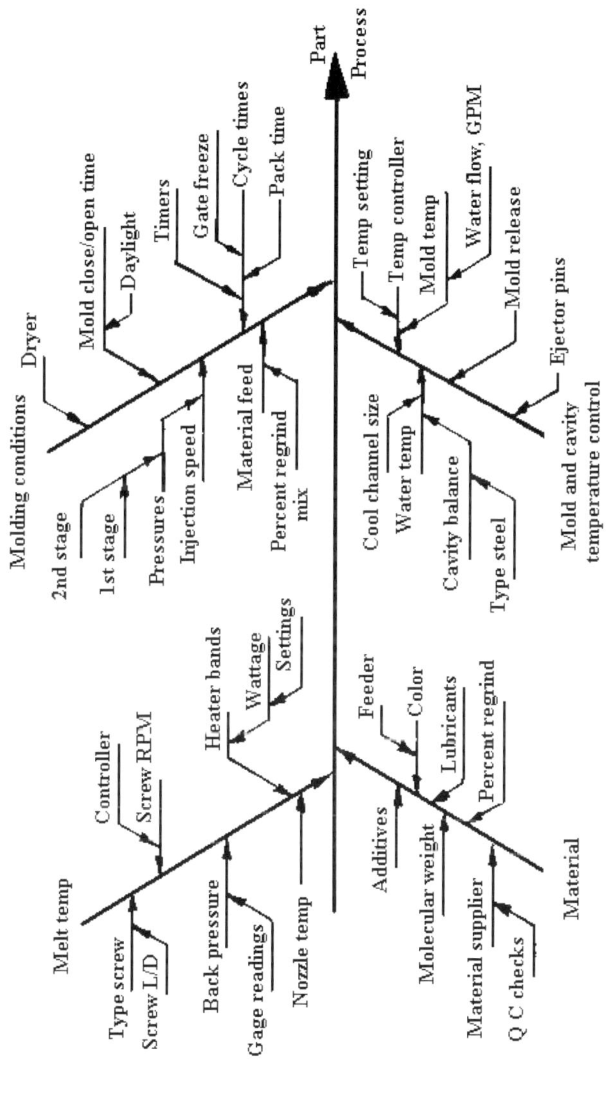

Figure 16. Ishikawa "fish bone" diagram.

becoming a black belt. The Six Sigma team members gain on-the-job experience and in-house black belts training sessions taught by the black belt or are sent separately to black belt training sessions to become black belt certified trained personnel. General Electric has adopted a strategy to train all of their personnel in the Six Sigma methods at their plants and other business operations to ensure positive results are always attainable in all their operations.

All Six Sigma team members must be encouraged to continue their educational and process improvement skills. These can include working for a degree in a business or engineering discipline. Attending specific topic seminars in program management, process, and quality control, business operations and problem solving seminars. These will assist all team members to grow in their profession and become a greater asset to the company and team. Selected team members should also attend, creditable, Six Sigma training.

Be aware some major companies: Motorola, General Electric, Honeywell, Boeing, and others offer training, at reduced rates, for suppliers who provide their products and services. These companies usually only require that their programs be the first Six Sigma programs developed for improvement and meeting cost of quality cost reductions.

Each team member should be interviewed for their current areas of knowledge and expertise plus what other educational needs they may acquire to be a greater contributor to the company and their team. This information is initially developed by the Self-survey forms the team members filled out. The Black Belt needs to discuss in depth the actual knowledge the personnel say they have and actually capable of producing. You do not want some one to begin gathering data and find out later it was not done correctly and the data is useless.

To achieve the greatest benefit, schedule training to always precede the next process to be implemented in the Six Sigma program. This ensures the knowledge will be fresh in the member's memory. This will assist the black belt in training other members of the team, often with new techniques and refreshing old knowledge skills that need updating to assist in the program.

Communications within the team must be in "Real Time". Usually introduced during the daily quality Six Sigma team meeting. All program information must be communicated to the black belt team members as soon as possible at it may affect their areas of the program. All information developed by the team must be documented in the program files with all members copied. A recorder is designated in the group and working with the black belt to summarize on a pre-determined schedule the data and information for the team and upper management. This summary can be short with bullet action and accomplishments with specific attached references for data collected, when more in-depth information is needed. Keep your points of information concise, direct, clean of extraneous information, and updated with new information as it becomes available. Archive your reports for later reference and for use in summarizing the final Six Sigma report for the program.

Green Belts are Six Sigma trainees in the process of working toward their black belt certification. The black belt can train team members to obtain their green belt in the process of becoming completely competent in process improvements.

BLACK BELTS EMBED SIX SIGMA METHODS INTO COMPANY CULTURE

Establishing the Six Sigma culture into a company begins with management. TQM (total quality management) received poor acceptance in some companies even after programs were developed and were successful. Management did not always understand what it could accomplish and how to successfully implement TQM objectives and goals. Other companies began TQM programs by sending management and employees for training, implemented sound programs that reaped the benefits from the TQM program. Then along came the more structured, consistent, and realistic ISO9000 followed by QS-9000 for the automotive industry developed by the major automotive companies integrating their existing quality methodology with ISO9000 requirements.

Other industry related quality programs including the prestigious Malcolm Baldridge National Quality Award, won by many large and small companies committed to continually achieve quality excellence.

Now the program of Six Sigma Process Improvement is the industry goal for improving processes and quality in business and manufacturing. Of interest, is that in 1924, Walter Shewart developed his Shewart, "Percent Defective Chart." He followed up in 1926 publishing the paper, "Quality Control Charts" in the *Bell Systems Technical Journal*. From this work the

"Control" word became synonymous with the lexicon of all quality engineers.

Shewarts ideas were presented in his book "Economic Control of Manufactured Products". He later teamed with W. Edward Deming on reinforcing his use of control charts in "Statistical Method from the Viewpoint of Quality Control". Since then, and up to Six Sigma, the control chart, has been modified to suit the data collection and presentation used to identify the repeatability of any process.

Control charts are the key method of recording and presenting data used to chart for control almost every process used, even with planned forethought of sound reasoning the results of sales.

The black belts job will never be completely concluded. New process improvements can and will always be developed. Keeping plant personnel thinking of new ways to improve an operation or process is a full time job. This idea must be continually reinforced at the company management level and with your suppliers who may not have Six Sigma programs. Remember, having a supplier ISO9000 certified does not, unfortunately, guarantee continued improvement with their products and services. The new ISO9000-2000 addresses this deficiency and within the next few years as companies have to recertify, will adopt and be re-certified to the revised ISO9000-2000 revisions.

With the new revision for ISO9000-2000 continued improvements, this will become a requirement within the standard. But, to what degree company's implement to this requirement remains yet to be realized. If your suppliers do not have Six Sigma programs, and where time allows, or need require it, offer the services of your black belt to assist them in the process and product improvements. Other major companies offer this service and it can be an excellent method for your company to reduce cost, improve supplier's product quality, and improve business relationships and better understand what is required of them to continue to be your supplier. They can also cover the cost of your black belt and improve their quality for a win-win program for each company.

The key to black belt and company culture reinforcement for Six Sigma is always striving, never letting up on better ways to improve the process, for business and manufacturing of the end use product or service.

The projected dollar value of the Six Sigma projects is as established in major corporations, in the $100,000.00 to $170,000.00 or higher cost savings. This is in line with what executive management would desire and expect.

SELECTION OF A CHAMPION FOR SIX SIGMA TEAM SUCCESS

For a Six Sigma black belt program to be successful requires a company management or executive person termed their "Champion". The champion is an executive of their company who has been sent to Six Sigma Champion training and will ensure their company's black belt(s) have no barriers to block their progress when solving a problem and/or improving a process. These vice presidents, presidents, CEO's or directors can ensure the black belt has the capital and support needed and any issues causing them a problem to be speedily removed. Black belts do not have the time or position to remove these obstacles or lobby for financial support to complete their Six Sigma programs.

Champions attend a short version of Six Sigma training, a few days, to obtain an overview of the Six Sigma program and problem solving methods. They learn what can be accomplished, time, and method requirements, how programs can and should be selected and how to guide, mentor and assist their black belts. They are typically at the start used to assist in selecting the programs for Six Sigma improvement, as they know the potential cost savings in selected areas of the company.

Champions are usually trained before the black belt so they can access company areas requiring problem solving and improvement techniques. They can then develop this list for black belt projects. Champions need to initially select the programs so there is ownership at the executive level. This will ensure the assets and support will be available for the program.

Should a black belt have a problem, they have 24 hours to resolve it before the "Champion" is notified. Then the champion removes the barriers, negotiates, and ensures the program moves ahead. Black belts are not permitted to become stuck with a company/personnel problem. The black belt is also full time on Six Sigma programs, and in major companies, so are all of their Six Sigma team members.

Project selection by the champion must focus on problem areas that can return a significant monetary reward for the company, originally targeted at $175,000.00. Major companies will not have a hard time identifying these projects. They, initially, must avoid the daily problems solved in the normal course of action by the production and quality personnel. Black belt projects should be selected as the ones you have tried to fix more than three or four times and failed and it is a major reoccurring problem with a high potential monetary savings and output improvement.

Coupled into the project selection process is a hard financial reward with management wanting the problem finally fixed for good. A methodology of selection was used for this selection process. Also, some companies tie-in management bonus for successful completion of long-term problem and processing areas.

Personnel Considerations in Six Sigma Operations

Hiring and retaining good personnel is very important for achieving and maintaining a high quality Six Sigma operation. Management must strive for employee buy-in for efficient, repeatable, and continued product and service operations.

Your champion and black belt are key to a successful implementation program. The personnel (who participate in the process) and later operate the improved system must be trained not only in the reasons for the change and how to maintain the control once established, but the why and benefits to them and the company.

Success has been produced by informing personnel of the improvement program and requesting input from personnel involved in the operation. Also, ask personnel who provide input both before and after the operation on how they perceive them and the operators will be affected by the program improvement, both from a positive and negative viewpoint.

Solving a problem often involves going through many layers of personnel and manufacturing methods and procedures to uncover the root cause effect of the problem or process. Also, by talking with informed personnel additional information can be learned to aid in a permanent solution and operation improvement.

Selection of team members, preferably Green Belts in training, who need not necessarily be all quality assurance personnel, are assembled and directed at solving a problem or improving a process.

A realistic job review is necessary for personnel to ensure the competent employee is challenged, informed, and rewarded for their success in a program of Six Sigma. When a solution is developed they must take ownership of the operation to keep it in control. Personnel must be retrained if necessary, procedures and work instructions updated, equipment modified where necessary, and evaluated by the team to ensure they know the who, what, and why of the improvement and how to accurately monitor a process and keep it in control.

It is important the Six Sigma team utilize the personnel outside of their group if necessary. Do not for lack of a team skill let the program languish. The champion can assist in these areas for supporting the team in developing solutions.

PERSONNEL REQUIREMENTS

Irrespective of where, who, or how the company Black Belt was trained the selection of team members must fit the Six Sigma problem or process improvement undertaken. Personalities of team members must be considered. A strong self-important personality is to be avoided. A team member must be a team player. Strong convictions can be a personality trait that can help or hinder a program. This is one area the Black Belt can assist to tone down dominant persons and to bring out the ideas of the less confident personality. One of the better black belts was a technician trained in the Six Sigma process who had an excellent working relationship with all employees of the company. He could easily interact with all levels of management due to his knowledge. Also, with his co-workers where he had gained their trust with the expertise and knowledge shown in his job performance.

What he was lacking in engineering and quality knowledge was made up by his ability to interact with all employees gaining there trust and ability to question them and get them to offer information critical to process improvements and problem solutions. He was liked, trusted, and non-threatening in his approach to personnel and knew how to interact, learn, and then translate, working in the team environment, his and their knowledge and information into workable solutions.

A good black belt will strive and want all team members to know as much as they can about a problem or process. This is accomplished by ensuring all information on how the Six Sigma methods work are shared by all team members. Only when all team members know as much as possible about a problem, and how to use Six Sigma quality methods can a program be successful.

SELECTING TEAM MEMBERS

Each member of the team brings their specific skills and personalities to the program. The team's selection should have the personnel with the knowledge to analyze the process and know how it can impact the organization or process. This must be considered both before and after the place in the operation that is being improved.

When improving a manufacturing or production process the Six Sigma team can consist of the following team members as listed in Table 7.

All may be potentially designated as team members but not have their time-dedicated full time to the program. The Black Belt determines how many members are required to be on the team. Personnel not selected as full time Six Sigma team members can be held in reserve for latter support. These personnel are used as support team members, to be called on when their knowledge and skills are required for problem or system analysis.

The Six Sigma team usually consists of four to five members, or enough team personnel to perform the required tasks without being so over burdened that they are unable to perform the assigned tasks on schedule. The black belt will guide their actions to ensure the program will be successful and completed as close to the time line as possible.

When the team is trained in problem solving methods, all are trained together. Be sure to use handouts on training material as frequent reminder of how it should be done are needed for team members to review.

Table 7. Team Member Selection.

1. process engineer (titles are left behind after team selection)
2. production supervisor
3. plant maintenance supervisor or technical
4. purchasing agent
5. product design engineer
6. tooling designer
7. sales representative
8. financial department representative
9. quality assurance
10. set up technician

DEFINE SIX SIGMA TEAM ROLES

Responsibilities and competence requirements for the Six Sigma team members is necessary to be discussed and decided before the team begins their program. Using the same personnel analysis for a process improvement program the core team could consist of personnel from the department or process under evaluation and selected others as:

process engineer
purchasing
setup technician
quality representative
plant maintenance
production/manufacturing engineer

Each team member has a vested interest in process improvement. Each are initially selected for their process, product, procedural, and problem solving abilities as known by their peers and supervisors.

These are also personnel who will be trained initially as Green Belts by the Black Belt during the initial Six Sigma process or problem solving program selected by the Black Belt, Champion and executives of the company.

The Six Sigma Program Can Start in a Number of Ways

1. A Black Belt is selected for team leader within the company and sent away for training in the Six Sigma basics. The company champion is also selected and sent to train and on their return from the first training session of the Black Belt, the program and Six Sigma team members are selected.
2. A trained Black Belt is hired from the outside and they begin the Six Sigma program process. After the executives of the company select their Champion and send him to Champion training the Black Belt selects his team members and begins the initial Six Sigma Green Belt team training. On the Champions return the Black Belt assists in program selection for Six Sigma improvement.
3. Six Sigma program has been initiated and some team members are certified Green Belts. The Champion selects a new program, Black Belt

selects a Green Belt to lead the program and the Green Belt selects the team members if one is not already setup.

The following forms are presented to assist the Six Sigma team leaders and members perform more successfully in the Six Sigma program. Each program will initially begin the same way but not always with the same leader and team members. Each problem will be different and process improvement a special case as the variables in each will vary even for the same type of operation.

Therefore, it is the charter of each quality team to be successful in their work. The challenges for quality improvement teams, especially Six Sigma teams, are very challenging. These teams are undertaking programs that have been termed difficult to unsolvable by others in their company. As a start the Six Sigma teams need a direction and clear understanding of what exactly they are to accomplish. It is extremely important that the team documents their progress to team members and others. The team must also ensure any changes they make are institutionalized, resulting in the process never reverting back to the old operation procedures over time, particularly when there is turnover of employees who held the knowledge of the process improvement efforts.

Experience has shown that programs not properly documented, procedures written, personnel trained, and the process audited on a rigid schedule, the process reverts back to its original condition.

The following forms can be used to assist in organizing, communicating, scheduling, gathering of data, reporting and finally ensuring the problem is really prevented from occurring or the process in "real time" repeatable process control. These forms can be storyboarded for the personnel of the company to see regarding progress of the Six Sigma teams program.

QUALITY TEAM CHARTER

The Six Sigma Quality Team Charter shown in Figures 17 and 18. These forms, Figure 17, first is presented as a formal listing of the teams name and who is the team Black Belt leader and the Champion for the team along with team members and their affiliation with the team.

The second form, Figure 18, of the charter lists the Six Sigma program in very simplistic terms. The form lists the process selected for improvement

Six Sigma Quality Team Charter

Team Name: ____________________ Date: ________

Chartered by: ____________________

Black Belt Leader: ____________________ Phone/e-mail: ____________________

Team Champion: ____________________ Phone/e-mail: ____________________

Team Members	Department	Phone/e-mail
1.		
2.		
3.		
4.		
5.		
6.		
7.		
8.		

Figure 17. Six Sigma team charter.

so list a short description of the process selected. Describe the products and or services related to the selected process and the customers who use these

Six Sigma Quality Team Charter (continued)

Six Sigma Process Selected for Improvement

Six Sigma process Improvement Goal(s)

Six Sigma Resources

Six Sigma Reporting Requirements

Six Sigma Suggested Program Timeline

Figure 18. Quality team charter.

products and/or services. What are the goals of the team, a percentage improvement, prevention of what problem? This will give more credence to the program and tie in others in the plant who possible had to deal with the problem or lack of parts if this is a productivity issue. What resources in team members, their time, funding for training and materials, assistance for other organizations, etc.

Coupled with this is the issue of reporting is very important to the team and the company executives. To whom do they report, how frequently, and in what format. How decisions for major change in the company/ departments will be made, team, executive, committee, and how it may impact on the department and personnel. A time line with action and decision points anticipated is desired even though it will change as the program progresses. Everyone wants and needs to know when the program is anticipated to be completed. Nothing is ever left open ended. Add a note that the information will be updated as the program progresses and new information is available.

The next Six Sigma team form lists the Six Sigma Team Composition, Figure 19. It is necessary to have these forms used so the company management and employees know who are involved in the Six Sigma programs. This is beneficial in that if employees have additional information to convey to the team they have a list of the team members they can contact. These forms can be storyboarded and retained within the team records for review as needed when additional assistance or information is needed by the team in the solution or improvement of the Six Sigma program. The team composition lists the team members and any comments on additional capability or training they may have had in their career that would be of assistance to the Black Belt and team.

The team member self assessment survey lists the training and knowledge the individual has had and their level of expertise. Each team member will fill out one of these forms and it will be retained and used by the Black Belt in selecting team assignments. It is also a good idea to have any person who may be considered for the Six Sigma team fill out this form for future consideration. These forms are in Appendix C.

The tally sheet for the team is then completed with each team members number for each item listed on the tally sheet. Once all members have had their numbers added to the tally sheet each section is totaled. The total number is circled in each section and this gives the Black Belt a good

Six Sigma Team Composition

Process: ______________________ **Date:** ___/___/___

Team member's name	Org./Unit/Dept.	Phone/Fax/e-mail	Comments

Figure 19. Six Sigma team members.

indication of the knowledge and expertise level of his team members for each of the listed skill areas. It also gives him a feel for the training necessary for the team to meet it's goal for completing the project or program. Feel free to add more skill areas as required. Any quality knowledge is helpful and if known should be listed for the Black Belt and team to know and draw on if required.

The competency of each team member is reviewed by the Black Belt to ensure their knowledge and prior training is as reported on their self assessment form and they are capable of performing the tasks assigned by the team. Then if a problem gathering the information occurs or a situation not anticipated, this is reported to the black belt immediately so additional training, personnel assistance, and knowledgeable program support can be obtained.

The Team Meeting and Action Plan, Figure 20, is the core form used to report on the Six Sigma programs progress to management. It is an abbreviated form listing only the critical information necessary to communicate the progress of the team. It is essential that all meetings have an agenda, time for discussion of open items and time for new items as they should occur.

All team members should always be present unless a very valid reason is presented and approved by the Black Belt. It is also the responsibility of team members to add new items to the agenda for the next meeting by informing the recorder of their intent to discuss an item during the meeting. It is the recorders responsibility before each meeting to send out this form with the current meeting items to be discussed. This will allow team members to analyze and prepare for the discussion should they have information to offer to the team.

It is important to note that each team member is responsible for their own reports and not the team recorder. The recorder is only responsible for the team meeting notes and distribution of meeting information. The Black Belt can arrange with the company personnel if typing and computing assistance is required for the team. Team members should not be their own typist unless the data and computations make it necessary.

There are additional data charting, data reporting, and team forms that can be used by the Six Sigma team members for documenting and reporting on the program, collecting data, and charting data, and for reporting the results of the program to the team, company personnel, and management. These are listed and found in the Appendix C for easy retrieval and use.

Six Sigma Team Meeting and Action Plan

Team Name: ____________________

Process: ____________________ Date: __/__/__

Team Black Belt: Leader ____________________ Start time: ______

Recorder: ____________________ End time: ______

Members	Members
1.	7.
2.	8.
3.	9.
4.	10.
5.	11.
6.	12.

Agenda	Topic time	Agenda	Topic time
1.		5.	
2.		6.	
3.		7.	
4.		8.	

Reports made

Subject: ____________________ Reported by: ______

Figure 20. Team meeting and action plan.

TEAM MEMBER TASK ASSIGNMENTS

Each team member is assigned an initial task to begin the analysis of the Six Sigma program. These are assigned by the Black Belt based on the type of program begun and the capability of the individual team member. Always, try and use your best assets, team knowledge, correctly and wisely to gather the necessary information required for the program.

Whenever possible after the initial team training program has been completed by the Black Belt, try and team up a knowledgeable person doing a specific task with a less experienced team member. This will assist each in increasing their skill and knowledge in performing the task on this and future Six Sigma programs. It is essential that the skill level of the team members always be increased as they are in training for their Green Belt and need to know as much information as possible on how to collect data and correctly.

COMPANY DEPARTMENTAL ORGANIZATION AND RESPONSIBILITY

With the establishment of the Six Sigma program and with teams formed for process improvement and problem prevention, it is important to know the method employed and personnel used to maintain quality. Up to this point these personnel had the responsibility for the company and department business, quality, and manufacturing operations. It is important to know what operations were established, methods employed, personnel used and what is the status of the established programs and were they successful.

There could be considerable information already documented, programs in place collecting the data, and personnel knowledgeable in the collection of the data and what it means to the process under study. Therefore, do not fail to investigate what is in place, if it is working, and who are involved.

Many companies are set up as shown in the "Company Departmental Organization Chart with Responsibilities" as shown in Figure 21. This organization and responsibilities chart maps out who and what department or manager has the prime and secondary responsibility for each of the listed area of responsibility action items. The Black Belt and management need to determine if the current responsibilities are correct, who is responsible prime, secondary, and how the information will be documented, stored, and who has the need to know and will have access to the data.

In any of the areas listed, in this example there are 23, the Black Belt with management input will select the areas of responsibility, personnel responsible, and they in turn will document the procedures they will use to

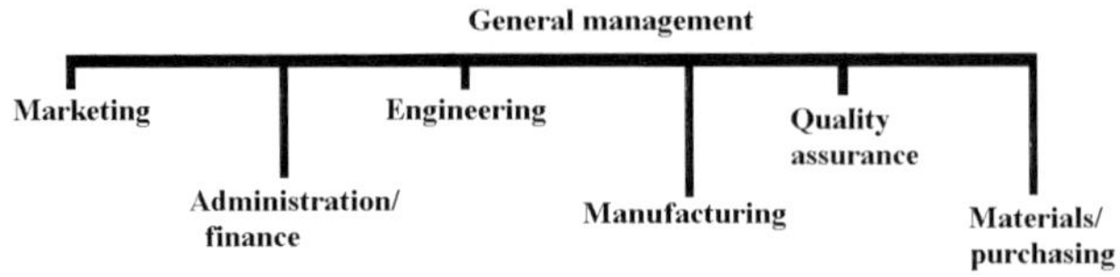

Department responsible for quality

GM = General management
MK = Marketing
AF = Administration/finance
EN = Engineering
MF = Manufacturing
QA = Quality assurance
MP = Materials/purchasing

Responsibility code:

R = Responsible
M = Must contribute
O = May contribute
I = Informed

Areas of responsibilities	GM	MK	AF	EN	MF	QA	MP
1. Establish business quality objectives	R	M		M	M	M	
2. Implement quality process control	I	I		M	R	M	
3. Plan quality systems	R	M		M	M	R	M
4. Determine customer requirements		R					
5. Establish product design specifications		M		R	O	O	
6. Establish manufacturing process control				M	R	M	M
7. Determine process capabilities				M	R	M	O
8. Quality vendors			M	I	I	M	R
9. Plan inspection and test procedures		O		M	M	R	M
10. Design inspection and test equipment				M	M		
11. Tooling reviews and purchases	O	I	I	R	M	M	M
12. Schedule production		I		R	M		
13. Material evaluation in-house						R	M
14. Produce products to specification	I	I		M	R	R	I
15. Quality feedback information	I			M	R	R	I
16. In-process quality audits	I				M	R	
17. Gather quality cost data	I	M	R	M	M	M	M
18. Analyze quality cost information			M		M	R	M
19. Final product quality	I	I		M	R	R	
20. Gather customer complaint data	I	R		M	I	M	
21. Analyze customer complaint data		M		M	M	R	M
22. Corrective action		M		O	R	R	O
23. Customer satisfaction	I	M		M	R	R	O

Figure 21. Company department organization chart and responsibilities. (Adapted from reference [1])

perform the duties assigned. These functions fall under the ISO9000 responsibilities and are often already specified and completed.

SIX SIGMA DATA COLLECTION

It is also the Black Belts responsibility to ensure that the information on how to do the tasks is available with enough copies for all team members. Also, the forms to collect the data on are critical for each task. Forms for this use are in the Appendix and others can be retrieved from the Internet.

An excellent source for information is available on the Internet with these sources found to be very up to date and informative in many quality areas:

http://www.qualityplasticconsult.com
http://www.qualitymag.com/articles/
http://www.controlmagazine.com
http://www.asq.org
http://www.aiag.com (Automotive Industry Action Group)

ESTABLISH DATA COLLECTION

Establish clear and precise data collection criteria with strict selection process to expedite a program. The team must know what major variables control the process and which ones can be improved to positively affect the process.

Preliminary studies of the process and discussions with the team and department personnel will identify the known variables to be investigated. But, as often occurs, some variables are more critical to producing quality parts and are better left as is. These should be identified and investigated to be certain changes in other variables will not cause a problem, or hinder the process. If unsure, run a DOE to evaluate what are the key controlling variables of the problem or process. Use a DOE when a decision point cannot be reached and the variables are too close in their influence on the process to select which one has the major influence on the problem. Then performing a DOE will show the variable or variables that really contribute to the end results of the problem or process.

For example with a manufacturing process where raising the temperature can be a possible solution a manufacturing and process decision must be made. This can be accomplished in many ways and some better controlled than others that have a higher degree of temperature variables. There are many in a process like injection molding where by increasing screw speed, changing the type of screw used, and replacing heater bands and other options variables that affect process temperature. There are both good and also not as effective ways to raise process temperature that can have a disruptive affect on the product and not positively affect the entire process. These could be too small a cavity gate opening, increased injection speed to fill the tool cavity faster, and excessive screw back pressure can degrade resins. Also, increasing injection speed may create excessive resin shear heat at the gate. Excessive shear heat can also cause the molecular weight of the polymer to be lowered, causing lower physical properties. The same can occur with glass-reinforced materials causing excessive glass breakage that will lower the materials physical properties. There are a lot of variable considerations as discussed for just controlling melt temperature. The variability of melt temperature has a major effect on the entire manufacturing process and product quality.

Creating a cause and effect matrix to discuss any variable change can illicit a good information exchange from the team. Any questions that have uncertain affect answers should be evaluated in depth before making a change. This is a form of the fishbone diagram and is very effective in determining all of the variables in a problem. Its use should always be considered first in the development of data for the solution of a problem or the improvement of a process.

Always verify your data collection equipment is in calibration. Any connection made at data gathering points, time, pressure, rate, temperature, etc. should be verified as tight and with good contact to ensure the data is accurate for the process measured. Discussing with your supplier how best to collect the data is recommended if the device is used for the first time. Also, ensure it is programmed and setup correct so valuable time is not wasted in collecting data of questionable nature. Also, the technician trained in the correct setup and use of the equipment.

Once the variable points are selected, collect data over the entire process run time and all shifts. Determining how long and how much data to collect on a process is determined on past process information. This means if equipment, for example a molding cell or machine dedicated to one material and process, may not have to be monitored the same length of time as a process with one that has multiple tool changes.

The goal would be to improve initially the long running process as it is constantly running the same material, producing the same product, and any changes, material lot-to-lot, and conditions in the plant and operators should be available. What may show up is wear in the equipment over time requiring periodic maintenance to be better scheduled especially when running abrasive and corrosive materials in your manufacturing process. Besides gathering useful data for process improvement, other valuable information can be learned if the team is made aware of these subtle changes that affect process improvements.

TRAINING OPTIONS

There are often options companies can select for training as off the shelf training programs versus training programs specifically targeted for the companies operating methods and products produced. There are benefits from each type of training. As discussed earlier, training in problem solving can be off the shelf type training using consultants who are able to train personnel in the fundamentals of problem solving. This training can be supplemented with in-house case studies of how problems were solved. Plus, the use of continued improvement processes using Six Sigma programs can be analyzed.

If the training source has experience in improving processes, business related for example, then similar examples of how to seek solutions using whatever tools are available for both the business and manufacturing communities. The processes are similar just different in seeking an improvement process. Good analytical logic and thought processes apply for each.

Six Sigma improvements are normally more analytical in their analysis. Data must show existing conditions that are used as a baseline so that improvements can be measured and plotted.

Also, monetary reductions in cost and resulting savings to the company can be gauged as improvements. This is in conjunction with defect reduction to a point of almost being unable to measure the final results when Six Sigma is attained. Imagine trying to calculate a loss of 3.4 defects in one million parts by collecting process data which has such a small deviation that it is almost impossible to plot other than as a straight line. This is one of the measurement problems when a process is in Six Sigma control. It is difficult to even try and call this a problem because the process is in such repeatable and continual control. But, the process to reach Six Sigma control will allow the plotting of data as when improving from three sigma to four sigma, etc. This information is covered later in a chapter on when and how to adjust the process.

To reach Six Sigma quality there must be specialized training of the team members. They are initially taught the basics of using the quality techniques. Which quality technique used depends on the process or business operation being analyzed. Training in basic statistical methods must be preceded with instructions in the quality methods for gathering the data, numerical, or procedural, as all will affect the improvement of the process.

Training not followed with immediate use is impractical and considered non-productive. Some companies spend time and money training their personnel without an immediate or intended use of the information. This results in a waste of valuable time and money teaching operators, supervisors, and engineers in new methods and never using the knowledge in practice on the manufacturing floor.

For training to become an asset to the company and employees, it must be used by personnel as soon as possible to reinforce and show it's worth to the company and personnel. Otherwise it is soon forgotten and never put to its intended use. This is what the Six Sigma training program does in the four months of supervised training. One week of training and back to the company to apply the knowledge with the team and then back for more advanced training. The cycle keeps repeating until the four training weeks of instruction are completed. Then the new black belt can continue with team member's support in implementing the training and knowledge to benefit the company's business and manufacturing processes.

There is a benefit to re-teaching, for example, statistical methods and analysis during and after the data is collected. This reinforces the team members knowledge in how it can best be used and applied to better understand how good or bad the process is in control. This training will be practical, applicable to the process and should yield the results desired to bring the program into Six Sigma control.

Remember, statistics are only a quality tool to show the degree of control a process is in and where it must go or remain after improvements to reach the final quality goal.

TECHNICAL SIX SIGMA KNOWLEDGE REQUIRED FOR QUALITY IMPROVEMENT

The black belt who has been trained in Six Sigma quality techniques, problem solving, variable analysis, decision-making, statistics, and decision-making must impart their knowledge and skills to the team members. Only when the team members are well trained and coached in the quality tools can improvements be achieved. This requires team action and interaction to analyze and make decisions for first improvement in a business and/or process and then continue the process to a Six Sigma level of accuracy when so specified by upper management. The black belt also needs to know when the process has reached it's potential and without major additional changes will not be able to move forward.

A process is said to be in good control when the calculated CpK capability is 1.33 or higher. Can this number be increased? Yes, but major changes will be required in equipment and support services in most cases. Depending on the business, customer needs, and requirements the black belt with management's approval may decide the process is as capable as need be for the customer base requirement and call the program closed. This technical Six Sigma knowledge is important to not continue a program unless the company actually needs the ultimate perfection to meet their business and profit goals.

To be the best you can be in your professional career is a worthy goal. To transfer black belt knowledge to others to continue these goals is even more worthy and a greater achievement. The black belt must always continue to improve their analysis, training, and business/manufacturing process improvement knowledge.

GOAL SETTING

Setting realistic goals is critical to a successful program. It is usually not necessary to actually reach the Six Sigma defect rate of 3.4 parts per million on every single business and manufacturing operation. To accomplish this goal, many more critical problem areas would be left waiting as assets were spent on customer acceptable areas. Six Sigma is not a numbers collection system to ensue all areas are Six Sigma capable. Some companies fail when they follow the single mindedness of only numbers. Chose wise projects that are big enough to be significant without being so large as to become unwieldy and not related to a workable solution. Six Sigma is based on fixing the root cause of a problem to successfully eliminate the daily problems that occur as a result of variability from a root cause effect.

The shotgun problem/process improvement approach is eliminated for the straight-line analytical method to fix the problem and then monitor and control the process so it does not happen again.

PLANNING AND IMPLEMENTATION PROCESSES

The majority of Six Sigma programs are focused and performed on manufacturing operations. These operations get the most visibility, easy to collect measurable data, establish a base line, analyze the problem, and work to find the root cause variables of the problem. A solution is then developed, verified, and controlled to eliminate the problem.

Eventually companies will approach the best they can be within a two year or less time frame with these current assets and equipment capability allocation necessary to meet their customer requirements. They must then decide if they want to actually spend the time and money to reach a true Six Sigma quality (3.4 defects per million parts) and to maintain it. The cost at this stage vs. the capability and maintenance of equipment and systems must be evaluated. These factors must be considered for cost to reach and maintain Six Sigma perfection. This is an area the champion and executive management team must decide if the cost justifies the end results.

REFERENCES

1. Harry, M. J. "The Quality Twilight Zone." Quality Progress February 2000: 68–71.
2. Harry, M. J. "Six Sigma Focuses on Improvement Rates." Quality Progress June 2000: 76–80.
3. Hunter, J. S. "Statistical Process Control." Quality Progress December 1999: 54–57.
4. Natarajan, R. et. al. "Applying QFD to Internal Service System Design." Quality Progress February 1999: 65–70.
5. Navy, Department of, "The Process Improvement Notebook (PIN)." Total Quality Leadership Office, Arlington, VA.

Chapter 3

Reasons for Implementing Six Sigma

Initially it was obvious that only high profile, money driven programs ($170,000) would be selected for black belt Six Sigma process improvement and problem solving programs. These were the programs major corporations wanted selected to improve their profitability. Not all programs were initially successful or yielded the desired results. With additional training and specifically applied solutions, latter programs were successful and companies have realized these savings and have moved into the billion dollar per year savings. Many companies can anticipate these high rates of savings for a few more years until their business and manufacturing processes are in as good control as the process and personnel are capable. Then as these programs are improved, and solved, the champion can select lower cost or less impacting programs for improvement. These lesser value programs are often classified as CTQ (contributor to quality) programs as illustrated in the following example.

CONTRIBUTOR TO QUALITY EXAMPLE

A situation involved the design engineer specifying thread-forming screws for the assembly of a polycarbonate-tooled cover. The cover had molded in bosses that met form, fit, and function. In the final assembly a stress cracking boss problem occurred in the field after only a short period of time for the customer.

Investigation found after an exhausting review of the material and molding parameters that purchasing had changed the original screw supplier. They bought an equivalent screw purchased at a slightly lower price from a new and approved supplier. The screw was said to be equivalent and met the same holding force but on analysis had a sharp thread form versus a tumbled, rounded, thread from the original supplier. The sharp edge created a higher stress concentration level in the polycarbonate boss. In conjunction with the same problem, the power screwdrivers used for assembly did not have torque-limiting settings to seat the screws. Also, the screws were not cleaned of cutting oils from this supplier that are a potential stress crack problem for polycarbonate resin. The combination of factors, sharp screw thread, no torque cut out limited on driver, and cutting oil on threads of the screws, resulted in the field failures. Another problem was discovered at incoming inspection, since a requirement was a verification of screw thread form and no oil or grease on the parts. This was not performed adequately and allowed the screws to be accepted for manufacture.

The solution was to return these screws and purchasing informed to only buy screws from the original supplier. The specification was sent to the original screws supplier with a notice to only supply screws to the specification for a rounded thread form and clean of all cutting oils and lubricants. Also, the screw cleaning solution used was to have been specified to the customer to make sure any oil and grease cleaning solution residue on the screws did not attack the polycarbonate resin. Plus, determine the required seating screw torque and buying power screwdrivers with this torque limiting capability.

Manufacturing also initiated a training program for the assembly team with a scheduled torque verification program to ensure the seating torque was maintained within engineering values for the seating of the screw in the boss. A certification for each lot of screws was to verify that they met the specification and incoming inspections work instructions were revised to ensure the inspection procedure met their requirements.

Engineering also became more specific in their bill of materials for these applications to only use a specific type of screw that allowed purchasing to buy at the best price a screw that met all of the requirements. One problem with four solutions for three problems not obvious at the start of the program that resulted in increased customer satisfaction. A typical Six Sigma program, of lower value, solved using the Six Sigma tools and methodology that in the long term would eliminate the screw assembly problem for their product line and customer product.

HOW TIGHT IS SIX SIGMA QUALITY

A question the company chairman asks is how close to Six Sigma need we get? Your management must answer this question. The ideal is Six Sigma for

Table 1. Significance of Sigma on Results.

Sigma numbers	Defects per million
1.5 σ	500,000
2.0 σ	308,300
2.5 σ	158,650
3.0 σ	67,000
3.5 σ	22,700
4.0 σ	6,220
4.5 σ	1,350
5.0 σ	233
5.5 σ	32
6.0 σ	3.4

(Adapted from reference [1]).

all operations. The customer need and additional cost may not require this degree of quality once major problem areas are corrected and brought within acceptable levels. The significance of sigma quality levels is shown in Table 1. The savings in defect prevention from three sigma to Six Sigma is 20,000 times better.

The goal then is to strive for a Six Sigma quality level and any change from three sigma to six sigma limits is a significant savings for the time and money spent. The reasons are obvious, reduced cost, defects, and rework to increase profitability, on time delivery, a quality product, and an increased lowering of customer risk.

INITIAL STATES OF IMPLEMENTATION OF SIX SIGMA FOR BUSINESS AND MANUFACTURING

Your management champion is selected and dedicated to implementing Six Sigma into your operations. The black belt leader is trained and the team members chosen.

Your first meeting with all Six Sigma team members is to discuss the program objectives, goals, cost savings anticipated, initial program activities, and to discuss the estimated time line to accomplish the selected task. Also, team members selected must discuss who will perform or team with to accomplish the actions and the necessary tools for solving the program objectives with anticipated completion dates. An examination of all possible areas exhibiting a loss of quality and incurred cost is like an iceberg, only 8% is showing, the rest is less obvious and not always easily visible in the Six Sigma selection process. Examples easily visible to management are shown in Table 2.

The example used in this text is common to all manufacturing operations; only the type of equipment, material and variables will vary. The hidden or less obvious business and quality items can vary in degree of severity and data must be collected to estimate its worth to begin a program.

The champion must initially look at all the possible company programs and their estimated pay back and savings value. This should be the first course of action for the Six Sigma team if a major program has not been selected by the champion, which almost never occurs. Management sees this

Table 2. Areas of Possible Loss of Quality.

rejects	warranty items	late production, increased manufacturing and quality costs
scrap	inspection problems	line down problems
rework	supplier product problems	customer complaints
The less obvious items usually known to only a few are:		
customer revisions	poor plant/equipment maintenance	late delivery
engineering change orders	Incorrect scheduling	excess inventory
long cycle times	schedule changes	unusable inventory
frequent setups	priority customer orders	damaged inventory
expediting costs	lost sales	unlocatable inventory
time value of money	excessive material orders/planning	incorrect inventory
working capital allocations	lost customer loyalty	lost supplier support
no increase in customer base		

(Adapted from reference [21]).

daily in their profit and loss sheets. Their problem is knowing the root cause of the loss and why it is still occurring.

Management must inform the managers and employees of the implementation of the Six Sigma program. Their participation, impact on quality and cost reductions and how improvements can enhance their employment, bonuses, and the company profitability.

Employee participation with the Six Sigma team is very important. Interaction may initially involve discussions on workflow, scheduling problems with equipment, personnel skill levels, training requirements, and output if quality or workflow problems exist. Should any personnel problems occur during the program they must be handled on a one on one discussion to avoid any wrong rumors being generated to cause a new problem and to verify if it is an actual or perceived production problem. A delicate situation that must be handled on a supervisory level with the workers and the output levels developed to meet customer delivery schedules.

The champion will, by the direction and input of upper management select the Six Sigma program. It is very important the black belt brings up all the possible situations that require improvement. Also, be aware the champion if a department manager, not the CEO, may show preference to their departments problems especially if a considerable saving can result. This is the main reason all interested parties agree on the best programs to work on first and then prioritize all remaining selected programs for later implementation.

Some companies have tied executive bonus and advancements into successful selection and implementation of Six Sigma programs. The company must consider the cost of training a black belt and team members, plus their time devoted at 100% to projects, the reward, must be substantial to justify the cost. Do not ever consider a Six Sigma program without a champion as too much is at stake to not fully support a program.

Pilot Projects, Problems and Solutions

There are many worthy quality cost saving programs to be considered when the first cut is performed for selecting Six Sigma programs. Selected is a manufacturing program that will typify a manufacturing situation almost all production personnel encounter in their daily work.

The example selected is:

1. Reduce/eliminate the high rejection rate for a tight tolerance and long running product. But, before presenting the actual Six Sigma program for reducing any defects created in processing, an overview of desired pre-production requirements that must be performed before the actual manufacture begins is developed.

Most business or manufacturing problems begin with a single reason for rejection of the product at a specific stage or period in manufacture. The root cause of any problem often was created earlier in the process. Six Sigma methodology is used to identify the root cause, where it first occurred, and to develop a lasting fix and solution to the problem. The root cause may have occurred at one of the listed areas in Table 3.

These areas of product production can be individually broken down into product or process variables that may have created or influenced the original problem.

In the design or manufacture of any product it is desired that all areas of the company involved in the product plus the customer, if appropriate, and material supplier and tool designers meet to discuss the product and input their ideas for producing a quality and profitable product the "first time."

It is critical to obtain agreement of all department participants in the early stage of design and manufacture. There are too many variables that can affect the risk factor in manufacture and end use reliability to not follow this operational plan. Check lists can assist in ensuring all questions are considered and answered early in any program. These same check lists can be used in program analysis to ensure all information was collected for the programs solution.

Table 3. Possible Root Cause of the Problem.

design	material selection	tool design
maintenance	purchasing	work instructions
tool building	scheduling	machine selection
plant equipment	plant systems	material handling
processing	machine controls	auxiliary equipment setup and operation
operator training	inspection methods/tools	procedures inadequate

If the product is not designed correctly for manufacture along with material selection it will impose many problems for both the tool designer and manufacturing as how the material behaves in the product cavity. Problems here can affect dimensions, appearance, physical properties, and manufacturing costs and defects. Each product is unique and the design should incorporate as many combined functions in the part such as assembly methods and dual functioning parts based on the materials properties.

For an injection molded plastic product the products layout in the tool to ensure good material flow and to fill thick part sections first and then thin sections. Plus control of the cavity-cooling layout in the tool to control the cavity temperature is very important for structural product integrity and good surface aesthetics. The ejection system must also ensure no distortion occurs on ejection from the tool cavity.

Also, a balanced runner system with cavity coolant temperature maintained within 2° F, of in and out coolant flow. Good venting, draft, and final tool surface finish polishing in the direction of part ejection from the cavity. Also, the base tool cavity steel for good heat transfer and wear characteristics should be selected and specified. This is just one method of manufacture. Each material and process are unique with cycle, temperature, pressure, etc., all unique for their type of manufacture plus the behavior of the material in the operation.

The material selected must have the physical properties for the product use and also the flow characteristics to fill and pack out the tool cavity in a minimal time cycle with controlled and repeatable melt flow properties. There are supplier and incoming material tests that can develop data on lot-to-lot material flow and molecular weight properties that influence melt flow in the tool.

Once the product and tool are designed as capable for manufacture, the plants manufacturing processes, auxiliary equipment, molding machine, and plant support services must be evaluated for capability and repeatable supply of required services.

These include cooling water for the injection equipment, floor chillers, tool temperature control, and any air operated or other support equipment for manufacture. Also, consider the other auxiliary equipment needs driers, feeders, grinders, etc. and their capability plus the plant environment as a variable in the manufacturing process.

When all of these services, equipment and material, are capable then the actual injection molding process can be evaluated for capability and repeatable manufacture. But, do not forget the operator who must be trained in the operation of the process as established by the setup technician or process engineer who specified the initial processing cycle parameters.

Then evaluating the process for capability will involve running a capability analysis of the molding machine and tool for repeatable operation for CP and CpK repeatable control of the entire molding operation.

Each injection molding machines capability is tied to the machines controls, valves, electronic timers, pressure valves, temperature control of the hydraulic fluid and other operating variables. The capabilities of the entire molding process for the product are these and the tool being run in the machine. Each tool will behave slightly differently based on the size and age of the molding machine it is run in. This means that every tool and machine should have an individual capability study performed to evaluate the repeatable performance of machine and tool. This is done each time to optimize the process and ensure the ability to produce a quality product each time the tool is setup in a new or prior run molding machine.

Each injection molding machines capability is tied to the material being processed and the type of screw in the barrel and its size. Some thermoplastics require a special screw design to correctly process the material. The amount of shear heat generated by the screw design turning at a specified RPM (revolutions per minute) must be considered. Also, shear heat created by the injection pressure and ram speed of injection of material through the runner system and cavity gate. The plant variables will affect these machine variables and it's control settings for repeatable operation cycle-to cycle.

In the injection molding industry this example involved only one tool, material, and molding operation. The information required to produce a Six Sigma product repeat ably is enormous to staggering for the time and money that must be expended to accomplish this goal.

This is one industry of over 15,000 injection molders both captive and custom with varying number of molding machines, from one to several hundred. They also come in all size ranges and age of machine and type of molding and material processed in the machine. If 1% of these molders have process control procedures for manufacturing for each tool it is a miracle. Only the well organized and Six Sigma program leaders will have this information. They will also be the only ones using this procedure instructions to setup and produce products for their customers.

The quality control department should also have FMEA's (failure mode and effects analysis) written so if a problem should occur a rapid trouble shooting of the process can be conducted and a rapid and permanent cure implemented.

Only after a process has been fully examined and the system brought up to full or as near to full capacity as possible, run and analysis of the systems repeatability should be conducted. This is true for all methods of business and manufacture. Procedures are necessary and need to be followed in all operations to ensure repeat able and quality operations occur for the business unit and their customers and suppliers.

The simple operation of determining a machines numerical CpK capability now becomes more involved as all contributing variables are identified. This is one of many bits of information that should be determined for improving and maintaining a quality system and improvement of an operation. The CP value must also be determined to verify its centeredness to the mean of the variables evaluated.

ANTICIPATED EARLY SIX SIGMA RESULTS

The key to success for a Six Sigma program is having selected a program where the results will be attainably noticed, and the reward or payback, substantial. This will convince upper management that the time and expense will generate a substantial payback within expectations.

What has to happen before this is realized is ensuring the Six Sigma team members are well trained in problem solving skill and the basic quality analysis tools for solving the problem and improving the operation?

SIX SIGMA TEAM SKILLS

The selection of the Six Sigma team with your black belt as the leader is initially based on the type of problem to be solved. What experience has proven is the team members may not know the essential steps in solving a problem. The black belt has learned these skills and now must, before the program begins, train the team in what, how, and when these skills are to be employed.

Six Sigma teams will have high visibility within the company and their success depends on their ability to make wise and knowing decisions to solve a complex problem that other more knowledgeable department personnel failed to be able to solve. The problem selected is often based on others past failures or non-rewarding results. The team needs a structured problem solving training period discussing the factors that influence problem solving and decision-making. There is essentially a seven-step model and procedures to use to improve the teams problem solving performance. This review of problem solving methods and techniques should be taught and reinforced each time a team begins a new program. A refresher course is beneficial to the entire team to review basics and reinforce information already known and to share knowledge with others who may be new members of the Six Sigma team.

Some team members will often come from an area not related to the problem and are there to add an outside pair of eyes to aid in a solution and to provide expertise in their area of involvement as required.

Problem Solving Involves Six Basic Processes

Problem definition:	Define what the problem actually is, be specific, may contain more variables than first thought possible or the solution is not in the area first identified.
Data collection:	What method(s) are to be used and are the personnel trained in setting up program to collect the data? What data is to be collected? How many variables are there in the problem? Are all contributing variables identified?
Solution generation:	Collect the data from all the processes, sources, suppliers, machines, etc. and have it entered into a resource system that is system friendly for the analysis of the data. Allow sufficient time to collect the data over a wide time frame if possible to determine if the cause is in these areas.

Action planning:	With data collected, make a decision on what the solution to the problem is. Evaluate the data to verify that the variables selected are the controlling variables for the process or program. Verify that a change will influence the process in a positive manner. Relay results to management for their information and decision to proceed to prevent the problem from reoccurring.
Implementation:	Make the change and ensure that the variables changed are in control, and the process is performing as predicted. Run a trial program for the length of time to verify it is in equilibrium and stable with repeatability.
Solution monitoring:	Record and monitor the data to verify the elimination of the problem was successful and the process will continue in a repeatable operation. Transfer the learning process to all areas of the facility that need this implementation.

These tools should be used each time a problem solution or process improvement is started. A log book of all the methods used and data collected should also be identified during the collection and data recording portion of the program. Be sure the entire program is documented during the program. Do not wait until the program is completed to finalize the documentation, it will never be accurate and complete.

The biggest problem with teams is understanding the nature of problems and how to use the available quality and problem solving methodologies to affect a solution. This is the responsibility of the black belt. The black belt will provide the breakdown and definition of the problem to be solved with assistance from a team member from the area or department where the problem surfaced. I have used the term "surfaced, in lieu of discovered" as the problem may have originated from an earlier operation or been created by a change in an operation before the point of discovery or recognition. The latter may be caused by a business or engineering change not communicated down the chain of the organization which caused the problem to occur. This can be as simple as not selecting the right person to be program lead resulting in poor direction and confusion, with many errors committed unknowingly. This is an upper management created problem resulting in blame being cast on quality as they should have known it could happen and taken the lead to prevent it.

Also, team members may be lacking in the knowledge of understanding how the people think (human information process factors) and how personal biases can affect the success of the problem solving techniques they employ. This is why understanding how personnel process information and listen to others as the team members discuss items related to the problem or process information that is important understanding and solving a problem.

Human Information Processing Factors

1. Individuals are only capable of retaining 5 to 9 items of information at one time. This limits team members holding onto facts when solving problems and making decisions. This must never become a point of contention within the group, as one never knows whose suggestion may point to the solution. Therefore, complete documentation and notes of all meetings are so very important. Meeting agenda's with action items and a secretary to record the data are always required to document what was discussed and decided.
2. Individuals filter the information received and can often distort it to suit their style of reasoning. This also applies to how they interpret and use the data generated during the metrics portion of the solution. Their filters correspond to their personality factors, cultural biases, values, and ways of processing information. The methodologies of an accountant versus an engineer who both use numbers have slightly different reactions to the numbers on the outcome to a solution of a problem. The accountant, must be exact, no variance. The engineer knows the number is an estimate with a degree of variability due to the material or a process never being totally exact and always in a mode of change. Who then is right? They both are in their own area of expertise. The Black Belt must communicate to the team the type of data that will be collected and how it will be used to extract a solution or improvement.
3. People tend to be over confident in their personal judgments and opinions. Studies have proven individuals overestimate what they know and the probability of an event occurring. Do not be held back by not

asking the question, "what if." This question can often be a break through question for a discussion of a problems variables leading to a sound solution.

4. Individuals, based on the available data, anchor their starting point for their decision-making on it, even if it is wrong. Care must be exercised when using existing information especially when the method of generating the data is not known or verifiable as being correct. They must question any and all available data if not specifically collected, documented, and known to be the true output or from a reliable source verified by data collection. Poor decisions can result leading to inaccurate conclusions. It cannot be said often enough to "Verify" your information and understand what it is telling you before making a hasty decision. Often the best thing you can do with a process out of control is to shut it down until a satisfactory analysis and solution is attained. This can save material from being wasted, reduce scrap loss, and give the team enough time to evaluate the problem before making a sound decision on what to do to the process.
5. Experience has proven people lack learning from past decisions if the decisions do not meet their individual needs. People often lack the skill of analyzing and revisiting past decisions and their outcomes. They want to solve the problem now and move on to the next one. Most do not realize the thought process of how they process information, think, and learn from past experience. Documentation of data may recall a missed point or path to now investigate.

In the initial team training for Six Sigma, the black belt must emphasize and explain the problem solving process to extract the best results from the team. This involves knowing how to learn, solve problems, what strategies to use, and how the team can best use the knowledge and information taught and generated in the problem solving process.

Studies acknowledge the more individuals know about the way to process and use information, the better they perform in decision-making, memory retention, and problem solving. In proactive problem solving this relates to how much a team consciously knows about the ways they will solve a problem. How they make or arrive at a decision and use the problem solving techniques successfully will assist the team to improve the quality of making Six Sigma type decisions.

PROBLEMS IDENTIFIED TO COMPLEXITY

Problems are classified as either simple or complex. The nature of most Six Sigma problems will be complex as they keep reoccurring after solutions were attempted and did not work or keep the process in control. They may also be classified as projects to reduce a problem and reach a higher quality level as in our example of determining CpK capability. This involved the injection molding machines and tools, while considering the effects on the manufacturing process from the plant environment, auxiliary equipment support, and control systems.

1. Simple problems have defined boundaries and definition with relevant data available and fit into straightforward manageable problem solving models. These are typically not classified as Six Sigma projects. They are classified as the daily solvable problems that frequently occur due to known and understandable reasons. They can be addressed later or if too frequent, solved with new procedures and work instructions or training of personnel. These are solvable if a problem, job, or machine program log book is kept for each operation. A reference to this book of records can easily solve the simple day-to-day problems experienced on the factory floor due to daily variability's in material, plant environment, etc. Then if these were documented in the log book the solution and fix should be easily obtained and implemented.
2. Complex problems have multiple and complex components with multi-variables with data often not readily available, or possibly questionable. There is no known solution that has worked and quality tools are required to collect, analyze and evaluate the data to arrive at a solution.

The use of metrics to collect the data is required and a team, Six Sigma approach for finding a solution is necessary. Time involved may extend out several months or more with possible extensive changes in more than one department of the company. The reward for a solution is counted in hundred of thousand of dollars saved and customers retained.

ANALYSIS OF COMPLEX PROBLEMS OR PROCESS IMPROVEMENTS

Step 1. Define the problem and effects on the organization
Step 2. Determine the processes to use to describe problem effects

Step 3. Gather information, new and historical, for establishing limits
Step 4. Analyze data for effects to determine a workable solution
Step 5. Develop an action plan to affect the solution
Step 6. Audit and evaluate the action plan for an effective decision and process
Step 7. Document solutions, establish monitoring and audit schedule, and transmit solution to company personnel

HOW A SIX SIGMA TEAM SOLVED A PROBLEM

Using the example to determine CP and CpK capability of the company's injection molding machines the following is a synopsis of the use of these 7 steps. The results obtained improved the entire molding operation of the plant and support systems.

Step 1. Define problem and effects on organization
Problem was defined as determining the CP and CpK value of a specific molding machine. It is dedicated to a long running product experiencing variable product quality levels, shift-to-shift and season-to-season. Goal: Establish and monitor a repeatable process control process for this system.

Step 2. Appropriate analytical software was selected to gather data on the repeatability of the molding machines processing variables, injection pressure, cycle times, injection speed, temperature profiles, etc. Other information was selected for collection during the analysis conducted over a period of several weeks, shifts, and possible personnel changes. The customer was contacted along with the material supplier, machinery technician, and maintenance personnel in the plant to gather additional information as required. A schedule of events were developed and discussions with involved parties were selected and assignments of personnel to conduct the gathering of data delegated. Consensus of the team members and management on events, methods, and processes to be used for the analysis was obtained.

Step 3. The black belt and team agreed to select a software program company for their assistance for the analysis of their molding operation. An Ishikawa or fishbone analysis was conducted for the following equipment, plant, and processing areas.

1. Molding machine-process, maintenance, variables, controls, and screw design.
2. Plastic material-supplier, purchasing, properties, tests.
3. Auxiliary equipment-dryers, chillers, feeders, maintenance.
4. Plant support equipment-H/AC, fans, filters, maintenance.
5. Plant involvement-airflow, temperature, humidity, air quality.
6. Tool-design and operation, steel type, cooling layout, temperature control, operation of tool.
7. Operator training and setup technical team.

The reason for developing the fishbone diagrams was to identify all variables related to the molding process even prior to the CP and CpK evaluation. A Cause and Effect Analysis could also be run for any problem that may surface during the analysis. The use of these two diagramming methods together can lead to a very detailed analysis of all the molding operations and what may influence each variable causing the problem.

Also employed was QFD (quality function deployment) for customer and supplier requirement and expectation analysis. Manufacturing engineering wrote a process control procedure for the molding cell and plants molding operation and had it implemented by training the operators and supervisors. A FMEA (failure mode effects analysis) was developed for the entire molding operation. Statistical data on temperature control, drying of material, plant temperature and tower water, tool and machine (hydraulic oil temperature) data was collected. Plus any other data considered to be a variable, data that can influence the manufacture of the molding process and affect the machine and molding operations CP and CpK values.

Employee, setup and operator training was evaluated for being capable for operation of the equipment, setup and control of the molding operation.

Product quality, customer complaints and any other problem items were Parato charted such as supplier problems, engineering changes, if any, and maintenance records of equipment and systems were analyzed. Incoming material records and material certifications, if used, specific lot testing data or typical lot analysis data was gathered, documented, and reviewed for involvement on the operation.

Step 4. The team was assigned responsibilities to get with the most knowledgeable person or persons to develop the fishbone diagrams.

A review of the process control procedures was conducted for accuracy and reevaluated to the operations functions. FMEA's were developed in a similar manner with production and quality assurance providing input.

Production setup with black belt assistance for the CP and CpK analysis and other team members setup monitoring equipment and recording sites to gather the other plant, equipment, process, and documented data for the program. The data collection required over several weeks to collect what occurred during a seasonal change. Also, during this time period other injection molding machines were monitored using the CpK software analysis so more than one machine and multiple tools were evaluated for capability.

The maintenance records for injection molding machines and tools were collected along with setup and process records for each product run along with quality records and any customer correspondence from sales and quality assurance.

Step 5. Analysis of the data required inputting it in SPC (Statistical Process Control) software program and plotting the information. The data output was initially plotted using Parato charts, histograms and respective SPC data charts showing if control of the variables occurred over the time period of data generation. All like data was reduced to the same horizontal and vertical scales using identical start and stop dates. Transparencies were made so similar data could be overlaid to inspect for any anomalies that may result during data collection. Likewise, product defect rates were plotted on the same horizontal scale so temperature, machine, and molding plant variables, for example, could be investigated for any changes in defect rates.

This was a monumental task with all data inputted into a master database for comparisons of like data that indicated a variation and could be overlaid for probable cause and effect analysis.

Cause and Effect analysis was then performed on any anomalies produced in the data and brainstorming, question and answers, were generated by the team with interactions traced use the fishbone diagrams as applicable.

Analysis required considerable time working to identify the multi-effects created by the over 40 variables affecting the molding process determined during the analysis of the process.

Any variable determined by the data to be stationary, non-variable, for control over the test period was removed from the analysis. Areas where minor deviation occurred but corrected by the operators was considered normal process variation and not a key factor in the solution. A record of the type, amount of change, and frequency was recorded and documented on the molding machines process records. Only one variable at a time could be changed by the operator and this required the molding shift supervisor's agreement for change and the amount. Sufficient time was required between any process change to ensure the change was implemented into the system and sufficient time had elapsed for the system to reach equilibrium and for the change to have an effect on the process.

Anytime a decision point occurred with multiple variables involved, a DOE was run to determine the variable having the major effect on the process and outcome.

During these changes the machine was being monitored for CP and CpK with the processing data and effects being logged by the software program. A sufficient amount of data was collected for analysis of the process. Always collect all the data within the same time period to eliminate data anomalies that cannot be backed up with other corresponding or cause and effect results.

One item not mentioned so far is the capability analysis software. The software used was from Hunkar Laboratories. Their capability analysis system requires for best results, using their "Gold Standard" data collection sensors specifically used on the molding machine. They were calibrated to ensure the data collected was as accurate as possible and would compare to the data source Hunkar Laboratories used for determining their machine capability standard.

We did not have or use "gold standard sensors" and had to rely on the output sensors that came with the equipment and any variability they may have in their output. Therefore, the data collected was representative of the accuracy of the sensors in the molding machines operation system that reported output to the control panel during operation. This did not mean our data was flawed, only that a true comparison of analytical results with the standard, established by Hunkar's tests, would not be a direct correlation to the standard of operation procedures in writing his software analysis program. It is recommended that the "Gold" standards he provides be purchased for a direct one on one correlation if possible for the best and repeatable results.

The Hunkar system also has a classification and problem solution software system built into the program. This program assists the operator to make quality adjustments based on output data once the highest machine classification is attained. Classification is based on Hunkar's evaluation system developed using their "Gold Sensors" on other molding machines.

The final analysis of all the data collected yielded the following information:

1. Tower water temperature varied shift-to shift by over 25° F.
   Required temperature 78° F ± 7° F.
   Central chiller system supply cooling water to the injection molding machine and tool support was in specification, ± 3° F.
2. Incorrect screw design used in one machine for material processing, general purpose screw vs. the required three zone, high compression and metering screw.
3. Inability to maintain packing pressure on one machine, probable bad check ring.
4. Excessive screw RPM's required to generate sufficient melt to fill tool on one machine, screw design problem.
5. Each machine and tool combination tested could not obtain a CpK higher than 0.55, with 1.0 considered a minimum and a CP of 0.50.
6. Cooing fans used in plant, blowing on operators and machines, caused temperature variations.
7. Maintenance of auxiliary and plant air particulate filters very poor to non-existent.
8. Tool and machine matching (scheduling) non-existent for melt/clamp capability, some to big, others too small, no insulation between tool and machine platens, tool temperature varied at in and out tool temperature sampling points.
9. Insufficient operator training for responsibility assigned, poor record keeping, and process changes made without direct supervision.

All of these items are major in regard to quality, repeatability of the molding cycle, and product rejection rate. Each item can be addressed separately but the first team task is to determine the ranking for solving the product problems.

It was decided that some items could be addressed by maintenance, some by production, and the others required engineering support. A report was drafted with their recommendations and a meeting was held with the champion and upper management to discuss future improvements in the system.

Recommendations for each items solution were determined and presented in their findings along with time, material, cost, and savings to be realized by corrective action. Also, the anticipated results when each item was completed.

Included with the report was a detailed analysis of the machine CP and CpK status for machines not identified originally in the findings. These included ways to raise the anticipated CpK value and keep the process centered, CP within the process limits after the major processing problem items were corrected.

Step 6. Teams from each department were selected and directed to make the changes recommended with minimal disruption to the business and production schedule. Meetings were scheduled with department and the Six Sigma team to review their recommendations, decide on an action plan, set a schedule, order the necessary equipment, and parts. All plans were now in place to implement the corrections with personnel assigned action items with completion dates.

The action plan, schedule, and parts delivery schedules were loaded in the company business systems so all departments and work correction teams could be updated in "Real Time". Some programs required a critical path schedule to ensure any system down time would not affect their customers critical production scheduled products.

Sales and marketing were also tied into the improvements so they could market the planned repeatable capability of their customer's part production. This would also lead to bidding on more competitive programs with customers here-to-fore they were not able to bid on due to pricing and output deficiencies and quality issues.

Step 7. Audit and evaluate the decisions and process improvements

The Six Sigma team ranked the items for completion with agreement from all departments in the following order as they were reported completed per their ranking item number.

Ranked 1, Item 2. Correct screw was used for processing the higher melt material, nylon. The general-purpose screw resulted in immediate improvement in establishing the new molding cycle, faster cycle by 40%.

Ranked 2, Item 3. Pulled screw and found worn check ring. Replaced with new check ring and part weight stabilized for another process benefit resulting in cycle reduction of 30%.

Ranked 3, Item 4. Measuring barrel and clearance over new screw flights resulted in determining barrel inside diameter was worn. Also, found metal fillings in melt. Ordered new barrel and liner to match the high compression screw.

Ranked 4, Item 6. Turned off fans directing airflow over operator area and away from machine barrels. Also, added barrel insulating blanket to enclose barrel for heat retention. Dramatic improvement in heat conductivity, and able to lower barrel temperature on heater bands and lower screw RPM's. Cycle reduced 20% and part weight stabilized.

Ranked 5, Item 7. All air cleaner, dryer, feed system, etc. Filters were either cleaned or replaced with a maintenance schedule established to monitor systems on a more frequent time schedule to determine rate of their requiring changing or cleaning.

Also, began monitoring dryer bed desiccant for activity to absorb moisture from resin during drying. Began testing drier bed capability with the water glass differential temperature test.

Ranked 6, Item 1. Conducted a heat study for cooling water requirements for each machine, chiller, tool, and any auxiliary equipment needing cooling water. Determined tower water system was undersized for providing sufficient cooling, within 85°, ±7° F temperature range. Central chiller was under utilized and had the capability to maintain tools and some molding machine hydraulic fluid cooling within machine operation specifications of ±3° F.

With some major and other minor plumbing routings the tower and chiller would work in series to provide more uniform cooling to the plant until a new higher capacity tower could be installed or current tower reworked back into specifications.

During the plumbing changes, it was found due to poor water treatment, the coolant piping was collecting mineral deposits inside the pipes, restricting the required water flow rate to the machines, chillers, and tools.

To eliminate this problem an ozone water treatment system was to be installed at the main water feed point to treat the entire plants water system. Critical sections in the water distribution system were replaced to attain the required piping flow rates to equipment and molding. The ozone treatment system will over time dissolve the mineral deposits; return piping to it's original inside diameter, and return flow rate to the piping original specifications, and machine requirements. The treatment will also decalcify the molding machines heat exchangers, which will eliminate core replacement and flushing with caustic cleaners for cleaning and eliminate production down time.

Ranked 7, Item 9. An operator-training program was established with classroom and floor training in the basic operator functions. Operators were also taught basis statistical methods to understand, collect, and plot any statistical data on their molding operations and to aid in accessing control and repeatability of their molding operation.

Setup personnel were also trained in quick mold change procedures and how to install tool and cool and heating lines correctly and test the system prior to startup. The setup supervisors were also used to train the operators in production control plus material suppliers gave seminars on the correct processing parameters for their resins. The training program would progress to having operators, setup personnel, and some production personnel becoming certified in their professions.

Training of other plant personnel was scheduled for their respective job functions and a review and audit was performed for accuracy of their work instructions in accordance with ISO9001 quality certification requirements.

Ranked 8, Item 8. Production was charged with determining their molding machine, tool, auxiliary equipment, and support equipment's needs. This included matching injection molding machines to tools, cooling requirements, routing of coolant lines, flow rate requirements, etc. Also, an insulation tool pad was installed to isolate tool and platen for better temperature control. This reduced the platen from drawing heat from the tool and maintaining the tool cavity temperature as intended with the cooling system.

A master schedule of all molding machines and tools were matched for capability. All associated equipment and tools needed to setup and process an order was developed along with equipment and process setup checklists. Equipment was identified, placed in staging areas with the setup crew responsible for issue and return of these items. Also, auxiliary equipment was placed on a maintenance schedule for ensuring any item required for a production run would be available and capable of performing the operation.

Production also began tracking setup times per tool and machine and who the set up personnel were and equipment required was staged for easy access and identified for use at each setup. This included interfacing with purchasing to ensure the correct material was delivered on time for production.

Quality assurance, working with engineering and production, were charged, among other duties, with verifying the correct material requirements was given to purchasing and the required certifications required and obtained from their suppliers. Also, incoming material was verified as correct by incoming inspection and tested if applicable and then stored in a known, secure, and protected area of their warehouse awaiting production drawing the material, drying it and delivering it to the production floor clean and dry.

Ranked 9, Item 5. Only after all these actions and improvements were completed could a new evaluation for each injection molding machine capability be conducted. There was also time allotted to evaluate specific machines while the changes were being competed. This was to evaluate how the completed changes had on the cycle repeatability and reduction of the reject rate.

This was good mid-program information to give to management to increase their confidence that the changes, time, and expense were working in their favor to lower cost and increase productivity.

As each of the nine items were completed a review of the teams improvement programs versus the work performed is documented and presented to management for their information on the programs success. The cost reductions were plotted and the entire plant made aware of the successes in productivity. Also, emphasize when done correctly and improvement made it is only a "one time" cost to maintain a much lower "overhead" charge for each program.

Often during an improvement program additional plant or process areas need improvement, as the ozone water treatment system. These areas are brought to the attention of the Six Sigma team who take them to management for their consideration for fixing or consideration later in the program. The ozone treatment system will cure a coolant flow and heat transfer problem in the plants equipment cooling systems. Water treatment like air flow in a plant, is often the last item considered.

If this problem had not been detected, all of the plants heat exchanger efficiency would have been decreased and very expensive repairs or replacement of equipment downtime heat exchanger cores would have been necessary.

During the plant cooling system modification flow rate meters were installed. The flow meters monitor the effect of the ozone treatment system. As the flow increases, monitored at control points in the system, the plants chilled water capability will be increased. This may save having to increase the systems capacity just by improving the flow of coolant in the system. Cycle repeatability and control is very dependent on the steady state coolant temperature in the plant

Monitoring sites for temperature measurement were identified with instructions posted with a schedule for collecting the data.

When all program items were completed, the tower and central chiller system was balanced for uniform temperature control for the entire plant. Each machine had it's own cooling circuit, not a series piping system form one machine to the next. Also, molding machines designated for tighter tolerance products had a separate cooling piping system installed from the central chiller to assist in temperature control of the tool and machine.

During the assessment of the tool/machine capability the use of quick mold change was considered. The quantity of tool changes were many plus it took to long and was judged very inefficient. This resulted in a program to reduce tool change time and investigate why the frequent tool changes in the schedule. This was a new program to be considered for the next stage of

Six Sigma improvement. Data was now being collected and all machines were scheduled to have their CP and CpK values recalculated.

Step 7. The program of data collection progressed for the next six months. During this period data was evaluated for improvements in the plant and manufacturing operations that resulted in more worthwhile programs.

The plants cooling system modification resulted in much tighter temperature control for tool and machine temperature control. The recommended temperature difference between tool coolant, in and out temperature is $\pm 3°$ F with a controlled water temperature tool cavity temperature was easier to control shift to shift and season to season. Some tools were not capable as the cooling system was a series flow from cavity to cavity. What was found was the difference of inlet water temperature versus outlet water temperature had been reduced by 20° F due to improved water temperature control and flow rate in the system and tool. A turbulent flow could be maintained for maximum heat transfer in the tool cavity.

Automated temperature recordings were established for the cooling system. These were later to be wired in a central recording system in the production office for continual monitoring with a warning system for temperature out of the required range for the plants tools and molding machines.

CP and CpK values all improved with the modified cooling system but it took several months. Production, with engineering assistance implemented planned maintenance, variable monitoring, and recording, and evaluation of setup change over times and operator efficiency and operation training. Also, with improved productivity of machines and systems, production scheduling of tooling for specific injection molding machines yielded reduced scrap and operating cost with increased output of higher quality products.

During the program it was determined that a more efficient lighting system in the plant would assist in aiding setup, production improvements, and personnel satisfaction. The attitude of the plant personnel also improved as they realized management was behind and leading improvements of product, their job security, and worker satisfaction. Everyone now considered themselves part of the company and customer quality Six Sigma improvement team.

REFERENCES

1. Harry, M. J. "The Quality Twilight Zone." Quality Progress February 2000: 68–71.
2. Harry, M. J. "Six Sigma Focuses on Improvement Rates." Quality Progress June 2000: 76–80.

Chapter 4

Design Your Operations for Six Sigma Manufacture

Designing for Six Sigma manufacture implies more than having four defects or less per million parts manufactured. Many manufacturing operations are for part quantities of lower volume. Also, the products manufacturing schedule is broken up into quantities delivered per order or built to meet week, month, or quarterly scheduled delivery orders. Often blanket product orders are built ahead to allow openings for other unscheduled orders that come in from their customers.

Custom molders and other suppliers schedule their operations to meet customer delivery dates. How they manage their business to complete a time periods product quantities is their production control department's decision. This then leaves open machine time to schedule in small production runs to keep their overhead cost down and to efficiently utilize their open machine time. This is also true for major part suppliers as their utility of manufacture can vary from product demand to seasonal changes and unscheduled customer requests for products. Therefore, with changes in tools, materials, personnel, and operation variables to meet a Six Sigma performance, meeting output demand with quality products is difficult unless Six Sigma principles are applied at the start of all product programs.

SIX SIGMA DESIGN MATRIX

The Design Matrix Diagram, Figure 1, follows the prescribed flow of input information through manufacture. Teaming with sales, design, production, quality assurance, your material suppliers, and the customer is the recommended procedure for a successful product launch.

Leaving even one of these groups out of the design program can cause a future problem some where in the down stream design and development for

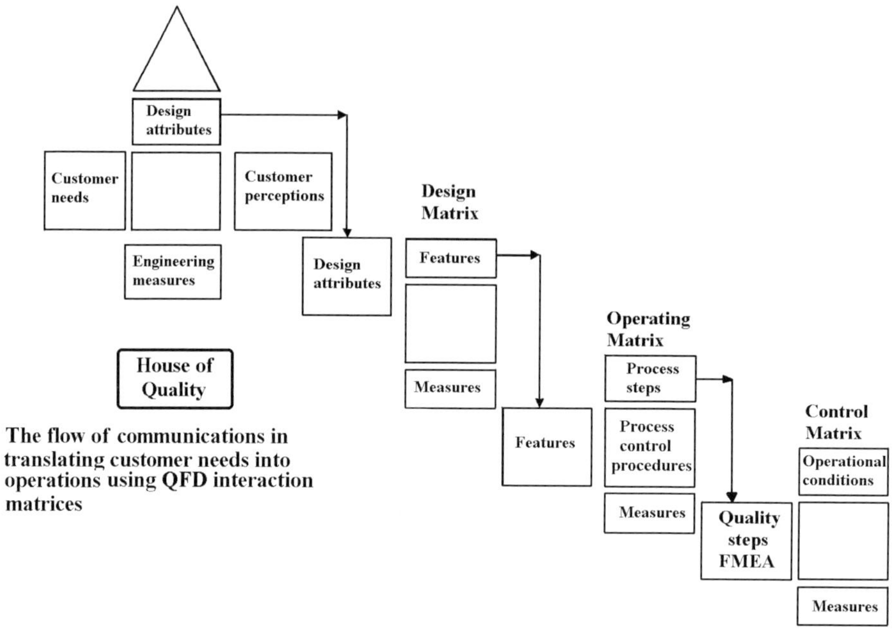

Figure 1. Flow of communications with QFD (quality function deployment). (Adapted from reference [7])

manufacture program. Where it may occur depends on the groups input if not all personnel are utilized at the start of a new program.

ELIMINATION OF SEVENTY TO EIGHTY PERCENT OF FINAL DESIGN PROBLEMS WITH SIX SIGMA

Eliminating customer design and manufacturing problems before the product design is finalized is the intent of using Six Sigma procedures. Often good design ideas and material selection does not have sufficient "up front" evaluation time by engineering and the evaluation of end use requirements of the customer.

Meetings with the customer and any of their outside service or product suppliers can be used to develop the product requirements that can assist the supplier in the manufacture of the product. All variables in the product must now be addressed and answered during this pre-design and production time period. The customer can also be your own engineering department meeting with sales and marketing to determine exactly what is required for the end use customers product. QFD can also be use for analysis of customer and product requirements in conjunction with specific check lists. A sample of the Design Check List is shown in Appendix A.

A list of product design requirements can be listed and a value of importance determined vs. the product, layout, aesthetics, and functional factors of interest to meet the customer requirements.

A list of product design requirements using the House of Quality format is shown in Figure 2, for a laptop computer housing. These answers can also assist the engineer in material selection for the product. Similar matrixes can be developed for all areas of design and manufacturing of the product including the tool, assembly, and decorating process control and quality limits.

After these matrixes are completed production can develop their process control procedures and with input from quality assurance can develop the

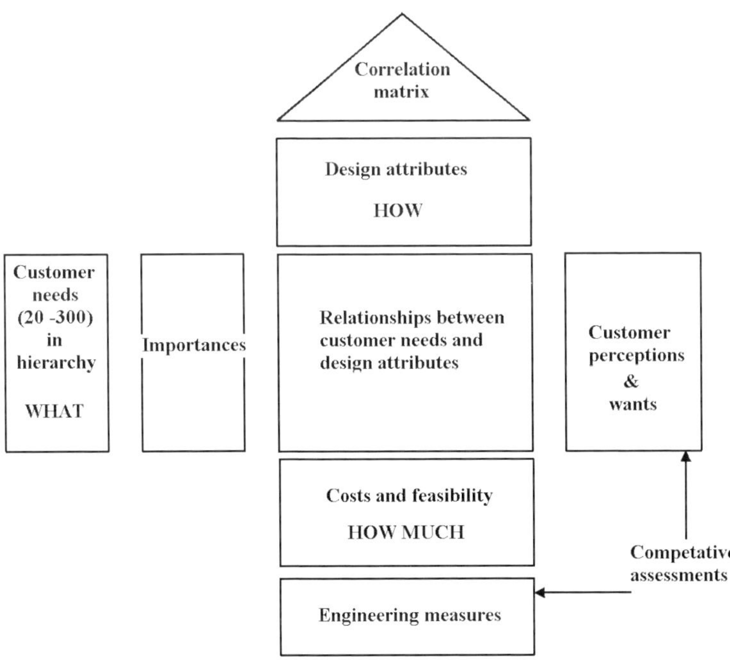

Figure 2. House of quality.
(Adapted from reference [7])

FMEA process sheets for mapping the entire manufacturing, decoration, assembly, testing, and packaging operations. An example of a typical Process FMEA form is shown in Figure 3, for the control of the quality for manufacturing where possible problems may occur and where required inspection points are for a laptop computer housing.

The FMEA is developed from the manufacturing process control procedures developed by the manufacturing engineering department of the company. When these are done in conjunction with each other, there can be a realized savings in both time and labor for each department's personnel.

Line Description: ____________________
Machine Description: ____________________
FMEA Number: ____________________
Core Team: ____________________

Potential equipment failure mode
Effects analysis process FMEA

Prepared by: ________________
FMEA Date (org.) ______ **(Rev.)** ______
Page ____ **of** ____

Risk priority number

Process Function Requirements	Potential Failure Mode	Potential Effects of Failure	Severity	Class	Potential Cause (s)/ Mechanism of Failure	Occurence	Current Process Controls	Detection	R.P.N.	Recom-mended Action(s)	Person respon-sible/Target Completion Date	Actions Taken	Severity	Occurence	Detection	R.P.N. *

Figure 3. Process FMEA form.
(Adapted from reference [2])

Input from the quality department may save duplication of efforts and vice versa with the manufacturing department if an inspection or other operation is performed in another department it may be eliminated in the new operation. Also, if done jointly or separately, always review the procedures with the departments key personnel to ensure the procedures are correct and follow the operations of the manufacturing or business process.

FMEA's are developed by quality assurance engineering. This is accomplished after the manufacturing or process control plans or procedures are written that are used to control the manufacturing process. These manufacturing process control plans ensure the flow of material, equipment, processes, and that all information is available and personnel trained to complete the manufacturing process accurately.

Integration With Other Quality Initiatives

The FMEA matrix is a form of QFD interaction that solicits input at the beginning of a program and throughout the development, manufacture, and customer acceptance of the product except it is supplier-manufacturing information based. FMEA's investigate risk points in the manufacturing processes in a step-by-step investigation process as the product proceeds, sequentially, during the manufacturing process.

Management must insist the FMEA interaction mapping is performed to identify and prevent problems during manufacture that need recognition to be avoided once identified. Agreement of where problems can occur and installing measures for their prevention by all departments will result in a successful product launch program.

The design matrix flow should proceed as follows for a new product or material replacement program. This is a typical flow for any new or redesigned product launch. The process follows these ten typical "design for production" steps.

DESIGN STEPS FOR SIX SIGMA PRODUCTION

1. Marketing/sales interface with the customer and bring ideas to design engineering. Design engineering seeks advice for customer acceptance from sales on product improvement and cost savings.

2. Management and the design team develop a matrix of operations with timeline and action items with a decision to proceed. Design develops the product based on a projected materials properties and customer and product design requirements. Product models are created using STL (sterioleothography) or other method for show and tell or testing to gain approval of the design to meet end use requirements. Material suppliers interface with design for their recommendations and cost studies are performed to meet cost objectives.
3. Production, molding, and material suppliers, tool designers, interface with design with input from quality assurance and the customer, using their design, manufacturing, purchasing, etc. check lists to develop answers to all product manufacturing questions. Tooling costs are estimated and a method of amortizing their cost over the life of the product are determined and often used to cover the molding cost in the products pricing.
4. The tool designers recommend part layout, cavity cooling, ejection system, and number of cavities for the tool to fit into productions injection molding equipment. Cost estimating is used to determine the 'piece part price' of the product. Input from the material supplier, tool builder, production for machine rates, and other related product departments is collected to calculate the estimated price of the product, even before any metal is cut for the tool.
5. Production can then finalize their production requirements; review machine schedules for open time, and complete the cost of manufacture estimate using machine hour rates established for the manufacturing machines. Quality assurance, based on customer requirements for part tolerances, work with design, the customer engineers, if applicable, molding, and production to finalize critical dimensions and inspection points required by the customers product to meet their specifications.
6. Any discussions of the number of critical dimensions, part colors, and inspection and control methods to verify product quality is maintained within manufacturing and customer specifications are now finalized. Any agency certification requirements are discussed and how data will be collected and sent with the products for their evaluation and re-certification.
7. A prototype model or tool is now usually built. It's layout and design should be as close as possible to the final tool design with the layout for the cooling, fill and cavity gating as will be specified in the final

production tool. Prototypes are made in the selected material with production producing parts as close to production parts as possible in a stabilized and repeatable cycle. This is very important as dimensions, part data, and appearance (aesthetics) factors will be obtained from these first sample parts, as may be cycle times for the initial cost estimate of the product.

8. Product is then assembled, tested, and any changes required are discussed by the team and product, prototype molding are modified to produce the final product for testing.
9. This can be a lengthy or short period of time depending on many factors such as field testing and acceptance. The goal is to produce a product, test it, and then with as high a degree of reliability and low manufacturing risk go into final production molding. This is often based on using the prototype data to cut initial product dimensions, tool cooling layout, and part ejection and finishing requirements.
10. At the completion of final product test and sign-off for production the final production tool is built and evaluated by production in the cycle selected and parts evaluated. Any tool dimensional changes are now made with additional trial production runs to see if any fine-tuning is required in the tool before final release to production. In many cases, if the material shrinkage is in doubt in certain areas of the tool due to gating or other reasons, the tool cavities are not cut to their final cavity dimensions until the production pretrial run is completed. Then based on the results the final dimensions and part finish is completed for each tool cavity.

During the preproduction evaluation, typically conducted on the selected for production injection-molding machine, the tool is trailed on the estimated production cycle. This is the period when final cycle time and process variables are finalized with production and quality assurance collecting data on dimensions and any other product and process variable requirements. This is also the time to run a CP and CpK evaluation of the process. This information will assist the production team in ensuring the tool, material, cycle, machine, and personnel are as capable as possible. The software evaluation system also evaluates the molding machine for any operation and machine wear deficiencies prior to running full production.

Also, incorporate in your analysis the service and repair possibilities of the product if applicable. The ease of service is very important for any product especially if self-service for replacement of expendables is to be required as in copiers and printers.

During the initiation of a new Six Sigma program the cost of manufacturing capital expenditures estimates must be considered. These are broken down into six life cycle phases of cost. These can assist the team and management define the cost of projects and doing business beyond the traditional accounting metrics when identifying potential costs and benefits.

**Six lifecycle phases in capital expenditure projects*

There are at least six lifecycle phases inherent in most manufacturing-based capital expenditure projects, according to "Business Justification of Open Architecture Control*, White Paper Version 1.0" by the open Modular Architecture Controls (OMAC) Users Group's Working Group on Business Justification of Open Architecture. The six phases can help project organizers think beyond traditional approaches when identifying potential costs and benefits. The six phases are:

1. Justification-concept development and initial project planning; examination of strategic business direction and definition of upgrade to address particular business needs; evaluate one-time and recurring costs and then benefits.
2. Design and development-time and costs of detailed design of machine or system, can be minimized if adequate planning and work is done in justification phase.
3. Machine acquisition, build, installation-time and costs of building, and installing capital equipment, such as control system, support equipment, test and evaluation, network and direct numerical control (DNC) capability, as well as training.
4. Product production, operation-includes day-to-day cost of supplies, labor, power, more training, data collection, and planning and scheduling.
5. Maintenance-costs of support personnel, training in maintenance and troubleshooting, downtime, spare parts, software maintenance and configuration control, and preventive and optimization maintenance.
6. Reconfiguration-improve/disassemble, phase out – includes costs of recovery or salvage, reclamation or disposal, overtime or outsourcing during changeover, and retrofit, rebuild, or buying new equipment.

There are a multitude of quality tools and methods that can be used. These will be described and shown how best to be utilized during the Six Sigma implementation programs. The selection of these will be the decision of the Black Belt team leader for the company improvement team.

FAILURE MODE AND EFFECTS ANALYSIS (FMEA)

The FMEA is a very analytical, informative, and supportive quality tool for both business and manufacturing. It can be replicated in many forms to suit all facets of a companies business, manufacturing, and service operations. The use of the FMEA is required for QS-9000 and for ISO9000-2000. Many companies have not used the FEMA to its full intent and potential for analyzing what problems can occur at specific points in the business and manufacturing operations. It is not always used to its full extent by knowledgeable quality personnel when pushed to find solutions to frequently occurring process and product problems.

It is the only metric used today other than the fishbone that examines a process, when applied by a knowledgeable quality person that will map out all of the operations and evaluate the multitude of variables of a specific process for potential problems. The FMEA is typically implemented after the manufacturing process control plan and procedures written for each manufacturing operation by the manufacturing engineering department. Quality engineering follow the manufacturing control procedures and looks for areas that could possibly cause a product or processing problem during manufacture.

The FMEA will provide a company when combined with the other quality tools as the Ishikawa, or fishbone, an analysis diagram or map of all steps and variables in an operation in a sequentially ordered manner. The fishbone diagram when used correctly will list the process variables where potential problems may occur. The fishbone evaluation method assists the evaluator to break down the functions operations, either business or manufacturing, for all potential sources of system variability. This shows information of any potential problem area that could cause a problem to be identified and then investigated for preventative solutions before manufacturing even begins.

The fishbone started at any point in the process can identify all variables in the process prior to this point and allow the engineers to ensure they are controllable and maintained in process control to ensure a problem does not occur at this or other points further into the process.

FMEA's can be used to map and analyze all of the companies operations from sales through service requirements. They are recognized and have become widely used and are required for QS-9000 certification. Used as a controlled map of all a company's operations for all business and manufacturing operations and in any area where personnel can anticipate at an operation where a potential problem may occur and what variable could have caused the problem. A well thought out and developed FMEA will follow sequentially all operations, business, manufacturing, and service.

To develop a useful FMEA requires a through understanding of the operations performed in a department and at each step in the process. It is important that a team is used to map out the FMEA flow plan to ensure no potential operation or problem point is not overlooked. The FMEA maps out all operations and steps performed for a single manufacturing operation. A FMEA is an off shoot of the manufacturing process control plan or a flow diagram of what functions and work operations are performed and what potential (failure) problems may occur. Data recorded includes station and operations plus potential problem types, their effects, severity, cause, occurrences, methods in place to control the operation, actions to take when they occur, who is responsible and results of actions taken to eliminate the problem from occurring. This was shown earlier as Figure 3 for a process FMEA.

To assist in reaching Six Sigma control the FMEA is an analysis tool used to map out in an ordered manner the process flow of an operation with all likely and potential problems identified, ranked, and should a problem occur, what operation personnel should do if a problem occurs during their daily business operations.

Like manufacturing process control plans for mapping manufacturing operations, FMEA's follow the same sequence of steps for each operation but in a different sense of analysis. They are developed in a sequential process like procedures and work instructions in ISO9001, they follow a preset implementation operation, not in as great a detail, but addressed to what operations are performed by department personnel in their manufacturing steps and work instructions.

The major advantages are they identify potential problem points if procedures and instructions are not followed. When this happens there is the probability that a process control variable or work operation has gone out of control. They also assist in knowing and deciding what to do if the problem should occur. This information can now be well developed with potential solutions and best of all preventative measurement to ensure it does not happen. But, if it does and solutions do not work then the only solution at that time is to shut down the process and seek help.

ISO9000 assumes everyone does their job and operation correctly and if done so, no problems result from personnel reasons. Of course, it does not take into effect material, processing, manufacturing variables that are never in a steady state operation and constant.

Therefore, in an imperfect environment, problems occur for many personnel, material, and equipment reasons. Just like keeping a documented account of machine settings and problems and how they were solved. A FMEA correctly written aids in the prevention and solution of future manufacturing problems.

They should be written and documented for all existing business and manufacturing operations. This includes both new and existing operations to locate and identify any problem areas in a business or manufacturing operation. Preventing problems is the point your quality dollars should be spent not in correcting reoccurring problems.

FMEA's focus on the "what if" and then how, and who, will respond and with what, preplanned actions. Of course all problems are not exactly alike and the FMEA is only a guide or road map for an operation. The analysis and question responses on the FMEA hope to identify, classify, and respond to a problem type that could occur and how it should be handled if it does. It is a starting point for preventing a problem and a recommended plan of action and responses to the problem.

USE OF CHECK LISTS FOR A FMEA

FMEA's can be used to also evaluate what information may be lacking at the workstations for supplementing the operators information and problem solving checklists. These check lists for problem solving are often provided by material suppliers or equipment manufacturers when used for trouble-shooting a problem for a manufacturing operation or any business or service process.

USE OF TROUBLE SHOOTING GUIDES FOR A FMEA

All major material suppliers have their "Trouble Shooting Guides" for solving processing problems with their materials for the most common problems they have identified. Check lists like this say, if the outcome of the process is this, to solve the problem try this or this to get back into control. Very helpful when combined with an experienced operator or setup technician to get the process in control. But, not all processing problems are listed and some can result from plant, equipment, tooling, and personnel problems. But the ultimate cause and correction may be out of their area of expertise and operation. They then need to communicate the problem, their attempts at a solution, and what variables (s) must be brought into tighter control to eliminate the problem from reoccurring. Teamwork is often required to solve very complex processing problems requiring personnel and supplier personnel performing fish bone or C&A analysis of the problem to reach the root cause of the problem. All of these items must be implemented, monitored for effect, and changed when necessary to achieve and then keep a process in Six Sigma control.

Developing Your FMEA Team and Form

Writing the FMEA is the responsibility of quality assurance, with assistance from manufacturing engineering from the department where the operation is performed. A manufacturing check list can be used to ensure all relevant questions are answered for each operation. The processing control plan and procedures for this operation, a manufacturing cell, are reviewed for the cells operational flow of work and correctness of the work instructions. A team is formed composed of personnel who have a stake in the process. In this example, a molding cell maintenance FMEA is developed. The team can be composed of the following categories of personnel with the team leader selected for their knowledge of the operation. The general FEMA form to use is shown with the heading for the operation to be examined is shown in Figure 4, and a completed FMEA for a maintenance operation is shown in Figure 5.

FMEA TEAM DEVELOPMENT

Maintenance manager – shift foreman, mechanics, electricians, pipe fitters, etc.

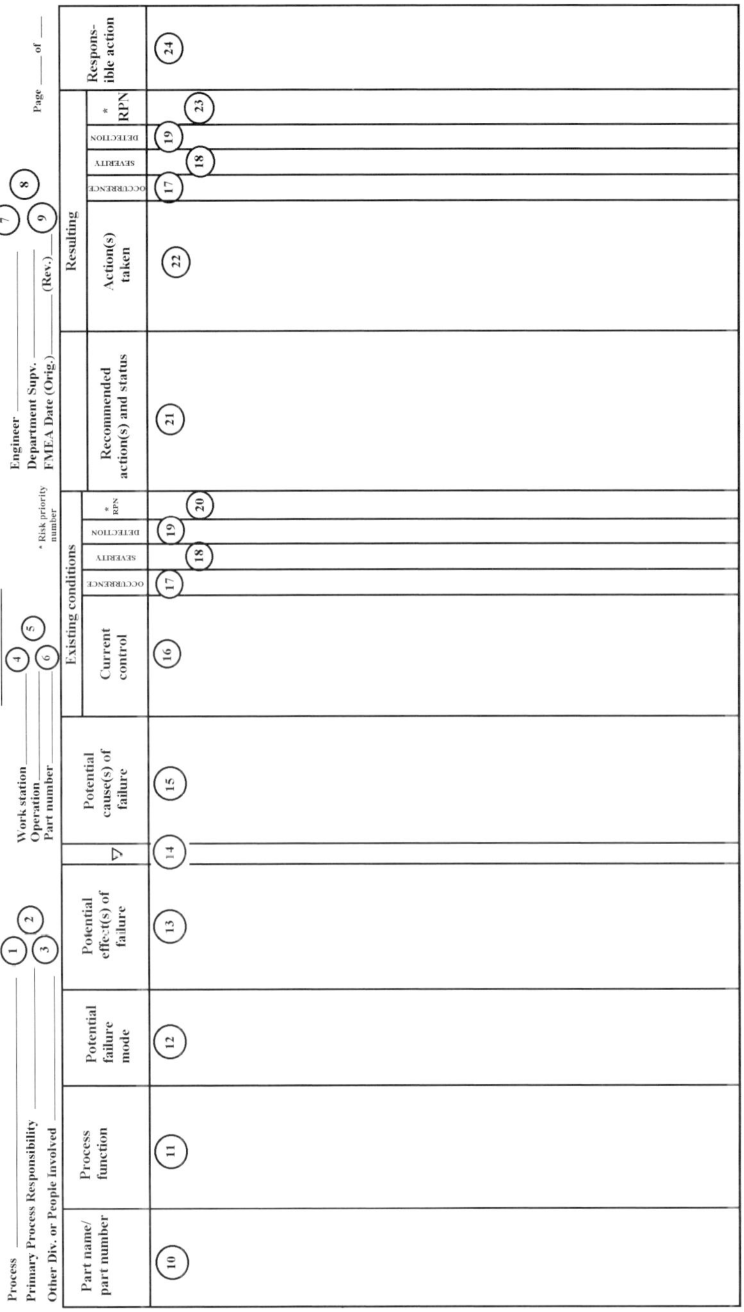

Figure 4. Process FMEA.
(Adapted from reference [4])

production manager – shift supervisor, senior operator, assistant production manager/supervisor

setup technician – mold technician, mold maintenance, mold room supervisor

plant engineer – purchasing representation, systems engineer, plant maintenance

quality assurance – quality assurance engineer, floor inspector, ISO/QS auditor or coordinator

Potential Equipment Failure Mode Effects Analysis Process FMEA

Line description: Semi-Automated Assembly Line
Machine description: Transfer Unit
FMEA Number: 64-5791
Prepared by: QM (Quality manager)
FMEA Date (Org.): 12-Nov. 2001 **(Rev.)** A **Page 1 of 1**
Core team: MM (Maintenance Manager), MB (Maintenance buyer), LHM (Lean hand mechanic), PS (Production supervisor), LO (Line operator), QM

Equipment Process function	Potential Failure Mode	Potential Effect(s) of Failure	Severity	Potential Cause(s) Mechanism of Failure	Occurrence	Current controls predictive methods	Detection	R.P.N.*	Recommended actions	Person responsible/ target completion date	Action Results Action taken	Severity	Occurrence	Detection	R.P.N.*
Transfer unit-main drive Drives the printer mechanism	• Chain breaks. P/N X 03-31-34	• Transfer unit down • Post-repair set-up problems	5	• Misses limit switch • Broken limit switch • Stretching due to normal wear	8	None	10	400	• Replaces limit switches with heavy duty type. • Install tension detector • Install cycle counter. • Monitor cycles. • Evaluate changes to improve accessibility. • Spares: chain & blocks mounting to chain. • Write repair procedure: describe tasks, change sprockets with chain, periodic lube and preventive replacement.	MM MM PS and MM MM and MB MB MM, QM, and PS					
Transfer unit lift Lifts the fixture support	• Chain breaks • Chain stretches. P/N Y03-31 Y03-30 2 each	• Unit down	2	• Wear • Misalignment • Lube	1	None	8	16	• Monitor cycles • Write and train repair procedure	PS and MM MM, MB, and QM					
Stroke control Controls the length of the stroke	• Chain breaks • Chain stretches P/N X-04-43	• Unit down Post-repair set-up problems	1	• Wear • Misalignment • Lube		None	8	8	• Monitor cycles • Repair and set-up procedure • Change chain with backup rolls	MM and PS MM, MB and LHM					
Turn-around drive Turns the fixture around and back to original position	• Sprocket wears • Sprocket breaks P/N 31-56	Sprocket becomes loose causing misalignment Causes chain to break	1	Set-screw loosens due to shock and vibration	8	None	10	80							
Frame Supports the fixture when it holds the part	Frame twists Frame bends P/N X-03-88 P/N Y-03-77 Two each P/N 03-03-99	Gradual decline in assembly quality Gradual decline in productivity	1	A failure which induces unusual shocks - chain breakage Improper storage	10	None	10	100	Check after a chain breakage Spares P/N 19 & 15						
Carriage Moves the fixture back and forth	Wear	Machine down	1	Friction	9	None	10	90	Periodic inspection and replacement as needed Spares	MM and LHM					

Figure 5. Examples of a maintenance FMEA.
(Adapted from reference [4])

ASSUMPTIONS FOR THE FMEA PROCESS

Prior to developing any assumptions, a fish bone layout of all maintenance, material, plant support and operation variables that control the molding cell are identified and mapped out for discussion and possible inclusion in the completed FMEA. From this a list of the molding cells machines, primary, secondary, and auxiliary is developed. Any plant systems affecting or interacting with the operation are listed as secondary units, even though they may not be considered as important for maintaining the process but could be a contributing variable to affect the quality and processability for the molding cells operation. Also, consider plant services, power, water, and environment in the plan, as it will affect the unit cell and quality of output.

A team decision is made as to how to consider these variables. As individual variables unique to a specific molding cell or in their total variability to the entire plants molding operation. In our FMEA, outside processing support equipment and functions will be considered separately, or as individual maintenance FMEA's for the entire plant. These support FMEA's will be referenced in the notes of the FMEA as they are used to support the cells molding operations. Too often these support FMEA's are forgotten for their importance and support of the manufacturing operation. Always consider all input from every contributing source to solving all types of problems. Only equipment or machines specifically located in the molding cell will be considered in the molding cell FMEA analysis. These are as follows:

MANUFACTURING CELL EQUIPMENT AND OPERATIONS

500 ton molding machine
auxiliary tool chiller
portable grinder for reclaiming material
hopper dryer/feeder
tool and setup at press
conveyors/robots support
operator actions required at press
secondary operations performed by operator
process setup instructions and data sheet
Quality monitoring process used and data to be recorded for process control

MANUFACTURING CELL EXTERNAL SUPPORT SYSTEMS

Systems outside the cell are for separate FMEA's are:

plant cooling water system, tower or central chiller
material drying system (central)
plant environment, temperature, humidity, and air movement and quality
support systems, air, power, material supply, etc.
tool maintenance
injection molding machine maintenance

Assumptions that can be made before writing the FMEA are:

1. Equipment meets process requirements.
2. Tool is capable with current updates and maintained separately.
3. Electronic controls to maintain system in control, calibrated and maintained by electronic/electrical shop in house, only minor support items in stock, fuses, etc.
4. Maintenance has responsibility for all items listed in manufacturing cell.
5. Backup items (replacement tool parts listed on parts list) are not stocked, rely on vendor for support.
6. If support equipment fails, a backup may be available but not always in the same capability range for the manufacturing cell. Could cause downtime until replacement available or unit fixed.

SUPPLIER AND CUSTOMER – PAST AND FUTURE REQUIREMENTS

In the past before beginning a move toward Six Sigma, the company's equipment was not always capable. Production had to be made and all resources were bought to bear to solve a problem, no matter what the cost. If a problem occurred to immediately solve the problem, the tool was moved to another available molding machine with out concern to its capability or size of fit for the material and tool. Defects and startup cost and material and product losses in scrap were an acceptable cost of doing business but now with increased sales; more demanding quality from their customers and being ISO/QS-9000 certified, down time cannot or will not be tolerated for dedicated molding cell production.

MAINTENANCE ONE OF THE KEYS TO SIX SIGMA QUALITY OPERATION

Management now realized they must improve their quality of maintenance for increasing business and maintaining quality. A new maintenance department philosophy was introduced:

1. Maintenance activities must be dedicated to ensuring all equipment and systems are capable of achieving a steady, efficient production rate of products to meet their customer requirements.

This resulted in reorganizing the companies maintenance program to be proactive instead of reactive. A supplier-client relationship resulted with maintenance, the service provider, production the customer, and the company the final benefactor of the maintenance services.

PREVENTATIVE MAINTENANCE THE COMPANY GOAL

A list of required items, to be incorporated in the program were initiated.

1. Maintenance activities must be dedicated to ensuring all equipment and systems are capable of achieving a steady, efficient production of products to meet their customer's requirements.
2. Develop a list of supplier recommended spare parts to be available for each long lead item.
3. Prepare a checklist and maintenance plan with a schedule of preventative maintenance operations to be performed at set, supplier recommended internals. The schedule is modified with in-house failure rate data for estimating actual service requirements.
4. Review equipment and system failure problems, Parato charting, actual down time, and equipment and services required to repair, get system back on line, and skill level of employee to diagnosis problem and implement repair.

5. If not in-place, develop individual equipment problem, repair and parts list of most frequent failure items. Review this with item two to ensure sufficient parts are always in stock or readily available overnight if to high a cost item to have in stock.
6. Review maintenance department skill and manning levels to ensure capability is available in-house to support maintenance requirements and repair when they occur on all shifts.
7. Develop rapid response repair teams with required equipment and tools for analysis of a problem. Repair stations are centrally located within the plant with equipment specific to the operations performed in the surrounding area available for use when necessary. Tool kits are identified in each maintenance shop for all shifts use with the supervisor responsible at the end of each shift that all tools are returned to the tool kits unless the repair is still being performed.
8. Establish estimated maintenance down time requirements along with personnel requirements to let production know effects it will have on their production schedule.
9. Provide work instructions and equipment manuals with part identification, for each piece of equipment in the maintenance program so that the correct replacement part is used and installed correctly.
10. Once the maintenance system is established, audit all first time maintenance procedures and work instructions to ensure they are accurate, complete, and within anticipated time requirements for service support.
11. Develop a communication system with production and plant personnel to inform them when scheduled maintenance is required, to coordinate maintenance with minimum disruption of their production schedule.
12. Document all problems for each machine or major piece of equipment in the 'Problem Solution Book' with the recommended method to use for developing and diagnosis a solution to a problem. In the Solution Book should be a list of commonly replaced parts and their part number for easy locating in your inventory. Also, if special personnel are required to correct the problem they should be listed with their home phone numbers if off shift and they are necessary to contact when a problem occurs. Also, any preventative solutions that have been developed to prevent or reduce the time interval between a wear or failure problem. Document the personnel required and time necessary to solve and repair the problem.

The results from developing a well coordinated and timely maintenance schedule will result in the following benefits to the company production schedule and product quality.

BENEFITS TO COMPANY PRODUCTION SCHEDULE

1. Reduce mean time between failures.
2. Ensure Six Sigma process control can be implemented and then maintained.
3. Reduce machine down time.
4. Ensure customer delivery of product will be on-time.
5. Reduce defects and problems during production.
6. Reduce maintenance cost, less overtime, and freight cost of emergency repair parts.
7. Provide company with improved JIT and lower inventory levels for all customer products.
8. Reduce cost of manufacturing.

The FMEA form used for analyzing a business or process are almost identical with minor changes in headings across the form to best describe the operations under analysis.

Differences are shown in Figure 6 and Figure 5, with a process versus maintenance FMEA column heading changes. Only two columns are changed for the maintenance FMEA along with eliminating the Class column. Process function/Requirements is renamed Equipment/Process Function and Current Process Controls to Predictive Methods /Current Controls.

USE OF THE MAINTENANCE FMEA

Of the 16 columns in the maintenance FMEA only five; equipment, severity, occurrence, detection and predicative methods are treated differently from the Process FMEA as now described.

P = Process

M = maintenance

Line Description:____________________ Machine Description:__________________ FMEA Number: ______________________						Potential Equipment Failure Mode Effects Analysis Process FMEA					Prepared by:_________________________ FMEA Date (Org.):________(Rev.)____ Page ____ of ____						
Process Function Require-ments	Potential Failure Mode	Potential Effect(s) of Failure	Severity	Class	Potential Cause(s)/ Mechanism of Failure	Occurrence	Current Process Controls	Detection	R.P.N.*	Recom-mended Action(s)	Person responsible/ Target completion Date	Action results Actions Taken	Severity	Occurrence	Detection	R.P.N.*

Figure 6. Headings for a process FMEA form.
(Adapted from reference [4])

Equipment
P-FMEA (process FMEA) describes process function under analysis
M-FMEA (maintenance FMEA) enter equipment, subassembly, or component being analyzed

Severity
P-FMEA, typical results in process down time or high defect rates being produced.

M-FMEA, down time, internal may be short or require extensive time affecting production schedule if replacement part cannot be interchanged to keep process on-line. Also, consider;

1. Available parts in stock or special order, with known supplier lead-time to deliver. Cost of critical spares in your inventory with established frequency of failure. Each equipment supplier should have an estimated MTBF (mean time between failure) rates on their parts.
2. Installation time for repair to be completed after receipt of part could also identify individual technician's repair times to affect the repair. May make a difference on how soon the system is running again. Also, the slower technician may require more intensive or special training.
3. Any special installation tools, supplier specified, and testing when needed to verify if part or fix was correctly installed. List the services of other maintenance support personnel when required and their equipment requirements.

Each company must establish their own severity rating system similar to Table 1. The rating should consider in-house stocked repair parts and lead

Table 1. Down Time Severity Ratings.

Duration of the Breakdown (in Days)		
Equal to or more than	Less than	Severity Rating
5	–	10
4	5	9
3	4	8
2	3	5
1	2	2
0	1	1

(Adapted from reference [2]).

time penalties for non-stocked special parts. A down time maximum limit can be established with the highest rating that may occur if special parts are required. From this number, the severity rating is decreased, as repair time is less.

To keep the lowest ratings the following pro-active actions would be necessary depending on the situation.

1. Carry replacement parts – in duplicate and when used, an immediate reorder is initiated. On-hand inventory depends on number of machines using similar parts and MTBF rates for replacement.
2. Bar code all replacement parts and at issue are scanned out of stock along with the repair technicians badge bar code. The transaction also alerts purchasing that a part has been taken out of stock and should be replaced at an interval previously established based on the MTBF records. This will automatically track inventoried parts used and by what technicians. Makes record keeping simpler and ensures used parts are replaced.
3. Periodic maintenance, inspection, and verification if a high wear or failure rate per item is identified.
4. Black box (pre-assembled) replacement modules for rapid replacement, typically analog or digital machine control units.

Only in the most sensitive customer relations would stocking the customers order in advance be justified as for JIT using JIC (just-in-case) inventory methods, which are expensive and not always the best use of the companies money. This could be very expensive if and when an engineering change order comes in and the JIC inventory is not usable. This is upper management's decision based on each individual customers requirements for delivery of their production units on time.

OCCURRENCE

Frequency of breakdowns is internally generated from past experience and the age and condition of equipment plus the maintenance reports of equipment capability and frequency of maintenance performed. In Six Sigma process control any wear of a critical control item or variable controlling item, may cause a process deviation large enough to be detected on the process control charts. Therefore, frequency does not have to imply failure; just an indicator of normal wear, which is prevalent in all manufacturing operations and only, varies in degree and rate. Adjustments in the process can often be used to bring the process back into control without a defect being created as long as the process is being monitored in "Real Time".

All manufacturing equipment wears and as an example an injection molding machine processing some corrosive resins and filled or reinforced materials will experience wear in the barrel, screw, and check ring assembly. Also, the tools gate will, over time wear and become enlarged. This occurs along with other areas as in the tool cavity where the resin impinges and flows during fill under fast injection speed and high pressure. All wear areas, components, and machine and mold systems must be considered even the hoses connected to the mold for cooling water. All items in the operation must be considered in any manufacturing and process evaluation.

Be sure to not mix failures of one specific cause with others. Keep each separate and if unknown, run an analysis to discover the root cause of which you should know already the effect on the process.

As a result you may develop more than one frequency of breakdown occurrence table as Failures and then Wear Replacement Required, etc. as shown in Table 2. This will be good data maintenance can use to schedule normal evaluations of a piece of equipment or system. It will also give them an idea of what else to look for in their maintenance of the piece of equipment.

Table 2. Occurrence Rating.

Frequency of the Breakdown (in Months)		
Equal to or more than	Less than	Occurrence Ratings
6	–	10
5	6	9
4	5	8
3	4	5
2	3	3
1	2	2
0	1	1

(Adapted from reference [2]).

DETECTION FOR PREVENTION OF PROBLEMS

The last statement is an aid to detecting problems before they occur. The best use of company funds is in preventative maintenance not repair or corrective action when a system has failed. Also, have in the maintenance instructions, for example, if a bearing or bushing is worn and to be replaced, the other bushings in the assembly have also been affected but to a lesser degree. Therefore, replace all of the related items to ensure the system is back to like new after the maintenance and repair. Failure to do so, with the new part now in place, will transfer higher loading to the older parts possibly causing them to wear out faster, requiring more down time for repair. Also, verify shaft wear or other related problems to replaced parts.

This is where good maintenance and repair documentation can answer questions as to this particular problem and its recurrence and frequency.

Even when a maintenance schedule is in place, there are usually warning signs, noises, process control and product shifts that will indicate a problem is occurring somewhere in the system.

Instruct personnel to report any and all signs of a potential problem to their supervisors when detected. Do not wait until the end of a shift or a break. These early warning signs are very important to the well being of the company with a detection rating system as shown in Table 3.

Table 3. Detection Rating.

Duration of Warning Period		
Equal to or more than	Less than	Detection Rating
0	1 day	10
1 day	1 week	9
1 week	1 month	8
1 month	2 months	5
2 months	6 months	2
6 months	–	1

(Adapted from reference [2]).

PREDICATIVE METHODS

With improvements in Six Sigma process control there are predictive methods that can be implemented and used to monitor wear or a system leading to failure. Wear measurements on barrel and screw and check rings during normal maintenance or cleaning should be made. There are specified machine clearance dimensions that should be followed. Knowing what a mold action slides original measurement were and comparing them to current dimensions, when broken down for service and cleaning, can eliminate a potential, production problem. When personnel know the minimum acceptable wear a part can experience before a problem occurs, they can eliminate it from happening. But, they need to know the required tolerances and what to do when the minimum are reached or will shortly. Measurements accurately taken and documented can show the technician a wear pattern and rate for the item. Then they can, with supervision interaction, if the next maintenance action on the item will still have the part in tolerance or must it be replaced immediately.

Maintenance is one of the keys to a successful Six Sigma process control program. Well developed maintenance and process FMEA's will create "Real Time" preventative problem operations within the company for continued process control of business and manufacturing operations.

CAUSE AND EFFECT ANALYSIS

Cause and Effect Analysis (CEA) when correctly performed will assist a company in solving a problem. All problems are initiated by a cause. It may occur at any time in an operation or process, business or manufacturing. Problems can occur at the beginning of a program based on a poor decision of design of the product or the wrong choice of a material or subcomponents. Problems can be created anywhere in an operation even during storage or shipping. Problems often occur when the customer is using the product. When and where a problem occurs usually was caused by an effect that was initiated and occurred prior to the point of failure.

The problem of failure analysis is determining the actual 'root cause' for the problem. The solution to a problem is to determine the root cause. This involves the use of the quality tools plus experience and knowledge of the failure analysis team. The teams greatest difficulty is tracing the effects back far enough to establish the root cause variables resulting in the cause of the effect. Therefore, sufficient time and analysis effort must be expended to collect and ensure the exact cause is determined. This requires a full and detailed analysis of the problem so a hasty solution is not implemented that only masks the real problem to be solved. A form for use in the analysis is shown in Figure 7, that is very similar to the fishbone diagram used to develop our operational variables.

Simply stated the Cause and Affect diagram was developed to assist in solving processing problems. The four main categories of process problems are; equipment, material, methods, and personnel. There are also other areas that may cause an affect and the team members are to add those categories. The form is well suited to present findings after a solution has been implemented. The simplest instructions on how to fill out the form are:

1. With the form turned horizontal, fill in the right hand box with the effect, output, or improvement that is being portrayed or desired.
2. Label the remaining boxes to show the categories of potential causes of quality.
3. On each of the four diagonal lines, draw smaller horizontal lines to represent subcategories, and indicate on these lines information ;that is thought to be related to the cause. Draw as many lines as are needed, making sure that the information is not too crowed and is legible.
4. Use the diagram as a discussion tool to help better understand how to proceed with process improvement efforts. The diagram can also be use to communicate the many potential causes of quality that impact on the effect/output/improvement goal.

EXAMPLE OF CAUSE AND EFFECT ANALYSIS

One very hard to solve problem was a 'molded in' stress problem as it usually only occurs after the part is assembled and often in the hands of the user. This problem usually causes a stress crack somewhere in the product. A high stress concentration is created by a high stress in the material of the product due to design or processing conditions at critical section in the product. During the processing operation these high stresses can occur at a sharp corner, where the material is forced to turn rapidly and especially at sharp, ninety degree, corners.

Then during normal use at these critical sections, a crack is initiated that causes the product to fail. Failures can also occur during assembly at a boss

Six Sigma Team Cause and Effect Diagram

Name: ______________________________

Process: ______________________________ Date: ___/___/___

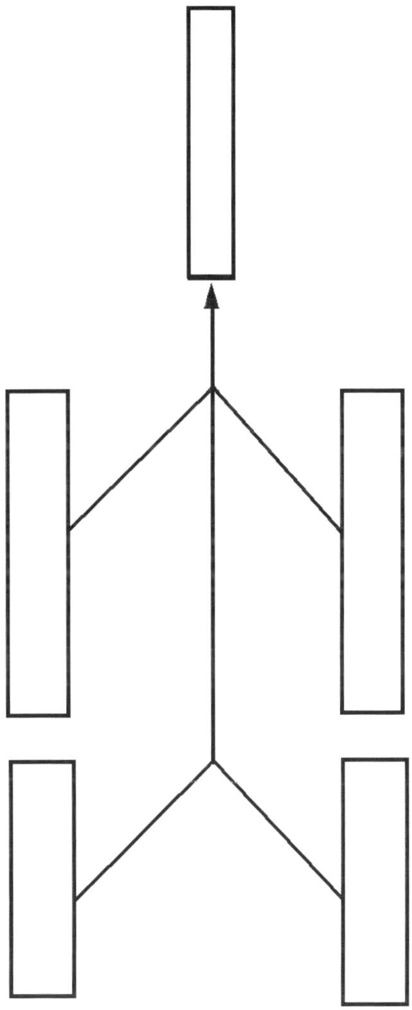

Figure 7. Cause and effect diagram.

when a screw is driven in, if the bosses' wall design was too thin or a material chosen with too low an elongation, then the stress caused by over expanding the material in the boss may cause it to crack. The sharp screw threads on a screw can also create a notch that causes a stress crack to form that cause a crack and the boss wall fails. High stress concentrations can initiate a crack in the part over time leading to failure.

The reasons for stress related problems are many and some may not always cause a problem until later in the products life when it is overstressed beyond the design criteria of the material. Plastic engineers have learned how to solve these problems by the corrective action of increasing the size of the radius at inside corners. During design reviews before a product is released for production preventive action ensures a minimum radius is specified by the design engineer and machined into the tool to reduce the high stress at critical stress points when in service.

This is also important when evaluating assembly methods to use the correct assembly screw thread type for a molded-in boss. Reducing weld lines at these high stress points also ensures there are no molded-in high stresses in the product to reduce the possibility of failure. Also, how to correctly process the material to ensure the product is well packed out without any molded in voids that cause molded in stress, and that the part section meets the required dimensions and has the correct part weight. Measuring the products molded part weight is a quality tool that is indicative of the product being thoroughly packed out without any internal voids in thick sections.

The time line for knowing the cause to when the actual affect is determined may be very long. A failure problem in the field during use can occur days to months after the product has been built. This is versus a processing problem that shows its effect, from a cause, almost immediately. But, processing problems may not show up until a later manufacturing operation that is why it is critical to ensure all processing and product monitoring variables are inspected and analyzed in "Real Time" to be sure the final product will meet the customers satisfaction.

A very important action that must be taken prior to even manufacture starting is maintaining tight part identification or good lot control for all products. Good lot control means that any product can be traced back, through excellent documentation and record keeping, to the original lot of material, date of manufacture, machine manufactured on, tool cavity number, processing conditions, quality inspection documentation with results of the product or lot of material, etc. Too many times when a failure occurs this information is not available. This is often because it was never recorded and the product cannot be traced back to the time and work order of manufacture.

This results in insufficient information to analyze the failure, as no traceability is available for an analysis to begin. Visual data can be obtained along with material identification and the possible reason for failure if this information was obtained from the customer. Whenever a failure occurs, always obtain the failed product for analysis of the possible reason for the failure. Examination of the product and some chemical testing may prove valuable to the solution of the failure problem.

The final solution to an affect on a problem may take minutes; to hours or longer to finally be resolved depending on how far back in the system it may have originated. It can also take extensive testing to show up long after the problem occurred. This is one reason for keeping accurate, up to date records of materials used, lot numbers, for what parts, on what machine, date of manufacture, etc. Trying to solve a problem after it has occurred days and weeks later without this information is almost impossible. It is made even more difficult especially if product lot numbers along with customer lot control has also been lost.

Customers are also liable to keep accurate records of supplier parts used and lot control in their factories if these parts were in sub-assemblies. But, this is not often the case as each customer has their own methods of controlling their product assembly and quality, some better than others.

EXAMPLE OF POOR LOT CONTROL

An example of poor lot control was a customer buying a capacitor. The supplier maintained good lot control through shipping until the part arrived at the customer's plant. Then the customer lost incoming lot control as soon as the part was assembled into their product, a military aircraft power supply. Then when a problem occurred with the capacitors, they were not able to provide their customer with any detailed information, other than a series of lot numbers purchased over a specific time period as the probable part creating the problem. After attempting to solve, unsuccessfully the problem, the customer was forced to maintain accurate lot control over what lot of materials went into each unit they manufactured. This simplified the problem and led in future situations a method to answer their questions as to why the capacitor failed or any other problem in their assembly.

In any manufacturing operation this is one reason why all tooling and work orders should be identified, and the manufacturing records noting all material lots used and in what time sequence the products were manufactured if a lot change occurred during the manufacturing process. Too often this information is never documented and as a result valuable information lost forever for solving a problem. This information must always be part of the process control plan for manufacture.

Another example of this problem is a blow molding bottle problem where highly active ingredients for flea and tick spray was bottled in a blow molded PVC bottle. Initial pre-marketing testing was done on clear PVC bottles and product that passed all tests. Prior to production, an organic pigment was added to the PVC resin compound in a very low concentration to only add a slight tint to the clear PVC bottle. After production started the bottles were filled and sent to the warehouse and some shipped to retail outlets.

At this stage it was discovered the bottles had become very brittle. The bottles tinted PVC body became very brittle and stress cracks developed in all of the PVC bottles. As a result with only minimal handling, while loading the individual boxes on a shipping pallet, the bottles easily cracked. Also, stacking boxes of bottles on top of other boxes of bottles, caused the bottles in the box underneath to also crack. This was just from the their weight and jolting during the loading of another box on top.

PROBLEM ANALYSIS, CAUSE AND AFFECT

1. Initial analysis by an expert in materials concluded poor quality process control of the blown bottles was the cause. His analysis stated the reason for failure was processing as the original bottles, blown from clear PVC resin, could be dropped, weeks after being filled without any failures. The attorney in this case for the customer of the blown bottles was not satisfied and sought out more assistance in the solution to the brittle bottle problem.

As a result I was asked for assistance in solving this problem. The same information was sent to me and after a review of the data concluded the only possible cause could be the tint additives. The dye or pigment of the composition was not known at this time except that it was used to tint the bottles a light blue in color during manufacture. This was the only documented change made since the initial testing of the clear PVC bottles filled with the solution. And since these bottles still experienced high elongation and toughness, ruled out the PVC resin processing conditions initially as the cause of the failure.

The blow molding operation was eliminated as the source of the problem as all dimensions of sample bottles were within specification and all met physically property specifications for toughness after molding, but before being filled with the solution.

FINAL ANALYSIS OF THE PROBLEM

It was determined after further investigation by a color and pigment expert that an ingredient in the pigment system combined with the solution attacked the base PVC compound and caused a rapid embrittlement of the PVC material. This was presumed to be caused by an interaction occurring with the colorant and the flea and tick spray components. The process was fast, acting very quickly to extract the plasticizer (the additive used by the compound supplier to add elongation and toughness to the PVC resin) from the resin in the bottles. This extraction of the plasticizer from the bottles material resulted in a very brittle thin walled PVC bottle.

Final chemical testing proved the pigment and solution caused the plasticizer in the PVC bottle to go into solution very quickly with the flea and tick liquid. This caused a rapid brittleness of the material with crazing occurring if the bottle received the slightest bump of the bottles side or base that caused a crack to occur causing the failures.

2. It was very difficult to obtain the required information that the tinted PVC bottle was tested with the solution before going into production. Documentation for testing and quality manufacturing records of the blow manufacturing process were poor to non-existent. Only some random physical measurements on wall thickness were actually recorded by the blow molded bottle supplier. These documented bottle measurements and processing records were never recorded on a repeatable or scheduled period of time during manufacture.

Therefore, with a catastrophic failure of all the molded tinted PVC bottles filled with flea and tick solution the only logical cause was chemical attack of the bottled solution on the PVC bottle. Further conclusive testing and input from a color consultant confirmed this analysis. Pigment (tint) additive, even in a few parts per million in this situation caused the plasticizer to leach out of the PVC causing almost immediate embrittlement of the bottles.

This conclusion was reached due to a prior experience that involved a similar situation only with a glass reinforced compound colored, type 6 nylon. A very small, less than one half of 1% of a pigment was substituted in the color formulation. The result was a loss of over 10% of the materials physical properties. This was a very dramatic effect for such a small change. Therefore, insist your suppliers always inform you if they want to modify or change any materials used in your product. Make this a main stipulation in your contracts and material purchasing orders and in any conversation with your suppliers.

CONCLUSION, THE PREVENTION OF A PROBLEM

When any change is made in a material the customer must be told, preferably with sufficient time to react if the material must be re-qualified in the product. It is the material suppliers and product supplier to the customer to act responsibly to both inform and test the material in sufficient time to ensure compatibility no matter how small the change as preventative action is the main issue to avoid a future problem. In this example the customer tested and approved after extensive testing, for both chemical and physical requirements, a PET (polyethylene terethylate) clear plastic bottle to market their new product. They did not trust the PVC compound, even though the PET bottle material was more expensive, and a clear PVC bottle would have been more than satisfactory, as the earlier non-tinted PVC bottle testing had proven.

DOCUMENTATION

Always document solutions to problems. This will lead to preventing the problem from reoccurring should the first solution not be satisfactory or conclusive in solving the problem. As personnel move, the solution of a problem often goes with that person. Therefore, documenting all problems in a problem solving book will it save time and expense should a similar problem occur. Cross-index the book so the problem can be easily identified, solution found, and implemented. It is also very critical to document process data. This situation revealed sporadic and essentially poor data recording with no continuity of repetitive information to show he process was in control. It was not sufficient for only documenting periodic measurements on the bottle to show the physical size was within specifications.

Moving toward Six Sigma will make solutions possible even more difficult to solve since control of all variable elements of the process are so tight. Therefore, careful problem solving methods must be developed and employed to discover the clues toward the solution of a problem.

Before problems even occur, FMEA's can predict suspect problem areas, leading to implementing preventative measures to prevent a problem from occurring. The goal is to analyze a process operation and document potential variables that reside at this operation so that appropriate action can be taken to prevent a problem in the present to prevent it from occurring in the future. We need to identify actions appropriate to help live with a potential problem that can be identified so it can be prevented or a solution developed to deal with the problem should it occur in the future.

This is shown in Figure 8, analysis through future orientation, anticipating, and eliminating likely causes of new and current problems even before they may happen in the manufacturing or business process.

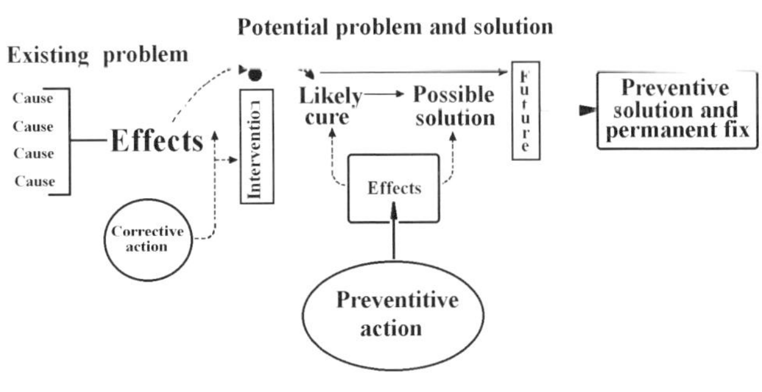

Figure 8. Cause-effect and corrective-preventable actions.
(Adapted from reference [10])

TAKING ACTION – WHAT AND HOW

When problems occur, management wants a solution now! We react and often make hasty inappropriate or ineffective decisions. These actions occur daily by operators and engineers making process adjustments. Often more than one process adjustment is made at a time that causes the process to go out of control. What results is no sensible idea of how to get the process back into control. When this happens the best solution is often shutting down the process until clear reasoning occurs and a review of the data causing the problem is completed. This involves information gathered from all sources for the solution of the problem being available and analyzed.

WHAT ACTION SHOULD BE TAKEN?

All actions taken in the solution of a problem should be well thought out and only 'one' variable at a time evaluated and adjusted. This is the opposite of a DOE. In a DOE a series of selected variables are evaluated that are considered key to the analysis of the problem or process. They are selected at their high and low values, randomly in a controlled testing program to determine the main variable acting on the process.

Begin by making one small adjustment at first to see if it has an affect on the problem. This process is designed to find the root cause variable of the problem, eliminate it and put measures in place to prevent it from occurring again. Corrective and adaptive action responses as shown in Figure 9, are one method initially employed to solve a problem.

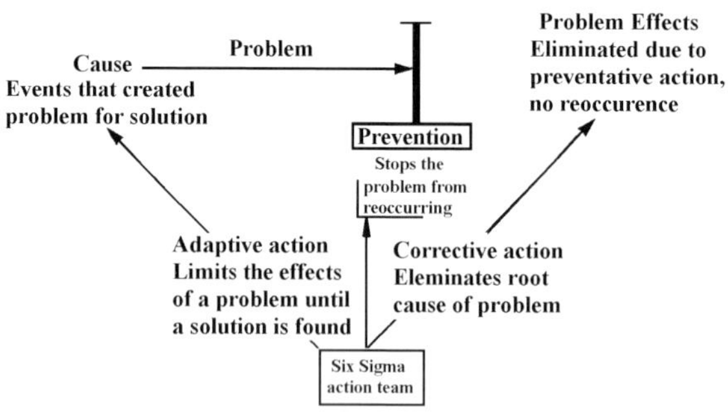

Figure 9. Corrective and adaptive actions.
(Adapted from reference [10])

CORRECTIVE ACTION

Corrective action is taken when a possible solution is developed. The solution is based on past experience and documentation that corrected the variable creating the problem. If this is a brand new problem, then the quality methodology analysis must be used to develop the information needed to solve the existing problem.

HOW IS ACTION TAKEN

The gathering of information is developed by questioning personnel and what they did to affect the process especially if any operator changes were made to the process. Often, it may have occurred naturally without any outside or known input from the operators. If the problem has occurred before, the 'problem solving' book can be used to assist in identifying the current problem and what is necessary to solve the problem. A review of the effects that lead up to the problem may also assist in solving the problem. But often if the parties involved in the problems solution did not document any changes made or do not understand the process, can easily fail to identify the root cause and how to prevent the problem.

An example of this was a company, sonic welding a fuel tank valve assembly. The problem occurred yearly and went away only after a set period of time. No one really knew why or how but it finally stopped. The next year it reappeared and no one solution seemed to prevent it from happening. Asking the right series of questions the problem was finally diagnosed and solved.

The problem occurred at their southern (South Carolina) plant when the humidity became high. Moisture forming on and in the products material at the surface, after molding nylon 6/6 parts was causing a very poor thermal sonic welding joint seal. The joint failed leak tests and exhibited brittleness and had spot leaks around the circumference of the joint. Changing all the welding variables did not affect the strength and even a new shear joint weld design did not solve the problem.

The solution proved to be very simple once the right questions were asked and a knowledgeable plastic materials person with past experience of the problem was brought into the conversation to solve the problem. The cure was simple, inexpensive, and proved very reliable after it was implemented.

The main goal was to keep the nylon parts DAM (dry as molded) prior to sonic welding. Moisture absorbed or present at the weld joint, even a shear joint, formed steam during the welding process and caused the joint to be brittle with spot leaks. Prevention was discussed with two possible solutions.

IMMEDIATE ACTION

1. Immediately after molding all hot plastic parts were put into a large airtight plastic bag. In each bag was a desiccant and after a specific number of parts were put in the bag it was evacuated and sealed. These parts for the entire fuel valve assembly are all molded and stored in this manner. The operator bagging the parts wore white cotton gloves to not contaminate the parts.

Gloves were worn to eliminate any hand oils contaminating the joint seal surface. The parts were now protected from any foreign contamination and stored in a DAM condition until sonic welded to assemble the parts. This is a major Six Sigma problem prevention method that must be enforced for a sonic welding operation. Any plastic part to be welded should only be handled with white cotton gloves to ensure hand oil or other contaminants do not get on the weld joint surface. This information was gathered by observing the welding operation, reading the suppliers literature on sonic welding, and constructing a fishbone diagram that uncovered all the many variables that would cause a leaking joint to happed during the welding operation. This evaluation should be conducted for any material and those with an affinity for moisture pickup after molding should always be vacuum bagged as soon as possible after molding. Hot parts will absorb moisture faster than cool parts and the intent is to prevent this from occurring, especially for hygroscopic materials as nylon that have a high affinity for moisture.

2. Also, due to subtle differences in each lot of material, manufacture all parts of the same assembly at the same time and from the same lot of material. Then after manufacture be sure all parts are identified as to their molding time, machine, tool, and lot of material. Then if possible, using the same operator at the manufacturing site, have the operator sonic weld the parts together. Parts can be assembled hot and allowed to cool down.

To do this depends on the companies use of their personnel and how they planned the operation.

If the welding is not to be performed at the manufacturing site, the parts can be conveyed to a central collection station where the sonic welder is located. There, another operator, wearing white cotton gloves, will load parts into the welding fixture, make the weld, and pass assembled parts to the next operator for leak testing the seal in just seconds.

3. It is not recommended to weld a just-as-molded hot part with a DAM cold part as the shrinkage differential, hot to cold would be to great and would cause a high stress at the weld line. Post mold shrinkage of engineering plastics is typically 15 to 25 mils per inch and can be more depending on wall thickness and manufacturing conditions, melt and mold temperature. This could cause the joint tolerances to be out of specification for a cold and hot just-as-molded part. Also, if the parts are dissimilar in size, the gating for each part is probably different and the parts will more likely be elliptical to some degree. Here a good tool designer can assist in determining the correct gate location, size and type. This is necessary to ensure the joint is as round as shrinkage, location, and molding conditions will allow by having the part fully packed out to maximum part weight on each manufacturing cycle.

Another fuel tank problem experienced was with an acetal part with too small a gate on the cap. This resulted in the part not being uniformly packed out, as illustrated by sectioning the part, voids were discovered. Another problem was the mating seal surface of the cap flange on the larger flange that made up the roll over fuel reservoir. The sealing surface for the cap was on a flange diameter over three times the diameter of the cap and much thicker. The seal cap seal joint was also not centered on the flange.

To further compound the problem the flange had only one edge gate and not spaced to ensure the cap seal surface would be as round as possible. This produced an elliptical shape for both cap and seal surface that were not elliptical in the same direction that the cap was to positioned in before welding. When both parts were measured each had a major degree of out of roundness. This further complicated the sealing process, as the stack up of joint dimensions was not compatible for the sonic shear joint interference requirements for a good seal. This meant that is a minimum dimensional situation, the amount of interference on one side of the shear joint was only 0.003″ and on the other side, 0.055″. Each was unacceptable for the type of sealing weld required for each assembly. Too much interference on one side would mean no welding depth and on the other, welding completed but questionable due to minimum interference.

After an analysis of the dimensions a solution was developed to solve this multi problem seal. The gate on the cap was relocated about 0.50″ to gate into the thickest section. A second gate was added to the flange to assist in control of the out of roundness of the flange seal surface.

Tolerances were analyzed and the shear gate redesigned for optimum interference and welding conditions. Based on the existing design, the amount of interference at the shear gate varied from a maximum of 0.054″ to 0.008″, that was greatly out of proportion for a suitable weld joint. The joint design used a step weld design so that the amount of shear interference was controlled with 0.010″ ± 0.002″.

Further investigation was performed as watching the welding operation. How the parts were handled and the care of keeping the joints free of any and all contaminants brought further changes to the operation. It was noted that the two operators involved with the welding operation regularly changed places during the welding and this caused additional problems.

Also, just after the joint was welded, the joint was tested for leaks. The sonic welding operation had a test station right next to the sonic welding machine that used water to test the welded parts for leaks. After each weld, the operator tested the parts by placing them into a test fixture that submerged the parts during testing in a water tank. After testing the operator removed the part and passed it along to the other operator who performed a similar but different type of leak test. The operator's hands were constantly wet and oil from the first test station further contaminated the operator's hands. This caused the operator loading the welder to have wet and oily hands due the immersion of the welded parts after welding for testing the joints tightness. Then based on the operator work instructions for performing the welding and testing, the hands of the two rotating welding operators, rotating with each other in their jobs every 5 to 7 welds, to have wet and oily hands because of the method used to test the parts after welding.

The Six Sigma permanent solution to their problem of over three years was then simple. After a redesign of the shear weld joint that implemented a step shear joint, required specific operator actions. The stepped shear joint gave the parts and operator a known starting point each weld cycle. Also, the amount of interference was the same for each weld and varied only with the tolerance in the tool.

Operators were also instructed to not change place or work function as frequently as before. The minimum position change time was each hour. At position change, each operator would wash and dry their hands and the welding operator would wear white cotton gloves. The gloves were used to keep the parts seal joint free of any foreign contamination, oil, water, and skin oils and lotions. Manufactured parts of the cap and flange were placed into dry and airtight sealed bags. To minimize the possibility of shrinkage variation, the cap and flange units were always molded from the same lot of material with date and time of manufacture. This was done so only like material and manufactured parts would only be assembled together for cap and flange roll over valve assembly.

Corrective action is taken when a known solution is developed based on past experience and documentation that corrected the cause creating the problem. Questioning personnel and what they did to affect process changes that lead to the immediate solution collects this information.

A Set of Examples Experienced

1. Clear acrylic parts suddenly becoming cloudy, the effect.
   Cause: dirty feed system prior to running acrylic, clear polycarbonate was in system. Poor cleaning, of feed system and/or grinder caused a small trace of polycarbonate to get into feed hopper that caused the cloudy acrylic material problem.
   Solution: Always break down the grinding system and vacuum out any traces of particle fine contamination from all spaces in the grinder system. Disassemble to ensure all areas are clean of the last material ground up. Vacuum out the system and never use high-pressure air as it spreads the contamination everywhere in and around the plant operating area.
2. Glass reinforced, bright yellow pigmented and compounded material turned a dull mustard color during initial processing.
   Cause: poor barrel and screw cleaning. Remains of a black reinforced nylon not properly purged from the barrel and the screw from a prior production run prior to startup of the yellow material.

Solution: 1. Repurge the system with a more abrasive purge compound, cast acrylic to scour the barrel and screw to remove all traces of black nylon. Then run clear polyethylene purging compound through the barrel to ensure the system is clean.
2. If still dirty, shut down, pull screw and clean.
3. If time permits, run clear polycarbonate scrap, purge the barrel dry, and shut down. When barrel and screw cool, the polycarbonate will separate, shrink away from the metal carrying all traces of black nylon with it. Startup next day with polyethylene purge compound to clean out the barrel of resin contamination. After cleaning the barrel and screw, the molded parts were a bright yellow.

If a material problem is suspect use adaptive actions. Material suppliers trouble-shooting guide and check lists are available to assist the operators to monitor the effect of their processing actions. Some actions may result in only a temporary or a Band-Aid fix with the problem coming back. Seek the root cause of the effect not a Band-Aid to temporally create a solution that will not keep the process in control. There is always an underlying cause as to why the problem occurred and a preventative solution can be found and implemented.

Material moisture problems occur when materials are dried and one bed of the dryer system is becoming inactive or has been contaminated. Dryer bed contamination happens by drying a resin at to high a temperature that caused additives in the resin to vaporize and contaminate the desiccant in one or more drying beds. The off-gassed volatiles plug up the openings in the desiccant and making it ineffective to absorbed moisture from the material being dried.

When one desiccant bed is saturated and the system switches to the other drying bed, the uncontaminated side, the materials are dried properly. When the system shifts drying beds again the contaminated bed is ineffective and wet material is fed into the system causing processing problems. Make sure your drying system has a moisture meter and alarm system to prevent this problem in your feed system. When the high moisture alarm is working correctly, it will warn the operator and the bed can be tested for absorption capability. Then if the bed is found contaminated the desiccant can be replaced.

This is a difficult problem to analyze as it can produce sporadic processing problems. Always test dryer beds on a regular schedule and make sure the drying temperature is correct for the resin being dried, as each is different. Test resin samples should be taken periodically and tested for moisture level. Samples should be drawn from the exit of the dryer and the entrance to the hopper. This will ensure that the material is dried correctly and when conveyed to the machine there are no openings in the transport system that allows moisture to enter.

If a separate, at the press, hopper dryer is used, then the sample is taken from the exit door at the base of the hopper for moisture testing. Each specific industry will have similar problems and processes to improve. The electronics industry has solder type and part placement problems in subminiature assembly devices. Cleaning is very critical so shorts do not result plus cold solder joints that are difficult to inspect. The examples describe here are for the plastic manufacturing. Learn from all industries as problems are similar only the solution is different and always remember to use the analysis methods and information in your respective industry.

PROBLEM SOLVING MATRIX

Preventative action is the goal to establish during Six Sigma programs. Analysis of your business and processing systems and implementing preventative measures can ensure manufacturing operations function as developed do not become a potential problem. The steps taken in this analysis are shown in Figure 10.

Two types of action plans

1. Preventative action.
2. Contingent action.

Preventative Action

Preventative action plans reduce the probability that a problem will occur by understanding the potential cause. Contingent actions reduce the seriousness of a future problem if it should occur.

These actions are based on the understanding of the probable effects such a problem will produce. An example of contingent action would be to

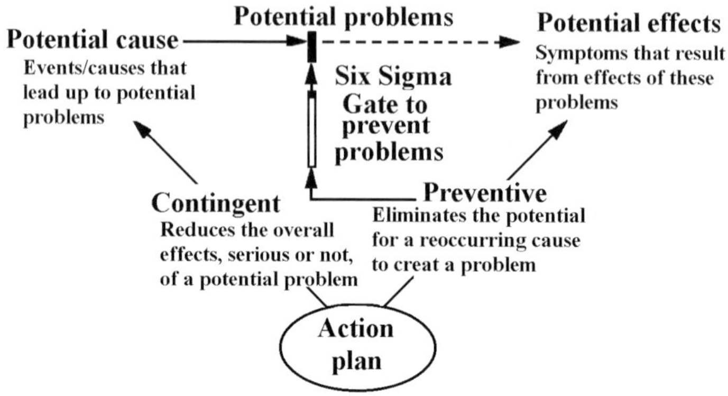

Figure 10. Preventative and contingent actions.
(Adapted from reference [10])

provide water splash shields around mold cooling hoses should one fail and spray water on the machines electrical system. A preventative action would be using reinforced armored hose with a replacement schedule implemented after a set number of hours of use. Periodic inspections for cracks, and degradation of the hose would be implemented for replacement with new hoses when maintenance deems it necessary.

The problem solving matrix (Table 4) needs some reference as to preferred actions, short or long term. The true focus for Six Sigma capability is to correct the problem permanently to ensure it does not reoccur. Adaptive action is used to only minimize the effects of a problem without a complete solution. Taking adaptive action to corrective action to final preventive action is the goal of Six Sigma.

Table 4. Problem Solving Action Matrix.

	Cause	Effect
Current problem or actual non-conformance	Corrective actions to eliminate root cause	Adaptive actions to minimize effects
Potential problem or possible non-conformance	Preventive actions to reduce probability of likely causes	Contingent actions to reduce seriousness of probable effects

(Adapted from reference [10]).

Contingent action does not prevent future problems, only to reduce or prevent their occurrence and their seriousness. Preventative action is knowing all potential sources of variability. Supplier problems can only be prevented if the supplier takes responsibility for their quality. Inform your suppliers through audits, open exchange of your requirements, and a dual QFD program so there is no excuse on either parties part for telling the supplier what you need and for the supplier not knowing and delivering to you what your require on time.

A proven preventative action program used by Lockheed Martin, for a launch of a space shuttle is a checklist. The checklist must be completed before each space shuttle launch covering over one million potential problem areas in the launch sequence. Check lists assist to ensure nothing is overlooked. Many people balk at using check lists but they do their job very effectively as a quality checkpoint for each step in a program or process.

The Lockheed check list was developed by creatively thinking through each step of the launch program. Each step in the launch sequence was analyzed and documented with checkpoints to evaluate information and system operations. This also includes any potential part of the system where a problem could jeopardize the success of the launch. Implementing a preventive action plan consists of developing a list of key thought generating questions or 'words' that open up the technical mind to a simple question such as, "What if?"

When all the "what ifs" are answered, a final release point is created to ensure answers to all the question are answered and satisfy all personnel the launch system is correct before proceeding to the next step in the launch sequence.

PREVENTIVE ACTIONS

The list of some basic characteristics or attributes of activity leading to identification of preventative actions is:

- periodic – time sequence sensitive
- planned – needs to occur before proceeding
- investigative – system check and verification completed
- analytical – data and readings are all in acceptable range
- deliberate – must occur before proceeding forward

• proactive	– are all steps implemented and thought through
• predictive	– what could happen if conditions (as is) continue
• no response or reaction	– a stopping point, must be corrected before proceeding
• involving knowledgeable personnel	– are the right people supplying information
• utilize quality data, reports, records, information	– are the measurement and control equipment and systems capable and correct
• based on the understanding of the quality system, processes, equipment, etc.	– are the right check points in place and data being monitored in "Real Time"
• management support is in agreement to implement	– the major hurdle to jump over for total program success

Preventative action involves "the use of appropriate sources of information such as processes and work operations that affect product quality, concessions, audit results, quality records, service reports, and customer complaints and corrective actions. These are used to detect, analyze, and eliminate potential causes of non-conformities."

Corning and Procter and Gamble use a formal potential problem analysis to help plan future business actions. P & G uses it when preparing for a scheduled downtime of their machines in their manufacturing areas. This saves time, money, and gets the lines back in as-new condition for longer production runs by thinking ahead in a systematic process.

Most problems never occur sequentially but come in bunches. Select teams of knowledgeable personnel, hopefully trained in problem analysis can be scheduled to work on each problem. Your black belt, who will direct each team assists in the solution and manages the team's solution and the final results to meet each group's problem solution requirements.

The use of Six Sigma analysis, design, manufacture, and inspection tools and methodology using these quality functions and operations will eliminate design, manufacture, and end use problems and strengthen the manufacturing and quality of the operation and products. These methods require management and personnel planning, scheduling, action, and implementation time. The goal of any procedure is to follow through using the technique to gather the correct information for always performing the process in "Real Time Control". The ultimate company goal is to reach for Six Sigma total organization and process control for all operations, business and manufacturing.

REFERENCES

1. Chase, N., asst. ed, "Uses and Abuses of SPC." Quality Magazine 1999: http://qualitymag.com/articles/1999/apr99/0499f7.html.
2. Cotnareanu, T. "Old Tools-New Uses: Equipment FMEA." Quality Progress December 1999: 48–52.
3. Deming Institute, W. E., Taccoma Partnership Notes, Project on family violence builds collaborative skills and new attitudes." Deming InterAction August 2000, Vol4, No. 2: 3–8.
4. Ford Manufacturing Staff, "Potential Failure Mode and Effects Analysis for Manufacturing and Assembly Processes, (Process FMEA) Instruction Manual." December 1983: (Preliminary), Attachment I and Attachment II.
5. Hunkar, D. B., "An Engineering Approach to Process Development and the Determination of Process Capability." Hunkar Laboratories, Cincinnati, Ohio, 1991, Document No. 228.
6. Knouse, S. et. al., "Getting Employee Buy-In to Quality Management." Quality Progress April 1999: 61–64.
7. Ohio State University, "Quality Function Deployment (QFD)."
8. Properties & Processing "Celanex" Thermoplastic Polyester, Engineering Plastics Division, Holchst Celanese, Bulletin J1A, February 1984: 1–62.
9. Smith, G. "Too Many Types of Quality Problems." Quality Progress April 2000: 43–49.
10. Wessel, D. "An Ounce of Prevention." Quality Progress December 1998: 33–36.

Chapter 5

Six Sigma Education and Using the Existing Quality Methods and Procedures

Every few years a new quality, or resurrected, quality method or technique is the champion method to reduce defects, cost, and business and manufacturing problems. It is then the buzzword in all quality circles and defined to be the key to a company's success in reducing problems and improving quality in all operations. Deming, Juran, and other quality champions developed these methods for companies to use to improve their quality. Companies in their hope of improving their output adopted these technologies but did not always understand or have the patience and agreement among upper management to fully pursue, master, and reap the rewards the new quality technology could deliver. The same is true for manufacturers to ensure quality products and operations occur to meet output and customer requirements.

Lean manufacturing is one program using small lot size, one or more units, which passes from station to station in a planned and timed material flow. Any defective unit is immediately reworked or rejected at point of recognition with the flow always continuous based on demand pull to fill the order. This system of manufacture, assembly line, is continuous, easily monitored, with problems easily identified for quality teams to focus on immediately to solve so the line does not shut down.

The system can be very vendor sensitive and previous work station dependent for each product quality item or operation to be in compliance for the product to pass on to the next station. When a major problem with a supplier occurred, it must be identified at its earliest point of entry into the flow of manufacture. If not, it can and does shut down the lean style of manufacturing until new product or process changes are made to correct the problem. This system also relies on suppliers being close and very reactive and supportive to any problem that occurs with a backup of known quality material or service support available to draw on should a shutdown problem develop.

As a result, the return on investment in manufacturing and supplier quality requirements did not always match expectations and manufacturing requirements. Fault must lie with company management as quality personnel could not measure quality into product. They were only capable of ensuring product was as good as suppliers could produce. What was needed in many situations was a rewriting of material specifications and obtaining supplier agreement that the specifications were attainable and could be delivered on a repeatable schedule. This required a better method of ensuring that quality was built in and delivered with the product at their plants. It required a series of suppliers being on the same level of quality for product manufacture. To meet this directive quality teams developed a new, common, and mutually agreed upon quality method that would initially attempt to bring manufacturers together to produce product of repeatable expectations.

To this end was developed the ISO9000 standard certification program. ISO9000 driven by European suppliers was finally adapted worldwide with some initial variations. The major corporations adopted ISO9000-1994 standards in their companies and expected their suppliers to also comply if they wanted their business. This was accomplished with some negativity and major foot dragging but accomplished by most companies in the mid 1990s. The only real negative comment was a company could meet the requirements and become ISO9000-1994 certified and still produce product of minimal quality.

To address this issue the revised ISO9000-2000 standard was developed to ensure more compliance to actual and auditable quality measurements, thirty four new requirements added. They were to be required the next time the companies had their ISO certification up for renewal. It is anticipated to prove successful in continually improving a company's quality of operations.

As a result companies wanting to participate in business with these major markets corporations were forced to comply. As more and more U.S. companies discovered ISO9000 did work for quality improvement, more and more became certified. When automotive adopted QS-9000 it became necessary for auto product suppliers to comply, as QS-9000 required adapting the twenty ISO9000 articles for certification into their standard certification requirements. We now have ISO9000-2000 and QS-9000 becoming IQS/TS16949 worldwide for all automotive product suppliers.

Six Sigma is now the quality "buzz word" with its conception by Motorola in 1986. Coupled with ISO9000 and the use of existing quality techniques and trained black belts an even greater savings and quality improvements can be realized. Six Sigma ties improvements to cost of quality and the companies balance sheets bottom line. Six Sigma is more capable to be monitored and assigned a monetary savings to the company. Remember, the initial criteria for a Six Sigma program was a project with a potential savings or improvement in the range of $170,000.00. That figure got management's attention. That plus the reported savings realized by Motorola, General Electric, Honeywell, and others in the million and billion dollar savings range of possibilities.

All companies can utilize Six Sigma quality techniques with only the savings to be realized factored down to their respective size. The outcome is that Six Sigma techniques work for all size of companies and type of business, service or manufacturing operations. This is the first "Real Time" quality system that has gained management notice and implementation.

What are these techniques used to obtain these monetary savings? We have already discussed some of them; Ishikiwa, fish bone diagrams, CP, CpK, QFD, and FMEA analysis. Design of experiments, DOE, and Cause and Effect, C&E, SPC, statistical process control, Kaizen, plus many others, GMP, CIM, PPAP, TQM, SQC, PFMEA, CR, cGMP and many other quality acronyms now being used by industry. Many of these acronyms refer to similar techniques only phrased to suit the author of their presentation or the industry used in for improvement. These techniques are not new, only in how they are faithfully applied to the improvement of business and product or service quality in their company for their customers.

THE NEED FOR A COMPANY CHAMPION

Six Sigma dictates a champion and one is always required or the program may never be successful. Educate your champion or better have them attend a short training course in the requirements and anticipated results attainable with Six Sigma. At the course they will learn how to understand variation and support your efforts in using process-behavior charts for interpretation of data.

CATEGORIZE AND ANALYZE QUALITY PROBLEMS

To fully realize the efforts for using one or more of these techniques the problem must be identified and defined for developing a lasting solution.

Classifying problems within the company at first seems to be a simple task. Each profession in a company from engineers, accountants, manufacturing personnel, sales, management, etc. classify problems as they appear within their area of operation that affect the performance for their area of influence and responsibility.

Each department in the company may use different problem solving techniques based on the problem identified within their operations. To avoid using the wrong problem solving technique, personnel must be able to categorize problems relevant to developing a permanent solution. This implies they have received training in problem classification. The majority of companies have implemented this training for their personnel.

This is termed problem 'taxonomy', where a formal classification system is used to sort and differentiate problems according to key characteristics. Using the wrong problem solving technique may lengthen the problem solving operation and not correctly identify the actual root cause of the problem.

Various methods have been proposed in both management and quality literature without wide spread use or acceptance. Frederick Nickols, as reported in *Quality Progress Magazine*, April 2000. page 43, "Too Many Types of Quality Problems", has proposed useful taxonomies recognizing the need to fit problem solving methods to the type of identified problem. Nickol's listing of problem types are;

Problem Types

1. Repair: To restore a malfunctioning system to its intended level of performance.
2. Improve: To improve a system so that performance goals are achieved.
3. Engineer: To design a new system or develop a solution that will satisfy anticipated goals.

Nickol's work was to promote personal well being when addressing problems of performance systems, structures, and processes. Consumer product manufacture, automobiles, electronic equipment, computers, air conditioners are performance systems, as are accounting and production operations in companies. How these operations are setup, operated, and respond to company business challenges must be considered. These operations can be differentiated as design, process, and performance problem solving areas when they occur. Nichols classifies repair tasks as performance and process problems with engineering tasks pertaining to system design. He also included a hybrid category (improve) that reflects the idea that performance improvement can be achieved by redesigning the current operation.

His research, published in a book *Quality Problem Solving* showed a distinct difference between performance and design problems. These were subdivided into 5 main quality problem categories. These are shown in Figure 1, showing the different types of quality problems.

It was found that almost every quality problem would fit these 5 categories. Note that not all organizational problems are direct quality problems such as decision problems, negotiation problems, and resource allocation problems among others. When companies investigate problems, invariably the problem will fit into one of these 5 categories.

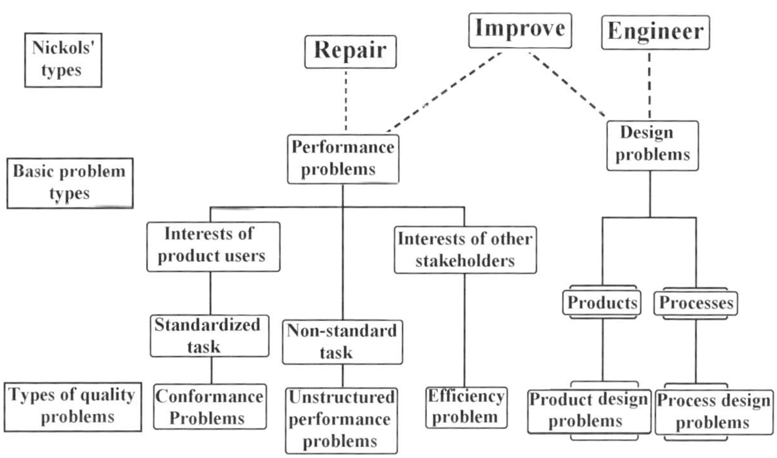

Figure 1. Toxonomy of quality problems.
(Adapted from reference [1])

PROBLEM SOLVING CATEGORIES

The five types of quality problems and their definition are shown in Table 1.

In brief: problems re-classified by defining these three characteristics:

1. Defining the problem characteristics.
2. Key problem solving tasks.
3. Strategies and techniques to solve the problem.

Part of problem solving after the problem is categorized by type is to select the problem solving technique and data gathering method to arrive at a solid and lasting solution. It is very important to document all information developed during the solving of a problem. This information must be filed in the department's "Problem Solving" notebook. To further assist the information should be divided into sub-categories as to the type of problem. This will ensure easy access and retrieval of information should a similar problem occur again. Then if it should, this information may aid in a quick and more lasting solution if additional time and resources are spent on discovering a permanent and lasting solution.

The Key to Problem Solving

1. Classify the problem.
2. Define its characteristics.
3. Talk with all personnel involved.
4. Collect data using correct metrology quality methods.
5. Discuss data and make a rational decision for a solution.
6. Implement the solution.
7. Collect data to verify the solution works.
8. Monitor solution for compliance.
9. Document the solution and file in department Problem Solution Book.

A short synopsis of each problem type will aid in correctly classifying a problem when they occur.

CONFORMANCE PROBLEMS

This problem is found in a highly structured system with standardized inputs, processes, and outputs. The system (an assembly line for example)

Table 1. Types of Quality Problems.

Problem type	Defining characteristics	Key problem solving tasks	Strategies and techniques
Conformance problems	Unsatisfactory performance by a well-specified system; users not happy with system outputs.	Diagnosis; determining why the system is not performing as intended.	Use statistical process control to identify problems cause and effect diagrams to diagnose causes.
Unstructured performance problems	Unsatisfactory performance by a poorly specified system.	Setting performance goals; diagnosis; generating viable solution alternatives.	Diagnostic methods; Use incentives to inspire improvement; develop expertise; add structure appropriately.
Efficiency problems	Unsatisfactory performance from the stand-point of system owners and operators.	Setting performance goals; localizing inefficiencies; devising cost effective solution alternatives and variety.	Use employees to identify problems; eliminate unnecessary activities; reduce input costs, errors.
Product design problems	Devising new products that satisfy user needs.	Determining user requirements; generating new product concepts and elaborating them into viable artifacts.	Quality function deployment translates user needs to product characteristics. Value analysis and design for methods support design activity.
Process design problems	Devising new processes or substantially revising existing processes.	Problem definition, including requirements determination; generating and elaborating new process alternatives.	Use flowcharts to represent processes, process analysis to improve existing processes, new processes, reengineering to devise new processes and benchmarking to adapt processes from others.

(Adapted from reference [1]).

is performing unacceptably with a higher than normal rate of rejects. On the business side, this could be the processing of a new order or flow of an engineering change or process order. All are standard, one operation after the other, in the flow of work through a department or company operation. It can be as simple as not achieving, "all department signoff and agreements" on the intended change of a procedure or operation.

A problem in communication occurs and it can be one of the most disruptive and more difficult problems to analyze. If the change order is not walked through the department or system, extensive scheduling, material, and delivery losses can occur. An important change order must always be walked through the organization to ensure all departments are aware of the change and are in agreement as to how, and when, it will occur in their department. The informed staff can ensure the change is made on time and correctly with the minimum down time and loss of productivity while maintaining the quality standard for the customer and operation.

The key feature for a conformance problem is there is a right and wrong way to perform the task. The process was working correctly before, but now is not. One or more steps in the input or processing activities have deviated from the specified requirement, so output is changed. The solution is finding where the deviation occurred, before input or during processing.

For the analysis of a conformance problem there are defined standards that can be verified for non-conformance. Any miss match is then documented and a reason for the change must be determined and where it occurred. Statistical process control methods are used in gathering data. The fishbone and FMEA analysis diagram can be used to map out or identify all variables in the operation if location is not known.

In a well-run company these diagrams should already have been completed and accessible to the problem solving team for their use in a problem solution. Cause and effect analysis may also assist in defining and locating the start of the problem. Using these two methods in conjunction with each other is a very strong analysis tool for solving a complex problem.

The process of finding the actual cause (Cause and Effect Analysis) is not easy as there may be hundreds of variables in the product or process and anyone person or variable could have created the problem. The process of diagnosis is performed by using data gathering with SPC, C&E; review of FMEA's, fishbone, and other problem analysis techniques. There is not one single technique that will always find the cause of the problem except good analysis, data gathering, and detective work followed up with documentation of the actual cause.

Since very rigid standards and specifications are used in processing, conformance problems are the easiest of the five problem areas to find solutions. But, finding the exact cause of a deviation in a very complex performance system may take, days, weeks, and even longer before a lasting solution can be implemented. Always remember when a supplier says they never change their product without notifying their customers, does not always hold true. The supplier contact may not have been informed of a small change made somewhere in the system that could have caused the problem. An example of this is as follows:

PRODUCT DEVIATION PROBLEM

A major plastic resin supplier, to save color pigment cost, changed a low percentage color pigment additive, about only 0.01% in a blue pigment for a glass reinforced nylon used in a power tool housing. As a result the part could have failed the mandatory UL (Underwriter Laboratory) 'six foot' drop test of the unit onto the floor. Fortunately this never happened or possibly the outside custom molder would have initially been blamed for poor manufacture. Testing proved the new pigment additive caused a 10% reduction in the physical properties of the base glass reinforce nylon material.

Always, have the material supplier carry out physical testing on each lot of product and send the certification of the test results with each lot of material to the companies doing the manufacture of products. This is not just an isolated case as it can happen and never rule out a material change as a possible source of a problem. Test materials at incoming, if properties are critical for the application. A test as simple as a drop weight test on a tooled product would have shown the material was weaker than requirements specified.

Only after the customer service representative informed the salesman that she was told by the color laboratory manager they had made a change in one of the base color pigments, to lower material costs, did the root cause become known. The lower cost organic pigment caused a loss of physical properties of more than 10% in property testing. This affected the drop impact strength in the existing product design. Unfortunately, the color manager and company did not have a procedure to follow when very minor modifications were made in an existing product. If they had, it would have listed a series of steps, requirements, and personnel to inform prior to making any change in a product.

The customer went to their alternate approved material supplier to ensure all parts would pass the UL drop tests. As this case testifies, without prior testing and being completely honest with the customer, no part of a color or material formulation should be changed. No matter how little, without thoroughly testing the material first and then obtaining the customers agreement before it is sent to the customer. In this case a change would have required new UL testing of the blue compound before being approved by UL, the customer and the color laboratory manager failed to complete this requirement. He assumed since the color remained the same the additive change did not require testing. A very poor decision on his part and the companies! Testing and customer approval is always required as the use of the product may have changed and any change in resin or processing could cause a new problem.

Conformance problems, as noted, are often the result of human error. Many reasons can cause this. Poorly trained, informed, personnel making changes without notifying others or even in the example, without thoroughly testing the material for property or process variances and informing the industry manager of the intended change that in this case, would have been denied! The ECR (engineering change request) process was not followed. To assist in avoiding this problem a sample form of an ECR is shown in Figure 2.

ENGINEERING CHANGE REQUEST (ECR)

Any business, engineering, quality, or manufacturing change must be analyzed for the effects it may cause once implemented. Failure to do this can cause procedural and work instructions to possibly change affecting other operations and possibly in a negative manner. For example, an ECR (engineering change request) from the customer must be reviewed in detail by all departments for effects it can cause. All changes must be evaluated before being made for effect on the products manufacture and end use application. The supplier has the responsibility to question and change if the effects can cause a problem further into the process. The customers change request may often not consider the process of the suppliers manufacture only the products end use requirements.

ECR NO.______________
1.DRAWING NO. ____________________________________
TITLE: ____
REQ. BY.______________
DATE:________________
ENG. MANAGER:______________

2. DESCRIPTION OF CHANGE
(DETAILED):____

REASON FOR CHANGE:____

3. INFORMED OR NEED APPROVAL OF CHANGE REQUEST (I =INFORMED A =APPROVAL)

__ __ MFG ENG MGR	__ __ I.E. MGR	__ __ Q.C. MGR	__ __ PROD MGR
____ MFG MGR	__ __MKT MGR	__ __ CONTROLLER	__ __ CONTRACTS

ESTABLISHED DATE:____________ DATE
4. ENG. ASSIGNED________________PROD. PLANNER__________OF RELEASE__________

5. ITEMS TO BE INVESTIGATED (INCLUDE W.O., P.O. & Q.C. TEST NOS. USED FOR INVESTIGATION
PRELIMINARY ECR REQUIRED YES:_____ NO:____
DESCRIBE DETAIL NEEDED: ____

6. APPROVALE FOR ECR RELEASE:
PUR MGR ____ COST IMPACT__________ CUSTOMER (Y/N) ________
MFG ENG MGR ____ I.E. MGR ____ Q.C. MGR ____ STD____ $/UNIT_____
MFG MGR ____ MKT MGR ____ CONTROLER ___ MATERIAL _______
LABOR _______

7. NOTE:____ TOTAL ______
ANNUALIZED
X VOLUME ______
TOOLING ______
SCRAP/
OBSOLETE ______
TOTAL COST ______

Figure 2. ECR (engineering change request).

Therefore, managers must carefully investigate the requested change on their operation, price, time, scheduling, and other operation changes not always evident in the ECR. It is also very important that all departments and personnel are informed of the change and when and how it is to be implemented. Status of any work in process and inventory of product or materials used for the product must be determined if still suitable for the products continued manufacture. If not it must be collected, segregated, and quarantined until an MRB (material review board) is held, with customer input, for its final disposition if needed. Also, program work instructions, if a procedure change is necessary, may require changing and retraining of personnel. If new equipment is required or existing machinery modified to meet the ECR a delay in time and schedule modifications must be approved by the customer and the contract renegotiated if necessary.

An ECR can generate a considerable amount of work and the cost may be covered in a price increase. If up front money is necessary to implement the change further negotiations with the customer is required. Also, if a contract item, the contract must be modified to reflect the change if necessary. The ECR is not a single piece of paper requiring department sign off but often a major work order to revise the manufacture and cost structure of the product.

Responsibility must be assigned in these cases to ensure the reasons changes were made are correct, all parties informed, and agreement from the customer is obtained so it does not cause a problem. A well-informed and trained staff can assist in reducing these types of problems. A method to reduce this problem is to have a material formulation check off sheet where all departments, sales, engineering, processing, quality assurance, etc. can review the changes, testing data and then sign off on the material or process change.

In maintaining control over all aspects of a product or process a Color Match Request form is shown in Figure 3. This form can be used for any item that is painted, colored or must fit up to or match a painted surface of another material. It is then sales and managements responsibility to offer the revised formulation to the customer and be prepared for their decision, to accept or reject the material for their product. Often, a change in a material requires the customer, if agency approved, to resubmit the material for the agency testing and it is expensive and time consuming. The customer will expect something in return if they are to go through these steps.

Customer:______________________________Contact:______________________________
Address:______________________________Position/Phone/e-mail__________________
____
Application:______________________________ (__) ______-__________
Product:______________________________ E-mail:__________@______
Material:______________________________Material source:__________________
Material replacement: Y/N, What:__________________ Prior material:________________
Part thickness:__________ Show surface: Y/N Class of finish required:______________
Estimated product weight: __________ pounds/kg
Color match: Exact: Y/N, color coding only: Y/N,
Color Target Standard Enclosed or Part to Match: Y/N
Color Sample required: polished, textured, type:____________, ribbed, what:____________
Date desired:____________________ Send to Whom?:______________________________
Molding sample for evaluation required: Y/N, Number of pounds required:__________
Must be compounded: Y/N, Salt and Pepper acceptable: Y/N, Other/what:______________

Color Match requirements:
An exact match requires a flat, glossy target standard. For critical colors use a Master Target Standard. If plastic masters are not available will supplier provide them? Actual parts available supply sample, adjacent parts or an assembled product and a pellet sample of one to ten pounds of current material.

Color Tolerances: Match Exact ( ) Close, or ( ) Wide
Parts have to match paint: Y/N Source available: Y/N Who:______________________
Matching parts
materials:____________________,__________________,________________
Are parts adjacent: Y/N, Separated: Y/N, How far away:________

Color match light sources customer will use for color approval: Mark 1 for primary, 2 for secondary light source.
( ) Macbeth Daylight ( ) Macbeth Horizon ( ) Tungsten
( ) Cool White Fluorescent ( ) Other ______________
Physical Property Requirements: Test results must accompany sample: Y/N
Mechanical requirements:____
Electrical/chemical/other:____
Pigment restrictions if any:____
Agency requirements:____
UV required: Y/N, percent of carbon black allowed:__________
Color Specifications:____
Other comments:____
____
____
Approval release:
Design team leader:______________________________Date:__________
Production manager:______________________________Date:__________
Purchasing Contact:______________________________Date:__________

Figure 3. Color match request form.

UNSTRUCTURED PERFORMANCE PROBLEMS

Unstructured performance is the exact opposite of the conformance problem. The tasks performed are not fully specified by procedures or requirements. This may be a new operation not yet completely developed with no written procedures and work instructions available. It can also be a service or sales situation where the ground rules are not defined. These and new systems and processes are unstructured and not specified exactly as to how to be handled in detail by rules or requirements.

Knowledgeable tasks cannot be standardized if an activity involves judgment and creativity. Many service and sales activities cannot be standardized because they must adapt to fit personalities, circumstance, and information received, and customer needs. What can be semi-standardized is the method and required type of data collected to document the business situation under discussion and what was agreed. These forms will also serve in gathering information for solving problems, establishing design requirements, or essentially any business or manufacturing operation. Establishing the known conditions and what actions are required for follow up is necessary for the operation of a company. An example of this information required for a business operation is the Sales and Contracts Check List in the Appendix.

Since these are mainly performance problems, the methods used initially can be a simple checklist for diagnosis to determine the reasons for the performance deficit.

Unstructured performance or system problems are very diverse with the most important tool for analyzing them is analysis, thinking carefully about the situation while documenting the information for analysis of the situation. Documentation is often forgotten in these "brain storming" sessions. Always have a note-taker to document team comments, recommendations, and agreements plus their data or ideas leading to a solution.

In some situations adding personal or company incentives can promote solutions to otherwise unsolvable problems. Solutions involving cargo containers not fully packed out when shipping overseas was a problem with one resin supplier. The cost of shipping the container was based on volume, cubic feet of container, not container gross weight. The supplier did their own container packing and with advanced foresight and utilizing known package size dimensions was able to increase product shipped capacity. This lowered current shipping costs by increasing the amount of product shipped per container by over 40%.

As a result of product dimensions versus current loadings directions, a procedure for packing was developed to utilize all available free space in the container. Guidelines were also issued to ensure materials were not damaged during shipment that also occurred due to prior packing methods. Inside container restraints and what could be packed on top of materials was also specified. The damaged good reports was not a first priority during the analysis, better packing methods further reduced customs damage claims and missed overseas customer shipments due to damage goods.

Once a procedure has proven it works, then fully implement it in your system. But, be careful to not limit the employee effectiveness in too rigid a procedure or work instructions for unstructured situations. Guidelines often can be used to ensure employees competence training make up for this lack of rigid instructions. Also, verify the effectiveness of the work before implementation at each end and in the system for effectiveness. The use of visual instructional material on how to handle routine and special situations can also prove helpful to guide the employee. Visual instructions and pictures with bullet identified directions at manufacturing work stations can substantially produce processing aids in problem recognition while improving workmanship of operators.

EFFICIENCY PROBLEMS

Efficiency problems involve material flow in the manufacturing operations, training of operation personnel or equipment rate output problems. While the quality to the customer is acceptable, the internal efficiency, cost and output may be below acceptable criteria to meet the customer's requirements. Time and rate studies can be performed to evaluate personnel and equipment. This will give management the initial data to make the required improvements. Lean and cell manufacturing sites may be solutions in output if lot size and equipment efficiency or test time causes bottlenecks in the manufacturing flow or product.

If shipments are late, schedules continually changed to supply key and job shop product with emergency shipments to customers, an evaluation of the manufacturing system is required. This is especially true when other work continues to fall behind schedule and output of the work force and moral suffers.

SIX SIGMA GOAL SETTING

Companies often use "goal setting" as a method for stimulating personnel for improving efficiency. The object is to tie in all stakeholders in the company to this goal, owners, upper management, and employees. To do this effectively all employees must be challenged, not threatened, into developing ideas and methods to improve efficiency. Many companies develop dedicated problem solving teams, some called SWAT teams in their departments to solve high reject problems by analyzing the problems and developing solutions to solve and prevent the problems. These teams focus on the highest profile problems obtained by Pareto charting the results of their manufacturing operations.

The SWAT teams were often assisted by quality assurance department personnel by assigning an engineer to assist the team members analyze the problem and use known quality methods to arrive at a solution. Then when a solution is obtained, further analyze the problem to ensure the solution prevents the problem and not just solves it for the time being. The only problem with this method is if the root cause should be in another department or area they do not control. The team must then get managements approval and the department manager to implement a change to correct their problem.

If a problem develops, the Champion is then called on to settle the problem. This can take many forms as improving material specifications, supplier performance, equipment maintenance, and upgrades and material flow and operator training. There are no single methods to solve problems permanently except a total analysis of the problem and working to find the root cause of the problem. This often results in finding solutions to other problems that were the cause of the problem identified and being solved by the SWAT team.

Often the level of the goal is not specified or set too high to be attainable within the time period required by management. This places an unrealistic burden on the team who must first establish a base line for determining the seriousness of the problem to be solved. Management often does not realize the time and assets required to solve what to them appears as a simple problem.

Until a study is performed, looking at manufacturing, equipment, personnel, and other problem issues. A realistic solvable goal to be achieved must be decided for the team. Therefore management must set an attainable and realistic goal with a time frame for all necessary and required problem solving analysis to be completed. They must provide the team the time to accomplish the data collection, purchase any assets needed in solving the problem, and ensure the personnel are trained and available to complete the work. This is required to be accomplished while production is maintained at the current production rate. This can be very difficult if personnel are not allowed the time and assets to do their problem solving.

Diagnosis of the problem after developing the preliminary data is necessary. Bottle necks can be identified, machine rates analyzed for intended throughput, operations investigated as to how, when, and where they are performed. Also, employee turnover and training has a direct impact on efficiency. Why the turnover, money, type of job, and how performed, supervision/management support, incentives, etc. all must be considered.

It is not always easy to identify promising opportunities for improvement if a baseline of problems or process improvement is not developed. Efficiencies must be evaluated for both before and after effects on the system. Often attempts are made to eliminate some operations. Evaluate the logic of these before implementing. An example of incorrect use was the elimination of verifying customer order entry data. This resulted in customer complaints for a variety of service, product, quantity ordered and where shipped, and delivery problems.

An effective use of idle or indirect time was to have machine operators perform secondary tasks at their work stations while waiting on the machine to complete its cycle. This eliminated down stream operators, made the operation more productive and saved the company money, while training an operator to perform additional tasks added greater value to the person. The operators due to long cycle times, had the time and equipment located for these secondary operations at their workstation. This little action helped improve employee moral as new job skills were learned enabling the person to grow in value to the company and in pay. Reward your people for their additional skills and operations. If necessary create a new pay grade to reward these personnel without creating problems with other personnel not doing this additional work.

Your employees are the best source for identifying efficiency problems in the plant. They perform the jobs and often, if they know what happens to the part after it leaves their station, can suggest new ideas for operation or product improvement.

Employees talk with one another and often discuss problem areas. I believe a new employee should be knowledgeable in as many operations performed in their plant as possible. Knowing what happens before and after can kindle an idea on how to do an operation better and more efficiently. This works in reducing problems and improving safety in the work area.

Problem solving evolves various strategies. Detailed analysis of all inputs, activities, and output. The use of fishbone, FMEA charts and checklists are the major quality tools used in these areas. This takes time and when changes are implemented must be managed so the change complement the before and after operations.

PRODUCT DESIGN PROBLEMS

If the product is not designed correctly for manufacture, assembly, decoration, and the end use of the customer, it can create a big problem at any of these steps in the operation. Simplistically stated, design problems can be avoided when all departments, business, finance, engineering, design, production, maintenance, quality, material suppliers, and the customer are involved. QFD should be used with the customer and internal operation analysis as an aid to ensuring the design meets customer requirements, is manufacturable, and able to meet quality and customer requirements.

Design of a product, or redesign, starts at the top with management involvement. The design and interaction of personnel is then broken down in sub-assemblies to the smallest component or operation. Reduction of parts, combining of functions into single parts, selecting materials to meet requirements and are the necessary and required steps of the design process. Agency and code requirements for safety and elimination of liability issues need to also be addressed. This places responsibility on the design team to ensure, even if the customer is not aware of these requirements for the product, must and should meet the manufacturing and product requirements for the products end use application. Always keep your customer current in agency and code requirements, should they lack this foresight of information.

Consider these ideas during the process of design; for methods analysis, design for value analysis, design for cost methods, design for manufacture, design for reliability and safety and any others that may apply for the product.

In the area of using plastics for metal replacement great inroads in bringing a product to market have been achieved. The use of die cast molding, existing metal platforms to use the plastic as a housing or support have reduced design to market times dramatically. These tools were used in the prototyping stage to make parts for testing and evaluation. The parts were then modified for their selected material and then final molding and procedures developed for manufacture adding features only possible with plastic materials. The quality movement has been a major factor in improving the design process by the utilization of their evaluation techniques and policies.

PROCESS DESIGN PROBLEMS

A process is designed to accomplish a designated task with a prescribed set of actions. Process design is therefore a task established to have a process achieve specified goals. Inherently if all processes were designed correctly there would be no problems. But, most processes always have a few inadequacies due to variables that cannot always be in control or tested to ensure they are always in specification. This is a matter of overall cost and just how much in control the company requires the process to remain, in control, during production. Therefore, performance and process design issues often blend together.

Quality organizations using TQM (total quality management) methods have begun addressing business organizational performance problems. ISO9000-2000 has implemented more requirements that require, SHALL changed to SHOULD, with companies to continually work to enhance how organizations perform their daily tasks requiring that procedures be developed and work instructions followed continuously. This has brought conformance of operations to business and manufacturing processes when strictly adhered to in all operations. But, as we all know this does not always continue as time and processes, personnel, and methods change.

Management did not realize, until problems occurred, that business operations could be standardized within specific boundaries so that operations, order entry, purchasing, accounting, and the like could do their tasks in a repetitive, repeatable, and quality efficient manner day to day. New situations would occur but these could be handled, in most cases, as other like or similar operations. This method causes less disruption and likelihood of a mistake being made. The base for organizational process and performance design is to ensure all operations are performed within the required business operational procedures of the company.

How business operations are performed can be structured using checklists, procedures, work instructions, and flow charts, with a form in Appendix C, similar to their use in manufacturing. Other quality techniques can be use, QFD, FMEA, fishbone, cause and effect, etc. to develop operation methods and assist in workflow problem solutions. Design of process flow charts and control plans are the most critical activity for solving process design problems. All processes have a start, middle, and ending with a specified or estimated time for completion.

Process design, like product quality can always be improved. A balance must be drawn when designing a process to ensure all requirements are implemented both old and any new requirements to meet product specifications or customer expectations. One should not overlook new methods as manufacturing equipment, station flow and layout, movement of material methods, and improved computer and software systems to track yield, rework, and scrap to increase efficiency, monitor product output, and quality. But, reinventing the wheel should not be employed if the current method has proven reliable and efficient.

The flow chart mapping of a process coupled with the fishbone diagram and known internal or customer requirements is an excellent set of process design methods to use initially for designing and establishing a process. There are existing quality tools called, process analysis, that can analyze and improve existing processes. It is not always necessary to reengineer a new process with often-untried technology when the existing system can perform within and meet company, procedure, and customer requirements. But, considering new flow or manufacturing systems as lean manufacturing, could improve and assist in output. Kaizen studies to redefine plant layout can also assist if used effectively in the organization.

Keep these reengineering ideas alive, develop a proposal with cost and engineering backup and present it to management for their consideration. In many cases, ideas from the proposal may be incorporated into the existing system for business and process improvements. Try a simple trial of the proposal to see if it solves or improves the process. This is less expensive and disruptive and should it not perform as planned, the line can be returned to its original operation with minimal loss and down time.

The areas of process solutions include process flow and layout, input screening, and control, handing, task assignment, and scheduling, setup, coordination, and consolidation of activities, process triggers and the handling of interruptions and delays. Always ensure if a problem with product is uncovered, an immediate MRB (material review board) is scheduled to evaluate the products quality. Also, if a problem occurred because of questionable material, workmanship, or testing, the material or product can be dispositioned immediately. If cause is determined, a change can be made immediately, material quarantined, or flow path changes made to continue manufacture or the line shut down until the problem is solved.

Problem solving should be a controlled and regimented process using the existing quality problem solving techniques that involves data collection, communication, and documentation of results. The final analysis then occurs with team support to reach a satisfactory conclusion. Implementation next proceeds with verification of results to substantiate if the solution was satisfactory to permanently eliminate the problem from reoccurring.

TOOLS FOR IMPLEMENTING SIX SIGMA

The black belt has been trained in the use of the quality methods to access the companies business and manufacturing capability. If your plant, equipment, and personnel are not capable of achieving Six Sigma these quality tools will show the areas where there are weaknesses in your business and manufacturing systems.

After these weak areas are known, a solution must be developed to eliminate these deficiencies. If equipment and plant services or supplier problems are discovered they will be identified so a fix can be implemented. If a personnel problem is the cause, it is often not as easy to fix if the individual is not following the procedure or work instruction as developed. Only by conducting "Real Time" audits can this be uncovered unless product quality problems result from not following the instructions. This is difficult to identify and often hard to verify for some operations. Retraining has worked and reemphasis of using the procedures and work instructions as written. Also, verification that they are correct and current for the operations performed by the operators.

Therefore, getting employees to adhere to the ISO/QS9000 and Six Sigma methods and any other quality program procedure is essential for its success. Quality is an employee-enhanced attribute of the product or service rendered. Your employees make it work. Management Buy-In and Support is crucial so employees can be empowered to perform the tasks to ensure quality is number one at the company. Recognition and rewards can assist to ensure their acceptance is made easier often they are trained in how and why they are to be followed as written.

PROCESS MAPPING FOR SIX SIGMA TEAM ACTIONS

All Six Sigma team process or problem improvements should be displayed to company management and all employees. All employees may not initially be involved in the Six Sigma program or even be aware a program is underway in their area of work. But, as time and actions proceed may be asked to participate in one or more solutions developed by the team.

Progress of the Six Sigma Team must be displayed and a good area is the cafeteria or main common gathering area of the plant. The information can also be displayed at all entrances and exits of the building. This will keep the employees informed of progress and better understand what is occurring and being accomplished. This method of displaying information is known as storyboarding. An example of how this is done is shown in Figure 4.

Storyboards are useful in communication process improvement-related information and events as they occur. They can also prepare employees for

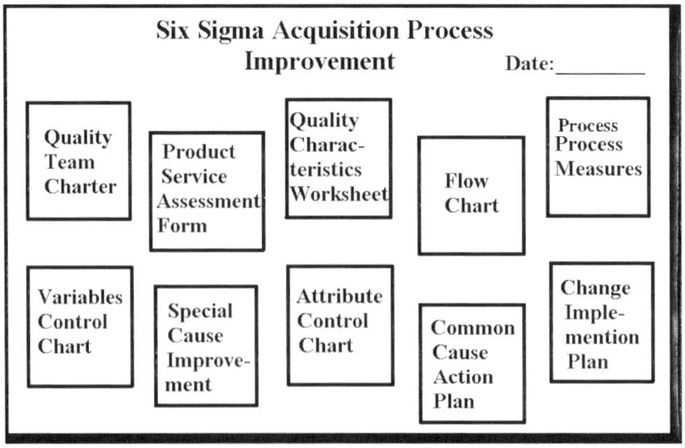

Figure 4. Storyboarding Six Sigma processes improvements.

any changes that may occur or training to be taught as a result of the changes in the system. Requests for input may also be an offshoot of the display requesting input from employees in affected areas.

Pictures, graphs, charts, and lists of improvement in output, quality, and savings are of interest and how the employee may benefit should be presented and outlined as a benefit of the Six Sigma program process. Ensure it is always a benefit to the employees, never a loss of a job but a chance to grow with the company.

PROFILE PROCESS IMPROVEMENT

All Six Sigma and all other business and process improvements should be profiled. This method is used to show anticipated and desired program results that are then posted in the company common area. This is the area where employees gather to eat and socialize during their breaks and where they can learn how their involvement is needed for a program to have a successful conclusion. Presenting the information of the Six Sigma Programs in storyboard manner will illustrate the steps taken by the Six Sigma Team members, decisions made, and the reduction in cost, scrap, and cycle times they have achieved.

These programs will become company success stories that can inspire new programs in other areas. Many employees know of other programs that could be developed and they just need the inspiration, knowledge the company is interested in their ideas, and willing to explore new challenges in continual quality improvements and cost reductions. A Six Sigma program suggestion form used in one company for obtaining business, manufacturing, product, and processing improvement and problem solving program suggestions from their employees is shown in Figure 5.

Many companies have Six Sigma program suggestion forms located throughout their plant. Similar to the original suggestion box that often lacked creditable employee feed back. Employees need to know if their suggestions were considered, initiator receive a response to their suggestion, and was their idea accepted or rejected and why. Suggestions considered should recognize and reward the employee for their careful thought that went into their filling out the improvement form. If this is not done correctly, the enthusiasm and thought process of the employees soon dries up. Some improvement programs can cause confusion if not thoroughly presented.

We are now involved in improving and preventing problems from reoccurring.
For this program to benefit the entire company and "you" our employee, we request your assistance in recommending cost savings and product and quality improvement programs.

These can be in both business and manufacturing areas as Six Sigma programs will work in either area of our business and will benefit everyone once implemented and working.

We are forming Six Sigma teams and have our new Black Belt, team leader now working with our company Champion, to select the program that will initially yield the greatest benefit to the company. This does not mean your department is not worthy, just that the initial program(s) will focus on the ongoing business, process, or manufacturing problems that have not yet been satisfactorily completed and improved.

Please list the programs you believe are worthy of being worked on and solved. Your input is needed as you are the one daily trying to solve or work around the problem you listed. Please fill out the form and put it in the collection box which will be collected weekly. After a review of the suggestions, our committee will get back to each individual to inform them of the status of their suggestion. This will be on a weekly basis. Each suggestion has merit and we want your input no matter how trivial or complex the problem appears to you.

We welcome more than one person turning in a form as the more employees aware of a problem the more possible solutions are available. THANK YOU.

PROBLEM:____
____
____
____
____
____
____

Submitted by:_____________________________________
____

Department: ___________________ Supervisor:___________________
Shift/phone: ___________/___________________

Date:_______________

Reviewed by:_____________________ On: ____________________
Recommendations:_________________________

Reply to Submitter:___________________ On:_________
By:_______________
Comments:____

Figure 5. Six Sigma program suggestion form.

Figure 6. A good quality aid poorly utilized.

Take the time to have a meeting to kick off the program and have knowledgeable personnel there to answer all questions from the employees.

Often, quality assurance or other departments put up posters and signs that can confuse the employees if not explained beforehand. One such example is shown in Figure 6, composed of (ACT, PLAN, DO, STUDY). When I first saw this figure, I believed I knew the thinking around the diagram. Talking with other employees I learned that some had similar beliefs but others found the figure confusing and wondered if there were to be a new quality initiative at the company. Since, nothing was forthcoming from the quality department, nothing was ever done with it and it soon became just another diagram on the wall without any meaning to the employees.

Since then, I found out Figure 6, is called "The Model For Improvement" developed by the Associates in Process Improvement – API. The model, developed for rapid cycle change, was used in the Community Collaborative on Family Violence. Mr. Lloyd Provost, of API provided volunteer mentoring to the "Tacoma Project", using an adaptation of the API learning collaborative methodology. The Tacoma project became a community learning collaborative, using rapid-cycle application of the improvement model for teams to become effective in making change.

The model for improvement asks these three questions, "What are we trying to accomplish?" next "How will we know that a change is an improvement?" and last "What change can we make that will result in improvement?"[1]

This sounds a lot like Six Sigma type problem solving. Identify the program and develop methods for gathering data, analyzing the data and acting on the results for improvement. This is stated very simply but means the same thing. The loss of this idea at the company I was with at the time is priceless and if only management knew what to do with the information.

When an improvement or quality plan is implemented, it needs to be explained to the work force and any information requiring employee actions fully explained.

Anyone who suggests a change wants recognition and when positively reinforced by the company, even a program that was initially not considered, deserves a positive response. This may even be in recognizing an employee for the number and types of improvements suggested and why or when it may or may not be utilized in a company program.

Some suggestions may not be of Six Sigma status, major savings in time, dollars, or reduction of defects, but acting on them could result in small ways to improve company operations and customer satisfaction. Keep your employees thinking, reward suggestions used, and all suggestions submitted. Who knows when there may be a golden opportunity that no one else has even considered. Also, rewards do not have to be monetary, plaques, certificates of recognition, and merit and even articles of clothing or coffee mugs can show others, this employee is thinking about new ways to improve the companies business of operations. Like safety and quality programs, posters and hats do not make a successful program the employees do this in their daily work habits and operations.

SIX SIGMA PROGRAM REQUIREMENT OVERVIEW

Ensure your black belt has an understanding of your industry for best initial results. The majority of black belts were selected from the major companies employees who were then trained internally by their own Six Sigma trainers. The major companies established a goal for quality and process savings and developed their quality methodology with their major savings projects in the initial conception of black belt training. Usually the black belts can quickly learn the variables affecting a new business and manufacturing operation. The training they received and how to translate it to any process savings are the same with only a different set of variables in a new operation. Therefore, the more a black belt knows about their industry and how the company operates and solves problems the fewer initial problems will be created. The new company black belt must read the companies business and quality procedures and work instructions to ensure they understand how the company operates.

Quality methods and procedures must be taught to all employees, managers, and owners down to the last responsible job in the company. Everyone has an affect on product quality even if they are not directly involved in it's manufacture.

SIX SIGMA TEAM TRAINING

Employees are the key to implementing Six Sigma after being trained by your black belt. The black belt is responsible for ensuring the Six Sigma team is capable of undertaking a Six Sigma program. When assistance is required the black belt ensures this is provided. If a personnel problem arises, the black belt cannot solve, the champion is called on to solve the problem.

Employees of the company comprising the Six Sigma teams must be told what their RJP (realistic job preview) consists and their anticipated goals. The RJP must show the teams a realistic expectation that leads to a higher job satisfaction and low turnover of personnel.

When the program is young, taping Six Sigma and problem solving training sessions is very helpful. These taped training sessions can be used to also train new employees in their job function and give the Six Sigma team a quick update as to how the current operation is performed. This in conjunction with working and watching how the operation actually runs will give the team a heads up as to are the employees actually doing the job as instructed. Should this not be as originally instructed why was a change necessary and never documented in the procedural process control system? The tapes can also be used for personnel to be trained in normal operations or special requirements for a job and if a point needs reinforcing or reviewed. These video training session tapes can be viewed again, hopefully explaining any misunderstood area of training.

Before the teams are formed the black belt should interview perspective team members to ensure the talent they bring to the Six Sigma program is beneficial. They are also interviewed to determine how good a "team member" they will make based on team members evaluation points listed in Table 2.

An employee team member check list may be developed to select the most knowledgeable candidates for the selected program. Key team traits for interaction must be explored. Keeping job titles out of the team mix at the

Table 2. Realistic Job Preview for a Quality Management Organization.

Topic	Positive features	Negative features
Empowerment to help customers and deal with peers	Creating customer satisfaction and agreement among peers	Dealing with angry customers and unhappy peers
Working in cross-functional teams	Member diversity producing new, innovative ideas	Member diversity producing arguments, conflict, disagreement non-solutions, blame, etc.
Using statistical tools	Getting direct feedback on actions	Dry and frustrating learning of statistical methods, no support, no understanding of requirements
Continuous training	Reward and recognition ceremony giving out training	Time spent on training both on the job and after work at home, no recognition for good work, criticism certificates, etc.
Culture of Quality	Good reputation, well-run organization	Low expectations for success, do not trust results, hasn't worked in past, negativity

(Adapted from reference [7]).

start is important. The black belt, no matter from what station they were chosen, is the absolute Six Sigma team leader. Even the president and upper management of the company must adhere to the black belts team's recommendations. They can agree to not abide by the teams recommendations or due to circumstances select a different solution to solve the problem. This is their final decision and they must take the responsibility of the program if they change the outcome of a program.

Team composition must be based on comparability of job skills, temperament, and ability to develop suggestions and speak their opinions. All team members are on the same level, with no leader other than their black belt.

The success in quality, product, and systems improvement and monetary savings are the result of the Six Sigma team and the black belt. The team and all support personnel are the responsible members for the programs success. Recognition of success is very important for team members. Their team success can make believers out of non-believers of a quality system. But, should a problems solution appear, not solvable, assistance from any and all sources both in and outside the company must be attained. Problem solutions are a team effort with no single person always capable of solving a problem. The company and all their employees, even calling on outside technical assistance when required are also a requirement of the Six Sigma team.

THE TOOLS FOR SIX SIGMA PROBLEM SOLVING AND PREVENTION

The quality tools already discussed, QFD, FMEA, DOE, scatter charts, fishbone, check lists, etc. are the primary tools plus others, such as SPC (statistical process control), to assist in collecting, sorting, and reviewing the data in it's basic plotted relationships to understand what has happened in the process to solve the problem in the company. A collection of data reporting forms, graphs, and charts are found in the Appendix for assistance in finding the right form for data collection.

The key to problem solving is employing the tools of quality in the correct manner. If not used as intended, or short cut, the necessary information may not become available, known, documented, and used in solving the problem completely in the shortest amount of time and effort.

One documented problem involved a resin supplier who required a uniform pellet size. This was always possible until one day the process, pellet size, went out of control. All variables in the feedstocks were evaluated and determined to be within specifications. The process did not appear to be out of control based on then known information. Discussions with operators and data collection of temperature, pressure, rate, and physical properties did not show any variance. Since the process was a batch process and pellet size was created in the reactor, with the solids (pellets) forming in suspension, it was determined the feedstocks must have changed as nothing physical in the process appeared to be out of specification.

Contacting all raw material suppliers resulted in feedstock quality testing good until the chemist on the Six Sigma team requested the suppliers test for any foreign impurity in their product. One supplier found a trace element never before detected in their material. It was later determined, a replacement coupling was of the wrong alloy and due to the leaching effect of the chemicals extracted foreign elements that contaminated the feedstock. Chemical analysis revealed this contaminant was the catalyst for the non-uniform pellet size problem. The supplier withdrew their product, corrected their problem to remove the trace element, and resubmitted material for reevaluation. The trial was successful and they were allowed again to be a supplier. This information was sent to other feed stock suppliers to ensure they would not cause a similar problem with their products.

As a result the other suppliers of a similar product also began testing for any foreign elements. This raised their product cost but not enough to justify a price increase. It also ensured their customers of continuing to receive a high quality product without their having to perform an incoming chemical analysis of their feed-stocks.

This particular problems solution indicated data showing the system was in control but an unknown variable was created that suggested a feed stock problem. The feedstock supplier's quality checks initially did not show a problem until a full spectral analysis of all components was requested by gas chromatography. Then, one supplier found the stray contaminant, in a very small concentration that during the high heat and pressure reaction process produced an affect on the materials rate of polymerization. The customer gained increased confidence with their suppliers and with the new tests, added assurance this type of problem should never occur again.

STATISTICAL PROCESS CONTROL

When collecting data for any operation, set a goal. The goal may be as simple as determining what defects are in your system using a Pareto chart to visually show type and frequency. Quality methods were developed to assist in mapping out problem areas plus showing capability of an operation, and the variance of selected items to show if they are in control.

When a problem occurs or with Six Sigma methods someone in authority says, "We must collect data to determine where and how bad are our problems". Fortunately, Six Sigma is turning this statement around to positively ask, "How good is our operations and let us see the results". The former statement is often said without any idea of what to monitor or how it is to be done and for how long a time.

Before any data is collected a plan must be developed which is well thought out for what the primary mission statement for corrective action or process/system improvement is to accomplish. The company champion sets the goal, the black belt and team determine what is required to accomplish the goal and do it.

SPC will be used in the analysis of the data collected. The Six Sigma team must be trained by the black belt in the correct methods to use to collect the data so it is not corrupted in the gathering. Data can be corrupted if not correctly gathered using calibrated equipment and recording methods. The team must know these good methods for the correct use of SPC data to correctly analyze the process or solve the problem under investigation. For any process under analysis the data must be presented in a manner to describe how the process is operating in "Real Time".

This information can be found in a quality textbook and the team can select from a menu of charting and presentation methods how the data will be presented for analysis. The list is almost endless and includes; Pareto, acceptable quality level (AQL), acceptance sampling, average outgoing quality (AOQ), assignable causes (of variation), attribute data (quality from go/no-go, pass/fail determinations), bell curves, bimodal distribution, binomial distribution (probability distribution), c-charts, capability of processes (CP, CR, CpK, and Z max/3, cause and affect, cell (of frequency distribution and/or histograms) chi-square (X^2), degrees of Freedom (DF), and on and on.

The goal is to select and use the data collection and presentation method that is the most capable to analyze and then present the data or the collection of data of the operation or problem under analysis.

When selecting the data to be collected, determine "how" the data is to be collected. This will depend on the process and the data collection method established for the operation. If data is already being collected the team must determine if the output actually describes the process under study. If not then how and what data is to be collected. In manufacturing there are analog and digital control units on most of the equipment and machines. The more modern machines come equipped, or can be purchased with data collection, storage, and output capabilities usually in a more exact digital format. Older machines may not have this ability and manual or external monitors will have to be employed to collect the necessary data. No matter what method is used, always be sure the equipment used; gages, valves, recorders, etc. are calibrated and in their manufacturers specifications.

SPC (statistical process control) is the most widely used process and product control method for controlling and monitoring the manufacturing process. Today, business systems and operations are finding statistical process control is very valuable for improving their business and manufacturing operations.

The term process control is used to collect statistically chartable data from all areas controlling the companies business and manufacturing processes. Without the use of statistical process control to monitor and control any manufacturing process it would be next to impossible to repeatably produce good products. SPC is used to control and monitor a process within preset limits determined by the processing conditions required for the material and manufacturing operations.

Monitoring the system ensures it remains within established control and process limits for the products manufactured. Most, if not all, new processing machines have process controls on the machine to regulate, control, and monitor the input variables operating on the machine during processing. Internal sensors on the machine report on "Real Time" machine variables that are time, pressure, temperature, speed, and other variable control settings for each cycle. Many new machines are built with controls to output the data using built in systems hardware and software programs. The system is then programmed to collect readings, average them, and then output them on the machines SPC reorder as how repeatable the cycle is progressing over time.

Process control requires an output to visually ensure the control limits are not exceeded during operation. Process control is defined as: "Maintaining the performance of a process at its manufactured capability level".

Process control can involve a range of activities such as gathering, reordering, and documenting process data, operation inputs from sampling the product of the process, charting its performance, determining causes of any major variation, and taking corrective actions when necessary to control the process, within established preset limits. SPC is not just collecting and charting process outputs, it is a visual "Real Time" map of what is happening in the process, cycle-to-cycle. SPC is used to measure the repeatable capability of any process or system used for manufacture.

Process capability is defined as: "The level of uniformity of product which a process is capable of producing". Process capability is expressed as a percentage of defective products, the range or standard deviation of product dimensions with measurements taken on critical dimensions. When the data is plotted it forms a bell shaped curve of the readings around the processes target mean with the data analyzed to its extremes and then determined to be within a known sigma limit of control. The data represented in Figure 7, shows that the process can be in control but not centered around the mean. The figure shows four situations with acceptable CpK process in control values but data indicates the readings are not centered.

Control (of a process) is said to be within statistical control if the process only has random variations grouped around the control mean and within the established control limits for the process. When control charts are used to monitor control of a process, a state of control exists when all points remain between established upper and lower control limits. This is shown ideally in Figure 7, for establishing the UCL (upper control limits) and LCL (lower control limits) that are shown as the tighter of the three. As shown, all limits are within the three-sigma limit.

There is sometimes confusion or a misunderstanding between drawing specification limits, manufacturing or process control limits and control limits as shown in Figure 8. This figure illustrates the three different types of control limits for a process. Control limits are tighter than the process limits that are used to control the manufacturing process. Process limits give the operators more variance during manufacture but are approaching the products actual permitted specification limits.

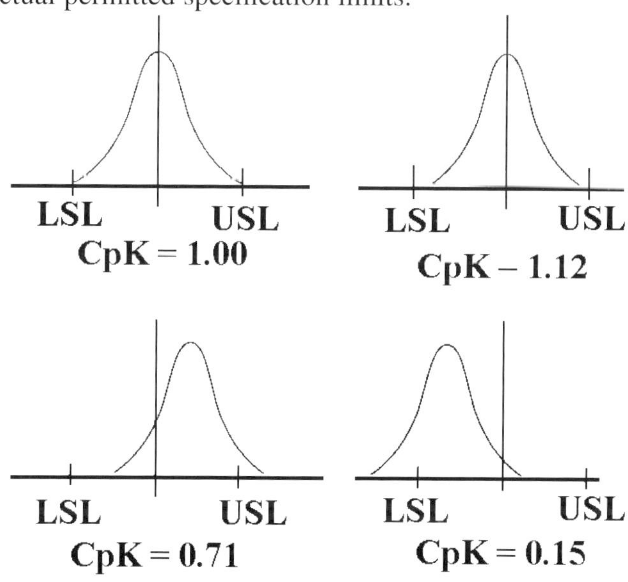

Figure 7. CpK relationships.

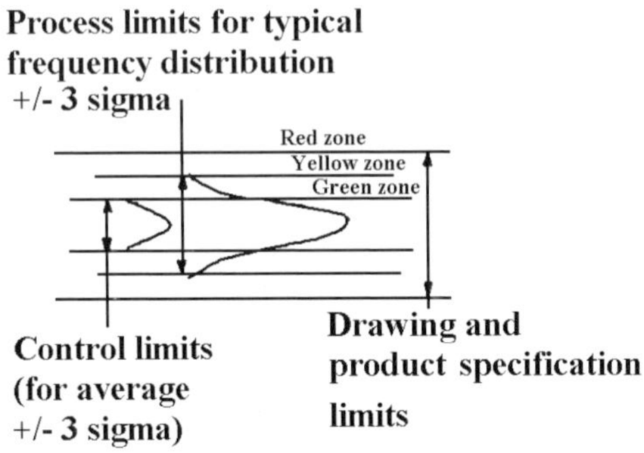

Figure 8. Manufacturing limits.

These limits are designated as green, yellow, and red. When all readings are in the green, excellent control is achieved. When some are yellow or have a walking trend to red, a signal is raised as a variable is out of control and has changed. Corrections are needed immediately in "Real Time" before the process goes into the red, the out of control process zone and out of product specification.

When the data points are in the yellow zone, the product is still within the customer's specifications. But, once the process goes into the red zone, the product will not meet the customer's specifications. When this occurs, product must be segregated and stored in a secured location for out of specification products. It is important that these do not become mixed in with good products. This is continued until the process is back in control, yellow or preferred green area on the control chart.

Once the process is back in control and verified as remaining in control an MRB (material review board) is then convened to determine if the out of specification product is acceptable for the customer or should be reworked or scrapped. This may involve the customer determining if the product is usable in their system. (Recall the example of the Black Belt and the housing, key way, problem that allowed the supplier to rework the part to accept a slight deeper key way. Discussions with the customer determined this slight modification did not affect the operation or quality of the product.) The supplier, if any doubt exists, must always defer this final product decision to the customer when applicable. Processes are non-stationary and over time, if a variable changes, will show a trend for going out of control.

Control charts are used to plot parameters of process performance selected as key indicators of process output. Data is collected at set time or number of cycles or other chronological variables either on the process or product produced.

Control limits are established based on product/machinery operation that when all data is between these control limits the process and product output is judged to be in process control.

Various parameters can be plotted and presented in a format to describe the processes normal variation throughout the manufacturing cycle. It is important to establish control limits that really record the processes true capability and control to produce the required product to repeatable specifications.

The control limits are established by selecting variable values or known to produce a product within specifications parameters or if a new process determined by trial to ensure the product or processes output stays within the customer's requirements and specifications. Control limits can be calculated based on known or process generated data. If the process strays out of these calculated or specified control limits, the process is out of control. This is a signal that action is required to bring the process back into control within the established limits.

PROCESS CONTROL CHARTING

Statistical process control data, to be of assistance, must be gathered, plotted, and used in "Real Time". Monitoring the process in "Real Time" will show the operator a trend unless a catastrophic failure of a component occurs. The ideal process control method is to check the system, monitor it frequently, preferably at predetermined set timed or cycle dependent intervals, and if a trend is developing, analyze the movement as to cause and if the trend continues bring the process back into control by identifying the offending variable and making adjustments.

Many of today's machines have statistical process control built in with installed software programs for control and analysis. A variety of control and analysis software is available. Closed loop continuous feedback systems are preferred for making minor adjustments once the system is in control. Control limits alarms can be set to signal a reading out of the norm and to alert the operator or process control team if a trend is developing.

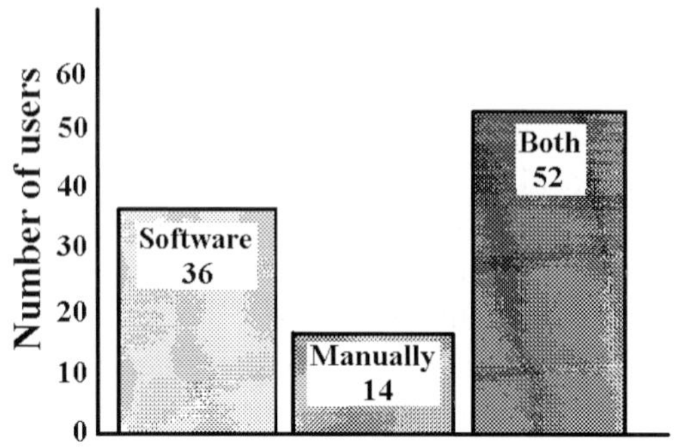

Figure 9. Creating SPC (statistical process control) charts.
(Adapted from reference [2])

A survey was conducted by Quality Magazine* with 102 respondents stating 36 used software base chart creation, 14 manually plotted the information, and 52 use both methods as shown in Figure 9, for those charting SPC data.

Another response in the same survey as reported in Figure 10, with 113 responses stated 103 used SPC in their company with 87 stating it was helpful with only 23 using it to its full potential. SPC is powerful when implemented as a culture, rather than as a statistical tool. For a quality culture change tools are not sufficient, it needs a change in thinking and working habits. Working habits, it is reported, are to some extent are controlled by management. I have seen both good and bad. The data

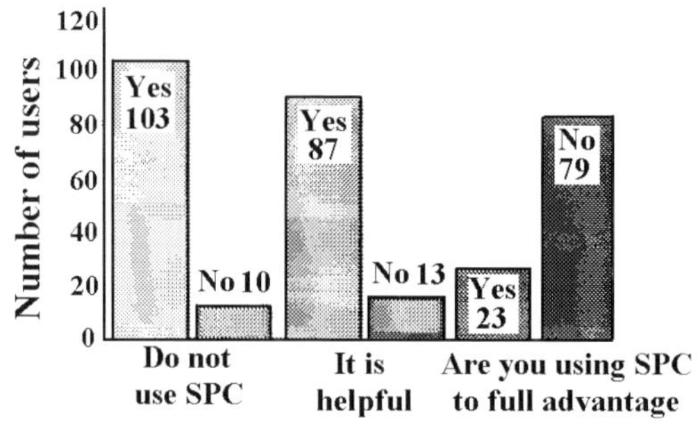

Figure 10. Use of SPC by companies.
(Adapted from reference [2])

collection system I have seen is obtaining electronic generated data and ensuring the collection and monitoring system are always calibrated. Also, the data is collected and used in "Real Time" for control of the process.

SPC should not be used to just create visual control charts so visitors and customers can see how good your processes are kept in control. SPC is for your setup technician and machine operators to use to establish and ensure the process stays within established control limits. QS-9000 requires that control process charts on the lots produced, accompany the products shipped into the customer plants to verify that the manufacture of these products was within control limits established between supplier and customer. This requirement for ISO9000 is very likely to be required in the latest revision ISO9000-2000. SPC is a process control and monitoring method to ensure the manufacturing process stays in control so only good and acceptable product is produced.

The creation of the SPC charts using operator, hand plotted, "Real Time" or SPC machine data collection and control software, also in "Real Time" does matter for accuracy of recording and then calculating the data points for plotting the results of the process. Using a good software program to collect and then calculate the data removes the potential problem of operator error. Also, the time factor must be considered. It is important the software program outputs in "Real Time" the data in either chart or numerical form for instant use by the engineers and operators monitoring the process.

Data accuracy and manipulation for ease of creating the process output charts must be considered. What is important, and often gets lost in the training is how the data is to be used in controlling the process and exactly what and how the process manufacturing variables controlling the process should be reported and in what form, numerical, bar graphs, or x-y charted.

Questions that must be addressed when charting variable data are:

1. Who determines the variables to monitor?
2. What should the variables mean value be and the UCL/LCL calculated values?
3. Where, how, and when should the data be collected?
4. Who should collect the data?
5. How should the data be presented for monitoring control?
6. How often and when should the data be collected?

7. How should the data be presented for monitoring control?
8. What is the allowable delay from collection to review of the data?
9. When and how often is the data reviewed?
10. How should the data be used?
11. Who uses the data and when are changes made and by whom?
12. What does one do with the data if high variability occurs and out of control happens?
13. Who makes the changes and what changes are allowed to be made?
14. How long do you wait to see the effect of a change?

These questions must be answered before any process is deemed worthy of charting for monitoring control of a process.

When new, unskilled, or current operators are trained to collect the data, sufficient time and training must be spent to emphasize the importance of accuracy in collection and recording the values. Also, the techniques to process the data to generate the output charts and information should be taught and understood when performed by individual operators and inspectors. The tools for ease of determining plotted values must be obtained and operators taught in the correct methods to calculate the data and then if by hand, plot it accurately on the correct chart.

CALIBRATION

A very important item often overlooked unless the company is ISO9000 compliant is the accuracy (calibration) of the process control instruments, temperature, pressure, time, rate, etc. of the equipment and manufacturing output systems. If these instruments that control the process and product are not in calibration, on all instruments and machines, your control settings and data output is deemed almost worthless for establishing process control and developing SPC data.

Therefore, ensure all sensing and data output and recording equipment are in calibration and on a scheduled calibration program. If unsure of any data sensor or recording equipment is not correct, have it calibrated immediately. This is also very important for portable measurement tools, calipers, dial gages, etc. used in verifying product measurements for data collection. The person collecting the data must also be trained in the correct method to take the measurement and the operation, maintenance, and setup for how to take reliable and repetitive readings of the product or process variables.

The purpose of training is to ensure all data collected is accurate and obtained in the same way each time by all personnel. There are special training organizations to train personnel and also to certify they have been trained to make specific type of measurements.

Statistical process control can be used for almost all business and manufacturing operations. Only in business operations, as marketing and sales, does its value become less traceable due to the nature of negotiations, price discussions, personalities, etc. which makes it harder to plot results of meaningful measurable data. Office operations of repeatable functions can be reported for analysis using SPC. The only requirement is determining what operations are of a repeatable nature or close enough to record data to measure the variable under study or requiring improvement. Manufacturing and service operations besides business operations required to meet customer specifications and requirements definitely need to be controlled and all are equally important to the success of a companies business operations.

In any process, who determines the variables to chart is usually the first question when control of a process is required. In a business operation it is less defined. The office processes information both verbal and then on paper or electronically in their computer as for customer orders, requests for information from department operations, and purchasing plus the accounting and business financial operations. These business operations are usually routine in their flow through the company. They can be tracked for problems, severity of errors (Pareto charts), and any influence on the company and customer relationships. It is the business manager's responsibility to ensure their operations perform as required and all tasks are completed on time and within the business operations of the company policies. In a product or service operation the company knows what the product or service is advertised to be capable of doing and now must ensure it is produced or the service rendered to meet this objective and the customers requirements and satisfaction.

OEM's (original equipment manufacturers) establish and design their products to meet the requirements of their customers. If the product is required to meet or obtain an agency "compliance" label for safety or other agency requirement, the specifications for the products testing must also pass their requirements. An agency such as UL (Underwriters Laboratory), test products to meet specific customer or product safety requirements before permitting their label to be used on the product. Consumers have come to rely on and insist these commercial and industrial products carry the UL label for consumer safety and reliable operation.

HOW TO DETERMINE THE VARIABLES TO MONITOR

In a new product design team program all departments should have had an influence in the products design, tool design, material selection, process equipment, required dimension, and final assembly and decorating areas. Each area of the product will have their own variable requirements. It is now the team responsibility to ensure the tool and manufacturing operation are capable of making the product, repeatable over the life of the product. Monitoring these variables will ensure if wear occurs, the tool, equipment, and process can be brought back to as like new condition before problems occur in the products manufacture.

The processing team selects the molding machine for the tool and material. Ensuring the melt capability, screw design, and size, and clamping force and platen size are within the tool and product manufacturing requirements as for injection molding a plastic product.

The material processing variables and molding cycle time, temperature, pressure, etc. are selected and the process is evaluated. Fine tuning of the process occurs after the process reaches equilibrium, usually within 30 minutes, depending on the size of the injection molding machine and the heat transfer qualities of the tool that can vary due to tool size and machine conditions.

When performing the initial trial evaluation, always reach equilibrium for all variables in the process, tool, material, cycle, plant variables, auxiliary equipment conditions. Only then can adjustments be made and variable data collected of meaningful value.

Whenever possible, evaluate the molding process on the machine to be used for production. During this period of tool trial, the process is evaluated to ensure the product specifications can be met within normal molding parameters. If special conditions must be established out of the normal processing range, the customer must be informed if not all dimensions are attainable or other problems occur to reach and maintain the critical product dimensions. This trial will establish the variable requirements for the process and you can evaluate others to see if the process is capable within the process and product requirements. This will ensure if the evaluation is successful, not requiring tool cavity adjustments, the process can be reestablished for production.

During this time period, the process team once machine and tool capability are known and with input from the design team, identify the controlling processing variables to monitor, along with ensuring the required product dimensions are obtained in the process.

ESTABLISHING CONTROL LIMITS FOR THE PROCESS

What should the variables mean values be and their processes UCL/LCL limit values. During the evaluation, the process itself will vary as the variables reach equilibrium. It is recommended that the variables selected are recorded even as the process moves to its equilibrium condition.

The early data collected will show effects on the product and may assist later in solving a problem when similar symptoms occur. Remember the temperature variable takes the longest to reach equilibrium in any manufacturing operation. Therefore, during startup, the cycle variables should be kept the same, producing product until the process team is confident temperature stability of the tool, material, and process temperature variables for the manufacturing operation are reached. Instruct the operator on how to determine when this equilibrium condition is reached, taking process measurements at timed interval such as product weight.

For injection molding, a constant part weight for the manufactured product is considered a very reliable measure of cycle and process repeatability. Once the molded products weight remains constant within a few tenths of a gram or less, just the part without the runner, is an indication of processing equilibrium being attained. Ensure this information is in the process and work instruction sheets to inform the operator how long to wait and how to make any adjustments so good parts are being produced in each molding cycle.

Preheating or cooling the tooling prior to starting up the manufacturing operation can save time and reduce product scrap in each startup situation. This should be a prerequisite in the process control startup procedure along with any other conditions in the operation and the plants environment that will affect a process variable.

For manufacturing and especially a Six Sigma process, equilibrium must be achieved before product is saved for the customer. There are established methods to determine what product and process variables must be to meet the customer's specifications and product requirements. A method often used is initially molding to maximum product weight. The cycles ram forward time is slowly decreased from a maximum determined time, to keep material flowing into the tool, until part weight begins to decrease. The ram forward time is then determined as the last cycle time before the first weight decrease. Other variables are then stabilized for the process using the best molding and product variable setup methods for repeatable production output.

DETERMINING THE CRITICAL PRODUCT DIMENSIONS

When determining the critical product tolerances do not use metal working, title block typical tolerances for plastic products. Use the tolerances from your material supplier and the SPI (The Society of the Plastic Industry) as shown in Figure 11.

The anticipated shrinkage tolerance listed for a series of product dimensions for a product made from the generic plastic material listed for the specific sheet. If tighter tolerances are required as in the electronic and medical industries, then the tolerances and temperature control of the tool, and number of cavities, balanced layout in the tool, must be much tighter than a normal plastic product. This may require metal tool cavity dimensions in the 0.001″ range or lower with tool temperature control, for example, cooling water entering and exiting in a parallel cavity circuit and controlled to $\pm 3°$ F or tighter. When determining what the critical dimensions should be, make sure the product and the customer agreed that the dimensions are actually required. Once these dimensions are determined, then the actual control limits can be established.

Based on the initial molding trials production will inform and work with the team on what variables to monitor for the product and process to ensure repeatability is achieved on each manufacturing cycle. These are normally the typical: pressure, temperature, rate and time variables controlling the process and tool temperature control. Remember the process limits and product specifications will vary with each product and method of manufacture.

Standards & Practices of Plastics Molders	**Material** Acrylonitrile Butadiene Styrene (ABS)

Note: The *Commercial* values shown below represent common production tolerances at the most economical level. The *Fine* values represent closer tolerances that can be held but at a greater cost. Any addition of fillers will compromise physical properties and alter dimensional stability. Please consult the manufacturer.

Drawing Code	Dimensions (Inches)		
	0.000		
	0.500		
A = Diameter (See note #1)	1.000		
	2.000		
B = Depth (See note #3)	3.000		
C = Height (See note #3)	4.000		
	5.000		
	6.000		

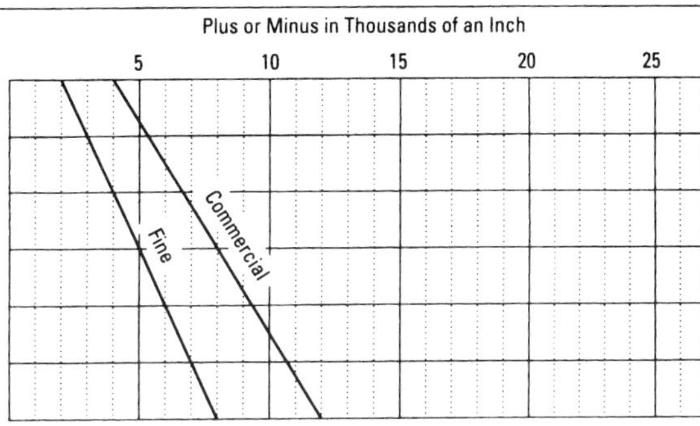

Drawing Code	Dimensions (Inches)	Comm. ±	Fine ±
	6.000 to 12.000 for each additional inch add (inches)	0.003	0.002
D = Bottom Wall	(See note #3)	0.004	0.002
E = Side Wall	(See note #4)	0.002	0.003
F = Hole Size Diameter (See note #1)	0.000 to 0.125	0.002	0.001
	0.126 to 0.250	0.002	0.001
	0.251 to 0.500	0.003	0.002
	0.501 & over	0.004	0.002
G = Hole Size Depth (See note #5)	0.000 to 0.250	0.003	0.002
	0.251 to 0.5000	0.004	0.002
	0.501–1.000	0.005	0.003
H = Corners, Ribs, Fillets	(See note #6)	0.027	0.017
Flatness	0.000 to 3.000	0.015	0.010
(See note #4)	3.001 to 6.000	0.030	0.020
Thread Size (Class)	Internal	1	2
	External	1	2
Concentricity	(See note #4) (F.I.M.)	0.009	0.005
Draft Allowance per side	(See note #5)	2.0°	1.0°
Surface finish	(See note #7)		
Color Stability	(See note #7)		

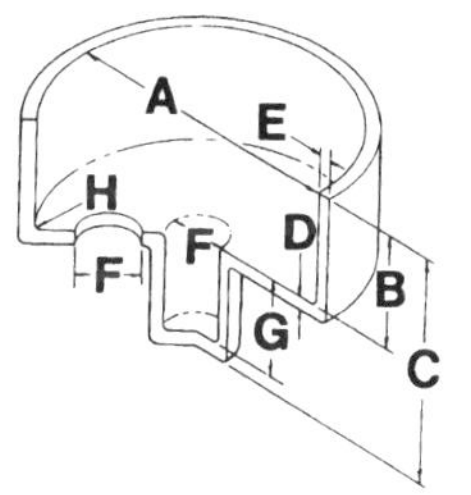

REFERENCE NOTES

1. These tolerances do not include allowance for aging characteristics of material
2. Tolerances are based on 0.125 inch wall section
3. Parting line must be taken into consideration.
4. Part design should maintain a wall thickness as nearly constant as possible. Complete uniformity in this dimension is sometimes impossible to achieve. Walls of non-uniform thickness should be gradually blended from thick to thin.
5. Care must be taken that the ratio of the depth of a cored hole to its diameter does not reach a point that will result in excessive pin damage.
6. These values should be increased whenever compatible with desired design and good molding techniques.
7. Customer-Molder understanding is necessary prior to tooling.

Figure 11. Anticipated material shrinkage tolerance for ABS.

PROCESS CONTROL CHARTING

During any manufacturing process, controlling the products quality is the most important function of production. The methods used to verify control must be established. The statistical tool most frequently recommended is the process control chart that was pioneered by Walter A. Shewhart. Control charting requires the understanding of only a few statistical fundamentals for practical applications. It can be easily implemented, and once in place, will show production a real time report card of progress of the process. The data generated can be given to customers to show compliance with their requirements, along with an indication of problem areas that can be corrected in real time.

There are many approaches used to determine the limits or specifications needed to keep a process in control. Some limits are determined by tests, some by experience, and others just selected because they were right for the process. Often control procedures are written in the form of manufacturing procedures or exist in the heads of veteran operating personnel. The latter is very dangerous for the process for if they are absent or leave the company, the process information goes with them.

Always, write down the specifications and manufacturing requirements and the variables that control the process. Also, their allowable range of drift or specification must be written down and followed during the manufacturing process. All manufacturing processes must have process control procedures written and then followed to ensure the product is made repeatably, cycle to cycle.

Experienced manufacturing personnel can separate manufactured product variations into usual and unusual categories. They make these distinctions within the limits of the process being discussed; if products fall outside these limits, then something unusual has occurred that must be corrected.

The control chart is a visual method of plotting and evaluating whether a product is in a state of "statistical control". It factually separates production personnel's distinction of variations into usual and unusual components. In essence, it compares actual production variations with the control limits established to make high quality products. When these limits have been computed and accepted for production, the control charts assume the visual role of aiding in the quality control of materials, products, and assemblies.

The economics of manufacture are the primary driving force in determining if the usual variation is within the specification limits. This is based on how tight the tolerances are set and how much the customer is willing to pay to have products produced to these specifications. In Six Sigma manufacturing the company determines if they want to control the process this tight so the amount of rejects and scrap is a minimal. There is a cost of manufacture to reach this point and management must determine the cost to justify the risk of producing out of specification and control product on a limited, some sigma limit basis.

There are typically three sets of manufacturing limits established on a manufacturing process.

1. Control limits – limits established to produce acceptable products.
2. Process limits – limits of process variables to produce acceptable products.
3. Specification limits – limits the process must fall within to make acceptable products.

Normally, the specification limits are the furthest apart of the three, followed by process and then control limits. Their relationship is shown in Figure 8.

Since it is risky to depend solely on the process information acquired by production people from years of experience, accurate record keeping should be supplemented by control charts. This information is valuable when new employees are brought on board or new supervisors are appointed. The control chart greatly reduces the operator's learning curve for process and product control.

Besides product variances, control charts show whether the process is in control. This is usually indicated by a bunching of points along the central tendency. Often, it may show wider fluctuations that may indicate the process is having control problems. Shifts or swings may also show a tendency away from control because of machinery or material problems, or other factors on the production floor.

CONTROL CHARTS

There are two basic types of control charts:

1. Measurements or "variables charts," that are used when actual readings are recorded, the so-called X, R, and s charts.
2. Attribute charts, that use visual or go, no/go data – the fraction or percentage defective charts called, P charts.

There are two different situations in which these charts are used:

1. When establishing a new process – either one that has undergone extensive changes or one that investigates on-going control after a preliminary frequency-distribution analysis has demonstrated initial control. Data is collected on product quality characteristics and control limits. And central tendency values are then calculated. Hence this condition is termed one of "no standard given".
2. When central tendency and spread values are initially established. This condition is known as "standard given". It assumes that the process is in control based on whatever data was used to establish the limits, arbitrary or real. These are based on production or service specifications and their requirements or on a target value established between the customer and supplier for the products.

The approach for calculating the control limits for these two charts is based on the laws of probability. The methods of calculations that vary between the measurement and percentage charts will be discussed in detail.

The process to follow in setting up the control limit chart is:

No Standard Given

1. Select the quality characteristics to be studied.
2. Record data on a required number of samples with an adequate number of units per sample. The minimum is five samples.
3. Determine control limits for the sample data.
4. Analyze the state of control in the process.
   (a) Too much variation.
   (b) Products move in and out of control.
   (c) Well-controlled process.

   When establishing control limits, several samples will often be out of control. In this case, trace down the problem in the process and repeat steps 2 and 3 until the process is in control.

Standard Given

1. Select the quality characteristic to be studied.
2. Establish the central tendency value and the spread to be used. All available data must be used to show that control exists.

3. Determine the control limits from these values.
4. Establish that these control limits are economical, practical, and required.
5. Establish the control limit values and plot them on graph paper.
6. With the manufacturing process in control, begin recording results from production samples selected at periodic intervals.
7. Take corrective action if the characteristics of the production samples exceed the control limits.

An example of the graph paper used to plot the data is shown in Figure 12.

MEASUREMENT PROCESS CONTROL CHART CALCULATIONS

The basic principles for computing measurement-process control chart limits are similar to those for calculating the frequency distribution of three sigma limits. The data required is as follows:

The *mean* (or average) is the most commonly used measure for central tendency. It is the sum of all the readings (X) divided by the total number of readings (n) for a specific trial run. It is denoted by $\bar{X}$.

$$\bar{X} = \frac{X_1 + X_2 + X_3 + \ldots Xn}{n}$$

More frequently, it is written using the Greek capital letter sigma (Σ) to denote the sum of the Xs as (ΣX).

$$\bar{X} = \frac{\Sigma X}{n}$$

For example if there were five readings – 4, 5, 6, 7, 8, – the average or mean would be:

$$\bar{X} = \frac{4+5+6+7+8}{5} = \frac{30}{5} = 6$$

The *range* (R) is the difference between the high and low values recorded for a specific trial run. For the example above, you have

$$R = X(HI) - X(LO) = 8 - 4 = 4$$

The *standard deviation* is a measure of the spread of individual readings for a trial run. It is usually computed for samples drawn from larger lots and in

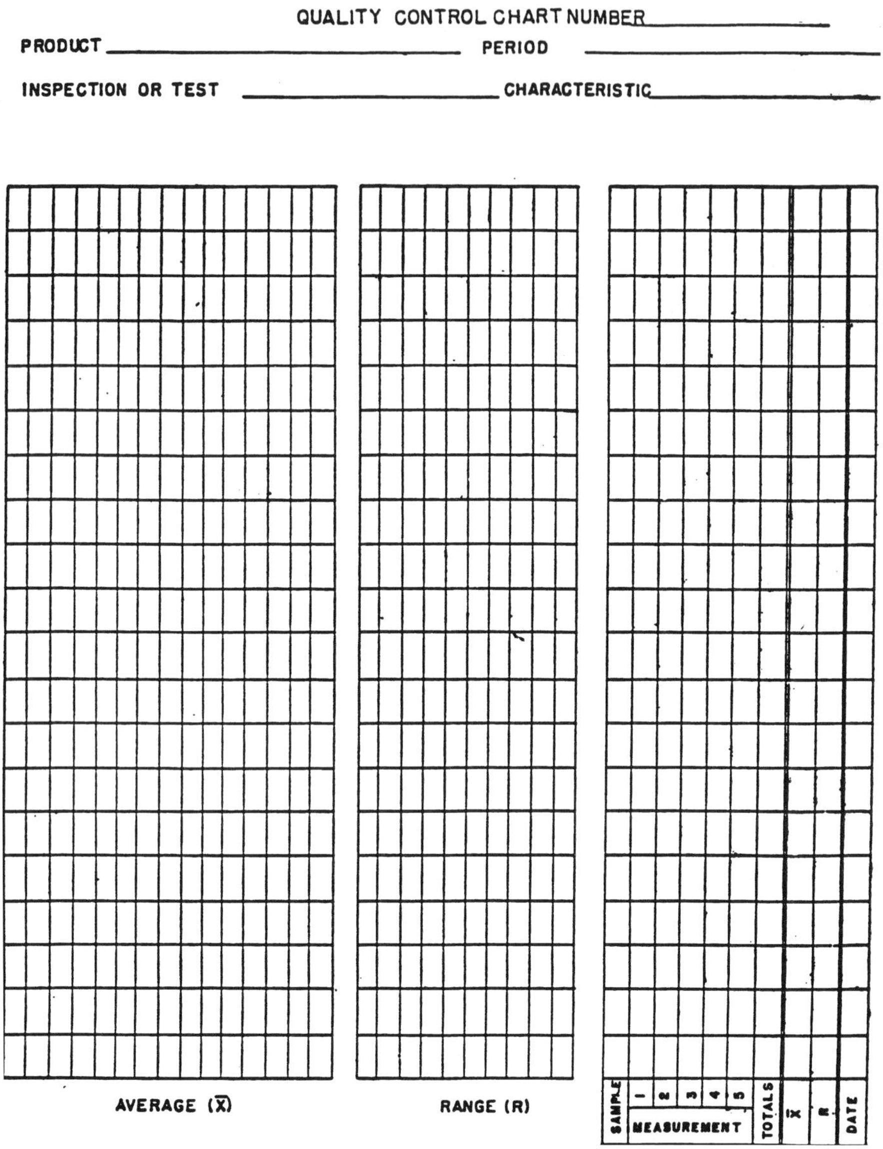

Figure 12. Control chart blank graph sheet.

these cases is known as the *sample standard deviation.* The sample standard deviation is the positive square root of the sum of the squared deviations of readings from their average, divided by one less than the number of readings:

$$S = \sqrt{\frac{(X_1 - \bar{X})^2 + (X_2 - \bar{X})^2 + \ldots + (Xn - \bar{X})^2}{n - 1}}$$

For lots with multiple readings.

$$S = \sqrt{\frac{\Sigma(\text{fx} - n\bar{X})^2}{n - 1}}$$

Where, fx = frequency of readings within each lot.

Thus, in the similar series for X:

$$S = \sqrt{\frac{(4 - 6)^2 + (5 - 6)^2 + (6 - 6)^2 + (7 - 6)^2 + (8 - 6)^2}{5 - 1}}$$

$$S = \sqrt{\frac{4 + 1 + 0 + 1 + 4}{4}} \qquad \sqrt{\frac{10}{4}} = 1.581$$

When comparing range versus standard deviation, it is readily seen that the range is very easily obtained – the difference between the high and the low values. But in samples larger than 15, it does not take into account the effect of multiple readings. With more samples, there is a greater probably of getting a wide-of-the-mark reading. With standard deviation, this effect is minimized by incorporating all the readings and balancing out the mavericks.

The primary difference is that a smaller number of readings is used to calculate and establish the central tendency and spread values. Since frequency distributions in industry tend toward normality, the three-sigma value for control charts has proven to be most useful and economical. The formulae and supporting chart data are shown in Figure 13 and Table 3.

As noted, two sets of calculations are given – one for range and the other for standard deviation. The choice of use is up to the individual. To calculate

No Standard Given:
When Range Is Used as Measure of Spread

Average:	Lower limit = $\bar{\bar{X}} - A_2\bar{R}$	(1A)
	Center line = $\bar{\bar{X}}$	
	Upper limit = $\bar{\bar{X}} + A_2\bar{R}$	(1B)
Range:	Lower limit = $D_3\bar{R}$	(2A)
	Center line = $\bar{R}$	
	Upper limit = $D_4\bar{R}$	(2B)

When Standard Deviation Is Used as a Measure of Spread

Average:	Lower limit = $\bar{\bar{X}} - A_3\bar{S}$	(3A)
	Center line = $\bar{\bar{X}}$	
	Upper limit = $\bar{\bar{X}} + A_3\bar{S}$	(3B)
Standard Deviation:	Lower limit = $B_3\bar{S}$	(4A)
	Center line = $\bar{S}$	
	Upper control limit = $B_4\bar{S}$	(4B)

Where X = grand average
R = average range
S = average sample standard deviation

Standard Given:
When Range Is Used as Measure of Spread

Average:	Lower limit = $\bar{X}o - Aao$	(5A)
	Center line = $\bar{X}$	
	Upper limit = $\bar{X} + Aao$	(5B)
Range:	Lower limit = D_1ao	(6A)
	Center line = Ro (or d_2ao)	
	Upper limit = D_2ao	(6B)

When Standard Deviation Is Used as a Measure of Spread

Average:	Lower limit = $\bar{X} - Aao$	(7A)
	Center line = $\bar{X}o$	
	Upper limit = $\bar{X}o + Aao$	(7B)
Standard Deviation:	Lower limit = B_5ao	(8A)
	Center line = So (or C_4ao)	
	Upper control limit = B_6ao	(8B)

Where Xo = value of the average adopted for computing the center line and control chart limits.
Ro = value of the range adopted for computing the center line and control chart limits.
So = value of the sample standard deviation adopted for computing the center line and the control chart limits.
ao = value of the lot or population standard deviation adopted for computing the center line and control chart limits.

When computing the lot or population standard deviation the lot size and number of measurements must be sufficient to categorize the lot with extreme values averaged out.

(Adapted from reference [3])

Figure 13. Formulae for calculating control values; UCL/LCL and range.

Table 3. Factors for Computing Central Lines and Three-sigma Control Limits for X, S and R Charts.

No.	Chart for Averages		Chart for Standard Deviation					Chart for Ranges					
	Factors for Control limits		Factors for Central line		Factors for control limits			Factors for Central line		Factors for Control Limits			
n	A	A_2	A_3	C_4	B_3	B_4	B_5	B_6	d_2	D_1	D_2	D_3	D_4
2	2.121	1.880	2.659	0.79790	0	3.267	0	2.606	1.128	0	3.686	0	3.267
3	1.732	1.023	1.954	0.88620	0	2.568	0	2.276	1.693	0	4.358	0	2.574
4	1.500	0.729	1.628	0.9213	0	2.266	0	2.088	2.059	0	4.698	0	2.282
5	1.342	0.577	1.427	0.9400	0	2.089	0	1.964	2.326	0	4.918	0	2.114
6	1.225	0.483	1.287	0.9515	0.030	1.970	0.029	1.874	2.534	0	5.078	0	2.004
7	1.134	0.419	1.182	0.9594	0.118	1.882	0.113	1.806	2.704	0.204	5.204	0.076	1.924
8	1.061	0.373	1.099	0.9650	0.185	1.815	0.179	1.751	2.847	0.388	5.306	0.136	1.864
9	1.000	0.337	1.032	0.9693	0.239	1.761	0.232	1.707	2.970	0.547	5.393	0.184	1.816
10	0.949	0.308	0.975	0.9727	0.284	1.716	0.276	1.669	3.078	0.687	5.469	0.223	1.777
11	0.905	0.285	0.927	0.9754	0.321	1.679	0.313	1.637	3.173	0.811	5.535	0.256	1.744
12	0.866	0.266	0.886	0.9776	0.354	1.646	0.346	1.610	3.258	0.922	5.594	0.283	1.717
13	0.832	0.249	0.850	0.9794	0.382	1.618	0.374	1.585	3.336	1.025	5.647	0.307	1.693
14	0.802	0.235	0.817	0.9810	0.406	1.594	0.399	1.563	3.407	1.118	5.696	0.328	1.672
15	0.775	0.223	0.789	0.9823	0.428	1.572	0.421	1.544	3.472	1.203	5.741	0.347	1.653
16	0.750	0.212	0.763	0.9835	0.448	1.552	0.440	1.526	3.532	1.282	5.782	0.363	1.637
17	0.728	0.203	0.739	0.9845	0.466	1.534	0.458	1.511	3.588	1.356	5.820	0.378	1.622
18	0.707	0.194	0.718	0.9854	0.482	1.518	0.475	1.496	3.640	1.424	5.856	0.391	1.608
19	0.608	0.187	0.698	0.9862	0.497	1.503	0.490	1.483	3.689	1.487	5.891	0.403	1.597
20	0.671	0.180	0.680	0.9869	0.510	1.490	0.504	1.470	3.735	1.549	5.921	0.415	1.585
21	0.655	0.173	0.663	0.9876	0.523	1.477	0.516	1.459	3.778	1.605	5.951	0.425	1.575
22	0.640	0.167	0.647	0.9882	0.534	1.466	0.528	1.448	3.819	1.659	5.979	0.434	1.566
23	0.625	0.162	0.633	0.9887	0.545	1.455	0.539	1.438	3.858	1.710	6.006	0.443	1.557
24	0.612	0.157	0.619	0.9892	0.555	1.445	0.549	1.429	3.895	1.759	6.031	0.451	1.548
25	0.600	0.135	0.606	0.9896	0.565	1.435	0.559	1.420	3.931	1.806	6.056	0.459	1.541

(Adapted from reference [6]).

the data, some production personnel prefer to use sample size of more than ten and calculate their control limits based on standard deviation, but the five-unit sample has become the industry norm.

In computing the control limits for a "no standard given" condition, the following eight steps establish process control. The range will be used as the spread, and formulae 1A, 1V, 2A, and 2B will be applied.

CONTROL CHART CALCULATION PROCEDURE

1. Select the quality characteristic to be controlled: length, thickness, warpage, injection pressure, time, packing pressure, etc.
2. Collect data by selecting an adequate number of lots or cycles based on a set time or frequency method. For this example a part dimension will be used. For a process, use successive cycle data in increments of five successive cycles. Therefore, use twenty lots as a start and select five successive samples to be measured for the specific characteristic from each lot. The lots should be selected at set internals every hour, half hour, or after 20 to 30 cycles, and the sample data should be recorded in successive order of selection.
3. Compute the average and range values for each of the twenty lots.
4. Compute the grand average X of the averages of the twenty lots as well as the average of the ranges R of the twenty lots.
5. Compute the control limits based on these lot averages and ranges.
6. Analyze the average and range values for each lot relative to these control limits. Determine if any factors require corrective action before the control limits are approved.
7. Determine if control limits are within economical limits for the molding cycle process.
8. Use the control limits in active production to b sure they produce parts within the limits established.

CONTROL LIMIT CALCULATIONS

1. The characteristic to be measured is pin length in an eight-cavity balanced runner tool on a 30-second cycle.

2. Assuming 240 studs are made every 15 minutes, randomly select 5 samples from this lot and measure them every 15 minutes to establish the data for the control limits.
3. Referring to sample lot No.1 in Figure 14, the five readings to compute the average are:

   0.498
   0.500
   0.505
   0.503
   0.503
   ______
   2.509

   The mean is:

   $$\bar{X} = \frac{\Sigma X}{R} = \frac{2.509}{5} = 0.5018$$

   The range is:

   $$R = [0.505 - 0.498] = 0.007$$

   Similar calculations are made for the other lots respectively.

4. Determine *grand average* or the mean and *average range*.

   $$\bar{X} = \frac{\bar{X}_1 + \bar{X}_2 + \ldots \bar{X}n}{n}$$

   $$\bar{\bar{X}} = \frac{\Sigma \bar{X}}{N} = \frac{10.0292}{20} = 0.50146$$

   Where N = number of lots measured.
   For the average range:

   $$\bar{R} = \frac{\text{Sum of lot ranges}}{\text{number of lots}} = \frac{\Sigma R}{N}$$

   $$\bar{R} = \frac{0.091}{20} = 0.00455$$

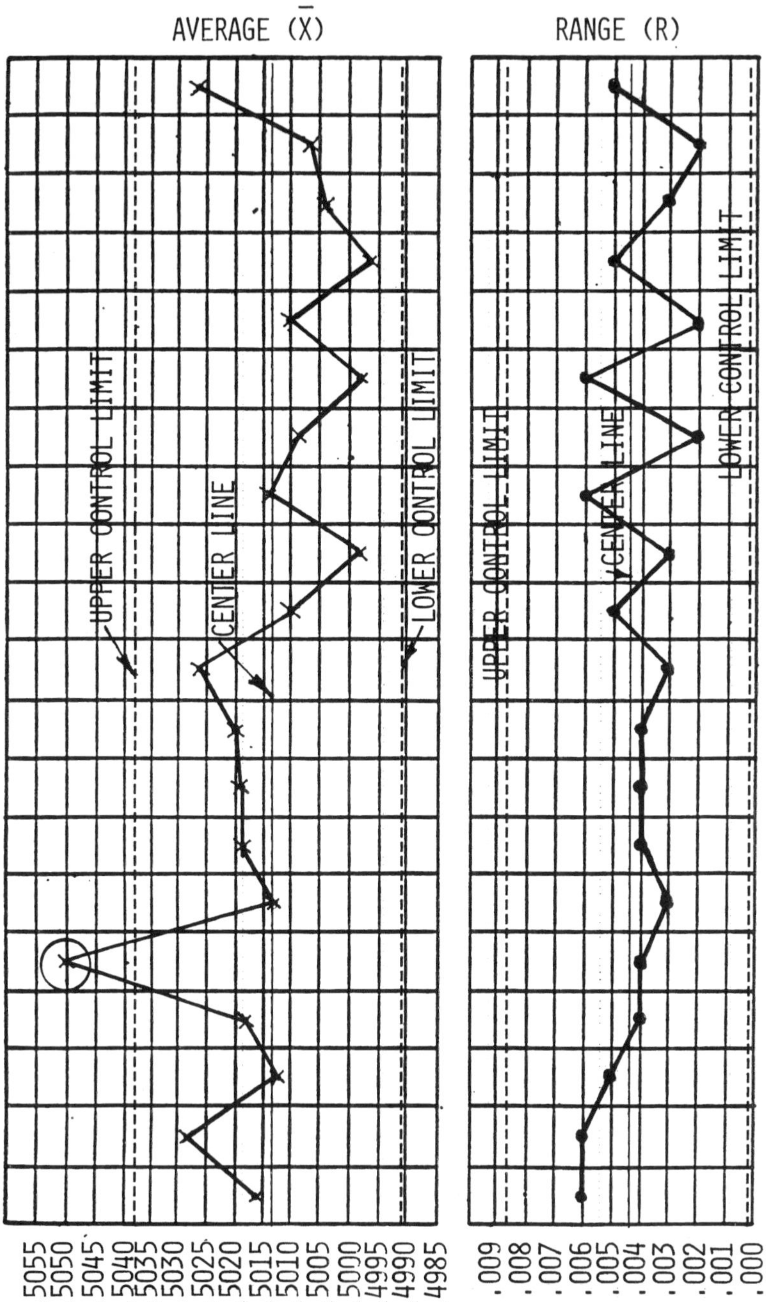

Figure 14. Plot of control limits.

SAMPLE		1	2	3	4	5	6	7	8	9	10	11	12	13	14	15	16	17	18	19	20
MEASUREMENT	1	.498	.504	.500	.499	.505	.503	.503	.502	.502	.504	.503	.500	.504	.501	.499	.502	.497	.499	.501	.505
	2	.502	.503	.500	.504	.507	.503	.502	.500	.503	.503	.500	.502	.504	.503	.504	.501	.500	.501	.501	.506
	3	.504	.505	.501	.502	.506	.500	.504	.502	.501	.501	.501	.499	.503	.500	.497	.501	.500	.502	.502	.500
	4	.501	.502	.501	.502	.501	.500	.500	.502	.501	.502	.500	.497	.498	.499	.500	.501	.501	.499	.499	.500
	5	.503	.500	.504	.502	.506	.501	.500	.503	.500	.503	.502	.501	.498	.501	.499	.500	.500	.501	.500	.502
TOTALS		2.508	2.514	2.506	2.509	2.525	2.507	2.509	2.509	2.510	2.513	2.506	2499	2.507	2.504	2.499	2.505	2.498	2.502	2.503	2.513
$\bar{X}$		.5016	.5028	.5012	.5018	.505	.5014	.5018	.5018	.502	.5026	.501	.4998	.5014	.5008	.4996	.501	.4996	.5004	.5006	.5026
R		.006	.005	.005	.004	.004	.003	.004	.004	.004	.003	.003	.003	.006	.002	.006	.002	.005	.003	.002	.005
DATE																					

Figure 14. Continued.

5. Computing the control limits.

Averages:
Lower Control Limit (LCL) = $\bar{\bar{X}} - A_2\bar{R}$
Control Center Line (CCL) = $\bar{\bar{X}}$
Upper Control Limit (UCL) = $\bar{\bar{X}} + A_2\bar{R}$

Ranges:
Lower Control Limit (LCL) = $D_3\bar{R}$
Center Control Limit.(CCL) = $\bar{R}$
Upper Control Limit (UCL) = $D_4\bar{R}$

For a sample size of five, the constants are taken from Table 3.

$A_2 = 0.577$
$D_3 = 0.0$
$D_4 = 2.114$

Substituting these data in the above formulae yields:

$$\text{Average (LCL)} = \bar{\bar{X}} - A_2\bar{R} = 0.50146 - (0.577)\,(0.00455) = 0.4988$$

$$\text{Average (UCL)} = \bar{\bar{X}} + A_2\bar{R} = 0.50146 + (0.577)\,(0.00455) = 0.5041$$

$$\text{Range (LCL)} = D_3\bar{R} = (0)\,(0.00455) = 0$$

$$\text{Range (UCL)} = D_4\bar{R} = 2.114\,(0.00455) = 0.0096$$

These control limits are then plotted on the graph in Figure 12. The center lines can also be plotted for visual clarity when reading the chart, although some leave these off.

When analyzing the data, be sure that out-of-control readings are not the results of human error. Such errors account for approximately three out of 1000 bad readings. Therefore, readings for this sample lot should be repeated as a control measure. The same person should take all of the readings with the same gauge. Until all questions are resolved, that specific lot should be set aside and properly identified. An example of using this technique for recording and charting part weight is shown in Figure 15.

With the newer process controllers, the control chart data output may be directly available. This output may be generated by internal software from signals pulled off the machines operations in real time. These signals can be fed into a local or central process control charting software program for generating "Real Time" output charts.

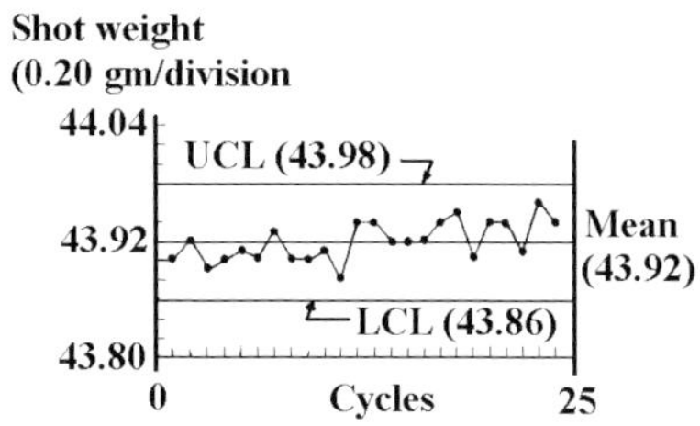

Figure 15. Part weight control chart.
(Adapted from reference [4])

This technique can also be used to generate machine process control limits, if not part of the program for monitoring the manufacturing machine's variables, such as:

1. Injection pressure and speed/time for the cycle such as:
   (a) first stage injection;
   (b) second stage – cut over to packing pressure.
2. Holding pressure on cavity.
3. Cavity pressure against core pins or sensors in the tool.
4. Mold cavity surface temperature.
5. Cycle times for machine operations.
6. Screw RPMs.
7. Barrel/nozzle temperatures.
8. Tool temperature control readings.

These process control readings will give the operator "Real Time" indicators of the manufacturing process and are very valuable in maintaining process control for the cycle and part manufacture.

DATA COLLECTION

Where, how and when should the data be collected. I believe in only "Real Time" data collection to ensure the process stays within control during the entire manufacturing process. With the product tolerances agreed on and knowing the manufacturing process is capable of achieving them, the process variables can be selected for monitoring manufacture. Knowledgeable process engineer and setup technicians can establish the value of the variables to monitor. The material supplier can recommend processing conditions and plant personnel can inform the team of the capability of their plants auxiliary and support systems.

Process condition variables are selected during the evaluation and recorded on process documentation records such as the Molding Data Record Sheet, Figure 16. Every time a variable is changed in the process. Failure to do this can result in a problem the next time the product is manufactured as the setup technician will not have a record of what process control change was made that affects the manufacturing process. These are the major control variable settings required to produce acceptable products.

Each supplier's material will have to have their processing variables adjusted to melt and flow in the process, injection molding machine and tool. The plant systems, temperature, air pressure, environment, cooling water, etc. are recorded and maintained for producing good products. This is true for "all" manufacturing processes and for each machine to make repeatable quality products to company and customer specifications.

The "where", variable data collection is determined such as injection pressure, temperature (machine hydraulic fluid, tool, cooling water), time (cycle times) etc. Older processing systems with outdated data sensing and reporting systems may require a person to physically record the data.

The "when" is never to frequent. Most plants monitor their processing conditions with computers to collect on each cycle, or at set internals the data variables in the molding process. Some collect all cycles and output an average, say after a minimum of 5 cycles. This data is stored at the machine or relayed to a control point for continual "Real Time" monitoring of the process by an engineer.

The "how" often determines the "when" data is collected. If automatic and electronically transmitted to the process control engineer or control room in "Real Time" this is the preferred method Only monitoring in "Real Time" can the process and Six Sigma capability be assured.

Data collected by roving inspectors and then plotted on graphs is too time consuming and liable to personal error in recording and manipulation of the data. But, if this is all one has, ensure the inspector has the best portable data entry and software to quickly calculate the data points for charting and documenting the process. Only in as close to "Real Time" recording will the process be monitored for 'true real time' process control.

If a variable changes, a trend of variation from the mean is often a pre-warning sign that a key variable is changing. When this occurs, the process

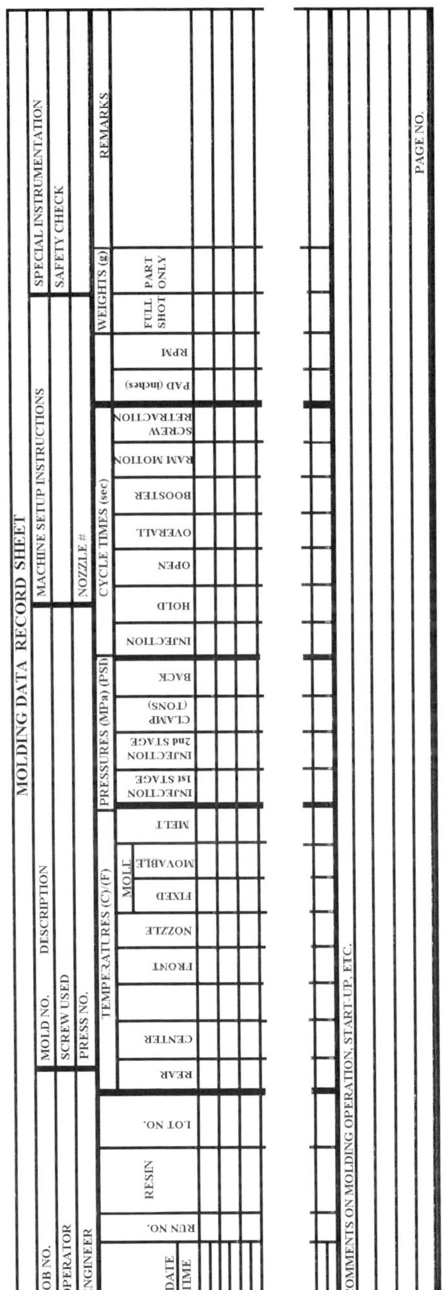

Figure 16. Molding data record sheet.

engineer/technician is called to determine the reason and make any adjustments if necessary to bring the process back into control before exceeding the product specifications. If the problem is more severe, not correctable by variable adjustment, then the processing team may be called on to solve the problem.

The data collected plus the manufacturing variable machine process record can be used in solving the problem along with the machines process, trouble shooting, and control log or material suppliers "Trouble Shooting Guide" for a specific product as shown in Figure 17.

WHO SHOULD COLLECT THE DATA

Electronic data collection in "Real Time" is preferred. When this is done, the software can be programmed to give an alarm if trends show an out of control conditions occur. Then if a trend for out of control continues, corrections can be made in the process before it goes out of control.

Control and documentation of lots of material, drying, feed, and blending data is also very important. Many plastic materials, resins, are made in batch lots as ABS and PVC compounds and will vary slightly in processing characteristics. You will not know this typically unless pre-testing of the material indicates a melt flow change or molecular shift as can occur with PE and PP and other types of resins. There should be available for each lot of material received, test information, supplied by their supplier on the materials composition, molecular weight, flow, and processing variables.

If there is only a slight variance in the data but the melt is still within processing parameters to produce good parts, this information must come to the plant floor, to inform the process control personnel of a material variation so an alarm does not indicate a real problem in the process. Again, Real Time information processing and alerts to communicate this information can ensure the process will stay in control. If a trend develops, at a lot change, but remains in process control nothing needs to be adjusted. But, if the trend for out of control continues, then an adjustment in processing variables is required.

It is recommended that all lot data used for a product become part of the products manufacturing documented data. Then should a problem ever occur, all data on the manufacturing and material will be available for analysis.

Trouble Shooting Guide for "Acetal" copolymer Molding Problems

PROBLEM	SUGGESTED CORRECTIVE ACTION
SHORT SHOTS, POOR SURFACE	1. Increase Feed 2. Increase Injection Pressure 3. Use Booster and Maximum Ram Speed 4. Decrease Cushion 5. Raise Material Temperature by Raising Cylinde 6. Raise Mold Temperature 7. Increase Overall Cycle 8. Check Shot Size vs. Rated Machine Shot Capacity; If Shot Size Exceeds 75% of Rated (Styrene) Shot Capacity, Move to Larger Machine. 9. Increase Size of Sprue and/or Gates
FLASHING	1. Lower Material Temperature by Lowering Cylinder Temperature 2. Decrease Injection Pressure 3. Decrease Overall Cycle 4. Decrease Plunger Forward Time 5. Check Mold Closure (Possible Obstruction on Parting Line Surface) 6. Improve Mold Venting 7. Check Press Platens for Parallelism 8. Move Mold to Larger (Clamp) Press
NOZZLE DROOL	1. Lower Nozzle Temperature 2. Lower Material Temperature by Lowering Cylinder Temperature 3. Decrease Residual Pressure in Cylinder by: a. Reducing Plunger Forward Time and/or Back Pressure b. Increasing "Decompress" Time (If Press Has This Control) 4. Decrease Die Open Time 5. Use Nozzle with Positive Shutoff Valve
NOZZLE FREEZE-OFF	1. Raise Nozzle Temperature 2. Decrease Cycle Time 3. Increase Injection Pressure 4. Raise Mold Temperature 5. Use Nozzle with Larger Orifice
DISCOLORATION	1. Purge Heating Cylinder 2. Lower Material Temperature by Lowering Cylinder Temperature 3. Lower Nozzle Temperature 4. Shorten Overall Cycle 5. Check Hopper and Feed Zone for Contaminants 6. Check Cylinder and Plunger or Screw Fit for Excessive Clearance 7. Provide Additional Vents in Mold 8. Move Mold to Smaller Shot Size Press
BURN MARKS	1. Decrease Plunger Speed 2. Decrease Booster Time 3. Decrease Injection Pressure 4. Improve Venting in Mold Cavity 5. Change Gate Location to Alter Flow Pattern
BRITTLENESS	1. Pre-Dry Material 2. Lower Melt Temperature and/or Residence Time 3. Lower Mold Temperature 4. Reduce Amount of Regrind

Figure 17. Example of trouble shooting guide page.
(Adapted from reference [5])

DATA PRESENTATION FOR MONITORING CONTROL

Visual charting is the best method as long as it is accurately documented. Numbers on the chart are only used to prepare the visual output. But, the numbers give the process engineer the degree of variation that can assist in solving a problem.

Accuracy in recording the data and its manipulation to present the visual results is critical and only trained personnel should be selected for this function. If the data is recorded incorrectly it may appear as a process problem except it actually is a data recording error. Therefore, the manual numbers recorded are very important in process control.

Electronic data collection is assumed to be more reliable but problems can also occur in their output to the data collection system. Therefore, a technician trained in process control (digital output predominately) is required to setup the output gathering and transmission to the computer software program to record, compute, and output the results. Ensure enough technician and engineers are trained to setup and adjust these controls for all shifts of operation. The program should also go through a system verification for accurate data recording and output of information. Your systems programmer can ensure this is performed correctly.

The use of user-friendly SPC software is recommended. There are considerable sources now available with easy cut and paste or, the best, a direct entry program with process output entered directly into the systems' spreadsheets to create charting visuals. Some software packages automatically store and generate the output as directed by the program. Keep it simple with as few operator involvements as possible to ensure the data is as accurate as recorded from the process variable sensors.

HOW SHOULD DATA BE USED

There is always the possibility that more data is produced or recorded than is useable by the Six Sigma team. This wastes time and ties up equipment that could be better used in other areas of manufacture. Select only the data useful to control the process.

There are over 40 variables that could be monitored in the injection molding process. Other manufacturing processes may have more or less variables that should be controlled and monitored. In many situations there is little information that informs the operator and/or manufacturing engineer what these variables are. Also, what they should be for producing a repeatable good product cycle to cycle. Therefore, if unsure, develop the process procedures for manufacture for the process and machine. Then analyze the items and determine what are necessary and required for ensuring the process stays in control. Then a determination can be made to see if all variables need to be monitored. This may not require all to be monitored as long as controls are in place to ensure variability is within product or typical process specifications. This information may be recorded separately as water temperature at the chiller, drying the resin information, etc. and then brought together at one collection site in the computer for the products manufacturing record.

All operations in the manufacture of a product, process control procedures for each operation and for all products and business operations should have a FMEA implemented. They will assist in determining where there may be weakness in the operation. Also each business operation should be evaluated in the establishing of process control procedures for the business and manufacturing operations of the company for a FMEA. Other

Table 4. Process Control and FMEA Schedule for Business and Manufacturing Operations.

	Date implemented	
Operation	Control Plan	FEMA
Manufacturing line(s)		
Receiving		
Purchasing		
Order entry		
Invoicing		
Incoming inspection and testing		
Auxiliary support operations		
Auxiliary equipment support and maintenance		
Equipment and tool maintenance		
Plant systems support and maintenance		
Process and tool setup		
Assembly and decoration operations		
Product packaging and product shipping		

company areas that may be considered are listed in Table 4 with a form for recording the data to ensure all operations are completed.

Essentially all operations in the company, business and manufacturing should have a process control procedure and FMEA generated to show the operation and any potential problem areas that could develop at specific points of the business and manufacturing areas.

This is initially a lot of work but once completed and adhered to, essentially the procedure tier 2 and work instructions, tier 3 of the ISO9000 certification requirements the company personnel can follow it daily and be assured of minimal problems. Injected into the operations I recommend 'check sheets' as guides to follow for only important operations, not the typical day-to-day business routines except for order entry, and contract reviews. 'Check sheets' can be used for machine and process setup to ensure all connections, tools, components; equipment, etc are available to perform the change within minimum time.

WHO USES THE DATA AND WHEN

Process, quality, and production engineers and technicians will review and use the process data to ensure the manufacturing process stays in control.

Should a product quality problem occur the data (historical) is available for review to identify any visible changes in the process or product during and preceding the problem occurring? If the check sheets were correctly filled out they may also aid in problem solving. For example, if a change was noted that may have inadvertently affected the process. Or normal wear and part failure in equipment or failure to service or clean a piece of equipment may have resulted in causing the problem.

It is very important to document all problems, their cause, and solution, then should they reoccur a solution should be speedily applied that may lead to solving the problem. A problem notebook and equipment record should stay with each machine and tool for documenting all changes, repairs, maintenance, and problems with each in a separate section. It is also important to document when a piece of equipment was last serviced or cleaned out as a grinder and is now ready for another job. This keeps all doubt out of the scheduling of equipment for all jobs.

WHAT TO DO WHEN OUT OF CONTROL OCCURS

A major cause of a process going out of control is caused by variety of variables changing as equipment wears and breaks or operators making adjustments when not necessary as during a shift change, new material lots used, changes in the plants environment as outside conditions, weather, power feed (brown outs at specific high power demands) etc.

WHO MAKES THE CHANGES AND WHAT CHANGES ARE MADE

Changes in any process are made by the operator/manager/supervisor that has responsibility for the action or operation. The authority to make the change must be delegated and making a process change usually is the technical person responsible to ensure the process remains in control. Operators usually do not have this authority unless senior and trained in the operation and control of the equipment for the process.

The changes that are made are always documented. The person allowed to make any changes must be instructed in the correct method to be used. Making a change to any process is important and must always be done correctly and within the agreed upon variable tolerance range. Only those variables that are deemed necessary to adjust the process to bring it into tolerance should be designated as the "what" to change. Other variables should not be changed unless they are agreed that they have an influence on the system and all personnel involved in the process improvement agree to the change and by what amount.

HOW LONG TO SEE THE EFFECTS OF A PROCESS CHANGE

Most changes in a process can be seen and evaluated immediately once the change in the manufacturing system has reached equilibrium. It may take a few cycles or in some processes, may take longer than an hour to see the effect of the change. It is very important the process reaches equilibrium before any new changes are evaluated before their effect on the adjusted process or problem.

A major chemical resin supplier has a continuous reaction process to make their resins. Control of the process was manual from the control room. Each shift change the process would vary, not enough to move out of process control but enough to see a variability in the resins physical properties each time a lot of material was tested and often in the same lot. Lot size was determined by manufacturing which could vary from 40,000 to a few hundred thousand pounds of material. Large continuous resin run lots were identified with daily run lot numbers determined by production. Also, output would vary from shift to shift on a daily basis and it was not attributed to the raw material feed stock, which in itself had a small degree of variability from their sister supplier plant.

A problem solving team was implemented looking for the variance. Raw feed stocks were checked, weather conditions, and operator experience plus the system itself for calibration and any worn pumps and control equipment not in calibration.

After a week, with no possible solution, the control of the reactor system was put on the computer for automatic control with control team members on each shift initially only observing the results and able to only adjust the system if required for control should it go out of the limits established for the computers control. When the newness of the control system became understood, it was noticed that the shift supervisor, after a shift change, made minor changes, interpreted in the control room process log book as required for process control, to the system as they had left it after their last shift.

These changes disturbed the systems equilibrium, which required more operator changes which affected rate and quality of the product. These effects were noticed in variations of the resins physical properties at testing and traced back to control room operator changes.

The solution was to retrain the control room operators, establish a process to achieve the quality and properties required and at an acceptable output. The computer control system was setup with the correct settings and locked out to the control room operators. The computers control system was only unlocked by the shift supervisor when the process control data indicated a process change was required to keep the system in control around the median variable value of the material. Also, all changes were to be written down in a control log as to why the change was required, responsibility assigned and changes documented for control of the process. This solved the material variation and once implemented throughout the entire plant, greater economies and better product was produced on all shifts.

Personnel caused problems, are the most difficult to solve as the variable behavior is often modified to avoid detection. Informing the personnel the solution to the problem was critical, so punishment was not an item, and the solution successfully implemented. Data doesn't lie, but data can be skewed to indicate other occurrences happening in the process that did not exist until an operation change was made.

Statistical process control is the best-known tool used to show control of the process and product. Select the controlling variables and monitor their variance. Using quality procedures keep it simple, easily understood, and show results graphically for the entire plant and company to see. This will show how department are progressing and awarded recognition to successful programs. Education of the work force is essential to continual improvement of the companies business and products.

Questions raised and their answers support my reasoning. The use of process-behavior charts to display data in a specified context is preferred. Using a cross-section of many measures can cause confusion and requires team analysis of how these variables interact on the process in their individual ways.

When monitoring a continuous process using $\bar{X}$ (x bar) charts and the process shows almost no further ways for improvement to concentrate on, there are other options. Use Behavior Charts, any variable outside the limit can be analyzed for assignable cause since it has an effect on the process. It can be economically wise to investigate these points and determine the cause for exceeding the limit. Here FMEA and fish bone charts and diagrams can assist in determining where the problem may have initiated and the reason the point on the chart went out of control. Always make sure an operator change in the process is documented so its effects can be analyzed for the solution of the operation. Solving these problems leads to continual improvement that is the goal of Six Sigma operations.

CONSIDER A DOE (DESIGN OF EXPERIMENTS)

If your process has peaked out with very minor or few points over the limits, then Six Sigma methods state you will have to fundamentally change the process. Here statistical DOE can be substituted to test each variable in a logical designed test program. In Appendix B is an example of a problem variable needing to be determined to control the effects of the process. A DOE was utilized to determine the variables most affecting the process for the product variable. The DOE was used to determine the main contributing variables, in this situation there were more than one, to improve the process for making an injection-tooled product and to control a critical dimension on the part.

Some processes have "trouble shooting guides" as shown in Figure 17 and supplied by the material supplier or generated in-house by technicians in their problem solving notebooks to diagnose and solve processing problems. Use them and make changes always ensuring the process is in equilibrium before analyzing the results of the output data. Also, maintain the change long enough before making another change to ensure the results of the change have taken affect and are recorded for the process. In making any process change, only make one change at a time except when running a DOE. This is extremely important for if the system improves and then changes again due to the other variable acting on the process, you do not know what variable was successful for the improvement.

When using DOE, a set of selected variables are set at their extremes, either high or low point, process value limits and through a set of organized random adjustments evaluated at one time. These process variables were chosen for analysis as suspected having the greatest affect on the process and product. This is just the opposite of making process adjustments or improvements, one variable at a time that is very time consuming. By using the DOE method it speeds up the variable evaluation process and the effects can be quickly analyzed and realized for the selected variables values affecting the process. What would have taken possible days or weeks to evaluate can usually be determined in hours using DOE. The method to be used is determined by the black belt for solving problems, locating the controlling variable, and improving the process.

When using DOE always remember to perform a fishbone first, then categorize the effecting variables, randomize them per the DOE procedures and then, first, test only variables that have a rapid, one to two cycles at most response in the process such as time, pressure, speed settings. Always perform tests requiring temperature changes last, as it takes longer for temperature changes to reach true equilibrium in the process. Always ensure the system is in temperature equilibrium before recording data.

Automatic control systems often re-compute control limits after each cycle or a series of cycles, (usually 5 or more cycles). If alarms keep signaling false alarms the problem can be in the computer software program. If this is suspected use hand charting and recalculate the process control limits. Most SPC books give standard formulas for computing control limits that usually inflate the limit vs. tightening them. This will be discussed later when discussing variables for tighter control of the process.

The solution may be incorrect calculation or a bad software program. Also, verify the controllers are in calibration and/or a faulty controller or value/relay is not dirty or sticking. Often the sensors relaying the reading to the controllers are well worn and may need service. Another reason for ensuring the maintenance and calibration of control equipment are always current, maintained, and reliable. Contacts should also be inspected as over time fumes from processing can coat the contacts and result in erroneous output to the recorders and controllers.

STATISTICAL PROCESS CONTROL FOR COMPLEX PROBLEMS

Questions asked were concerns over using SPC in complex situations. Process-behavior charts can be used in both simple and very complex situations. Good training, following the intended methods, good output and measuring methods (technicians trained in how to measure if performed manually is very important). Personnel not correctly trained and certified can create as many false readings as good readings. Equipment maintenance and calibration is also very critical. Be sure your personnel, recording equipment, and software program are capable when data is recorded.

Complex problems need to be analyzed to the point of monitoring the variable with the highest probability of causing the problem or for process improvement. In molding cycle improvements the variable often overlooked is the cooling water used for temperature control of the tool cavity and molding machine. If inlet and outlet water temperature varies by more than $\pm 3°$ part dimensional control can vary, cycle-to-cycle. Flow rate, parallel cooling circuits, tower and chiller water temperature control, plus contamination in cooling lines and condenser coils, etc. can be the cause of many molding and process control improvement programs for the machine cooling, it affects hydraulic fluid viscosity that can cause erratic operations of systems using the fluid for it's operations.

SPC is a way of thinking with attached tools for collecting and reporting output. To implement SPC without the methodology backing you and the process up, will only lead to busy work and marginal utility of a useful process and quality tool. Know what you want to measure, measure it correctly, plot the results, and then analyze the results based on the results of the product or operations performance. I have used SPC for the manufacture of only one product per day with excellent success plus in a high volume plant making 5,000,000 parts per week for over 350 separate job orders. The SPC process established was implemented to make collection and plotting of data very easy for the operators. It presented essential data the operator could use to monitor the output critical to the customer and our manufacturing requirements.

The results initially resulted in a 40% reduction in defects with the major effect and benefit resulting in repeatable product manufacture. SPC software and MS (Microsoft) Excel spread sheets are often limited and not always user friendly. Learn the basics for charting and select a SPC software program to meet your current and future business and manufacturing requirements. Remember the output, what our process is capable of performing, continuously, is the intent of SPC charting of variables.

A process said to be in quality process control is defined as "on-target within UCL/LCL values minimum variance". Conforming to a customer's requirement in establishing the company's specifications is just the starting point; the finish line is the repeatable manufacture quality goal to reach.

With more companies becoming ISO9000 certified their sales and marketing personnel market their use of SPC for monitoring to maintain control of their manufacturing process. If they use SPC correctly, using SPC as a cultural implement to continually seek improvement, it is an asset. Using SPC, as only a statistical tool to gather data is not it's true intent. A change in the thinking and use of SPC by management and operations personnel is required to always seek excellence and continued improvement in business and manufacturing operations.

STATISTICAL PROCESS CONTROL CHARTS

Before moving into Six Sigma methodology, more questions on SPC control-charts need to be addressed. In a company new to SPC who have not adopted the culture, management and employees must understand and know the benefits to be realized by monitoring their processes. This includes keeping it simple, understandable, and how it can benefit the process, employees, and company. I have experienced management (CEO) wanting rejects eliminated but resistance from department managers on implementing SPC. These managers did not understand the value or rewards of successful programs to improve efficiency and output. A culture, training, and knowledge of the use of SPC was the major problem. The break through resulted in showing successes, very simple at first, and then more complex and time consuming but with good results on how well their operations were actually performing with potential improvement discovered from the data. As a result improvements resulted with increased output through the elimination of defects.

The use of this data aided in assisting plant personnel in solving problems and resulted in improved business and manufacturing operations. Ensure charts and results are visible for all management and company employees to see as success breeds interest and more success.

Remember a state of statistical control is always predictable. A process exhibits a state of statistical control when it is performing as consistently as it can, either good or poor. The process-behavior chart then can show the results of variable changes on the process.

The data generated will be used to map out changes in variables that can improve the process. The visible SPC chart shows the effects on the process. These series of behavior-charts act as a report card for improvements. Charts show where you have been-not where you are going; only a trend in a specific direction at the time the data was recorded. Changes can be anticipated when solving a problem or improving a Six Sigma process. Some effects may result from unintended changes, often a surprise when a variable change is made. Or it may indicate a fluctuation in equipment operation, molding change, or plant system improvements or variations. Always look at points on a chart as an indicator of a variable change known and identified.

PROCESS LIMITS

Product specifications are determined based on the requirements of the product. Process and Control limits are calculated based on the products finished requirements and used to control the manufacturing operation so product is manufactured repeatably each cycle. Control limits are the desired process operation limits, with all products or operations maintained within an acceptable range. Process limits are defined as in control but close to exceeding the customer's specification or requirement limits which defines an-out-of-control process or product.

Process behavior is a separate issue from specification limits. Customer product requirements (specifications) will always be met when the process is maintained within its process control and specification limits. This means the process is operating in a predictable operation with cycle-to-cycle repeatability. The minimum limits a process should run in is three sigma process control. A move to Six Sigma control limits, the ultimate attainable goal, reduces process and product defects by 20,000 times.

Process and control limits must always be recalculated, once the process is in equilibrium. This is done by the collection of current data to be used to calculate the limits, termed soft limits, with as few as 15 new data points. These initial process control limits will change and must be recalculated when more data is available, 25 to 30 data points. The tighter the limits the more control is maintained on the process and product.

A process-behavior chart for low volume runs, prototypes runs specifically, can use as few as 8 data points with a good reliability of knowing the run was within process control. This aids in ensuring fill rate, tool cooling, and machine and material variables were in control or equilibrium to produce parts for fit-up and testing if required for the program, as is usually required.

I have witnessed short prototype runs of fewer than 30 parts produced where process equilibrium was never reached or monitored and part dimensions varied so much they were deemed useless for anything but show and tell. A total waste of time, material, and equipment resources resulting from the program manager not realizing what the prototype parts were to be used for assembly and/or functional testing.

Process limits once established and deemed in control should only be recalculated when a major process variable change is made such as the introduction of a new lot of resin or equipment modifications or adjustments to the process. The object is to take the right action rather than obtaining the right numbers. As long as the product stays in control by adjusting the process should new process limits be recalculated to tighten up the process control for the product and process. Only recalculate or revise the process limits based on sensible reasons and how the process is behaving. Remember Deming's saying, "If the process and product are in control, leave the process alone."

New process limits should be established and well documented each time a production run is made. This includes any time a tool is setup on the same or a different injection-molding machine in the plant. This serves several purposes. It establishes a set of variable settings that can be used to startup the tool each time it is run. It tells the setup technician what settings are or need to be set for the product to be made successfully. It aids in quickly establishing process control for the run to ensure not only good parts are produced but within the quoted price to the customer. A tool will have a machine setup log of process limits for each machine it is setup and run on.

DATA SUBGROUPS

Subgroups of data collection usually state, a minimum of five values collected, to evaluate a process variable. This is used to ensure enough readings are collected to show the true variance of the variable during the conditions the variable was evaluated. Conditions dictate there will always be variation, cycle-to-cycle, even under ideal conditions and a minimum of five readings will typically show this condition. No process is ever stationary; it is always in a state of change and variability.

Readings within sub-groups are called rational sub-grouping. These are readings taken as a part of a subgroup and considered to be within essentially the same conditions. But, should the cycle be interrupted, rational subgroup data can be compromised and may show up in subsequent readings, or subgroups as out of control. Continue the process, not using the data for charting or leaving it in but always noting on the chart the reason for this variance. Too often these conditions occur and later when the data is reviewed no explanation for the variation was documented. Was it a natural occurrence or a known adjustment or condition occurring on the process? Documentation is key when working with any process or problem solving.

Consecutive recording of data in subgroups is not a requirement for a process-behavior chart. What is important is that the readings, when taken every 5 to 10 or more cycles, they are rational and statistically consistent over time with other readings in the subgroup. Process behavior is based on data repeatability within allowable, process variation guides. The data stays within control limits or specification limit groupings and does not show a trend to exceed the products specification limits.

As discussed, process specification limits can be recalculated anytime with the process engineer's goal to always stay within the established process control limits during manufacture to produce acceptable product.

Use of pre-control is not to be used to control a process. Pre-control is based on controlling the process based on the specification limits. This allows too much variability in the process that is then reflected in the products dimensions. In essence the limits are too far apart to guarantee process control of the operation.

OUT OF CONTROL SITUATIONS

Should the process exhibit chaos, it not only is unpredictable it is classified as out of control. When this occurs the first step is to identify a reason for this condition such as an unknown variable change. Do not be concerned of the outcome of the process at this time, only find the reason and correct it. The goal is to return the process to predictable results.

When chaos occurs, use the trouble shooting guide, machine and tool problem notebook, check lists on process problem solving, and operator tools, pyrometer, visual checks of motion and travel, etc. to define the cause. Again, once the cause is found, document what happened to the process, how the cause was discovered, and how solved. Put this in the plant, program, and machine problem-solving notebook. This will be invaluable information should the problem occur again at a later time and date. Remember, if a solution cannot be found after extensive research, shut the process down and review the data for the probable cause of the out of control problem.

A process does not have to be in predictable control to calculate a reliable process-behavior chart. Based on how control limits are calculated, good process control limits can be developed from "questionable" data, not repeatable data. The robust calculations allow the technique to be sensitive to the occurring conditions.

The process-behavior chart is used to characterize a process for being either predictable or unpredictable. As a result, you do not have to wait for "good" repeatable data. Only remove "data spikes", totally out of control temporary data points, usually singular or at most 3, from the data as shown in Figure 18. But, if the data remains out of control, repeated spiking, a serious process control variable is out of control and chaos can occur in

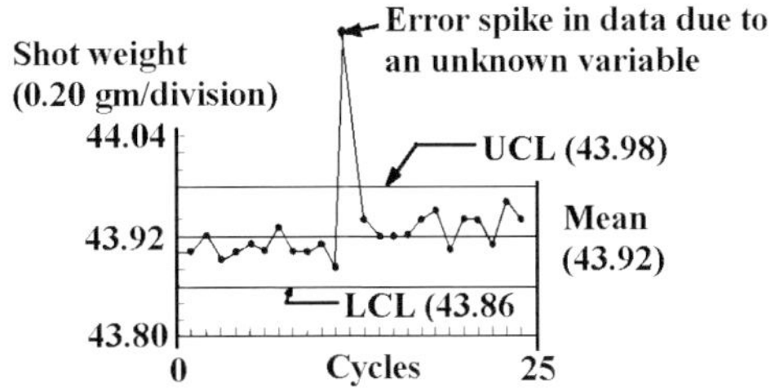

Figure 18. Chart of out of control data spikes.

the process. Removing known variable data changes is acceptable until the process stabilizes from the new change.

Process-behavior charts are easy to construct, visibly showing results and easy for all to use and interpret. They combine simplicity with sophistication for the user. When all changes are noted (documented) on the chart they allow the routine process variation in a predictable process to be separated from the exceptional variation of a unpredictable process. This allows for looking for the assignable causes of an exceptional variation, thereby improving the process and achieving repeatability, cycle-to-cycle.

REFERENCES

1. Nickols, F., "Too Many Types of Quality Problems." Quality Progress April 2000: 43–49.
2. "Uses and Abuses of Charting SPC Data" Quality Magazine April 1999: http:/ /qualitymag.com/articles/1999/apr99/0499f7.html
3. Feigenbaum, A. V., "Total Quality Control". New York: McGraw-Hill, 1983.
4. Hertzer, R. A., Green, M. W. "Closed-Loop Servohydraulics: Boon to Automation." PM & E Nov. 1988: 38–39.
5. Celanex, Properties and, Processing, Engineering Plastics Division, Holchst Celanese, Bulletin J1A, February 1984, 61–62 10M/1283 3M/885.
6. ASTM STP 15D from American Society for Testing and Materials.
7. Knouse, S., "Getting Employee Buy-In to Quality Management." Quality Progress April 1999: 61–64.

Chapter 6

Achieving an Effective Six Sigma Deployment Plan

Success depends on a well-planned program from the start to finish. The black belt with input from your company champion discusses the intent of the selected program. The monetary return to the company in business or manufacturing improvements is the key driving force for the program. The programs success must be visible to the employees as a key to improving process and profitability plus customer satisfaction. This can be displayed as discussed earlier as storyboard material in the common area of the company.

Some companies may have difficulty initially in selecting, the program to implement Six Sigma. Therefore, discuss the details with your black belt to determine a time line and number and type of key team personnel needed for the program. This is important for if key personnel are not readily available, able to delegate their daily duties to their subordinates the team will not obtain the full resources of this person. Six Sigma requires, by definition the full time and commitment of team personnel. This full time commitment means exactly that; no, daily tasks or meeting must interrupt the personnel of the Six Sigma team members. Others take up their duties as they are fully committed to only Six Sigma programs.

SELECTING THE PROGRAM FOR SIX SIGMA

The first test for program acceptance is: are the assets available, monitoring equipment, sufficient computers for analysis, and measurement equipment for processes etc., within the company or must be purchased, to perform the analysis of the program? This may not be known until the team meets and discusses the plan of action. Other department personnel, not team members, may have to supply these answers as how their data is generated and documented.

Management must also supply the monetary assets to obtain key equipment or services necessary for the program. Without these assets the program will not be successful. It is interesting to note what key analysis equipment is often lacking even for companies who have been in business for decades. They never considered it necessary to solve problems, analyze the root cause of the problem and know what was required to be used in solving their product problems over the years. This included writing meaningful process control procedure and manufacturing specifications for buying products. Also, obtaining support from outside the company to assist them in reducing their high scrap rates. This is necessary when they were so out of control a fast solution was required by upper management! This fast turn around of the company problems and processes resulted due to a rapid change in upper management forcing the department managers to look more closely at their bottom line profitability and not accept such high losses.

Next a plan of team actions must be developed with responsibilities delegated for data gathering without disrupting the normal business and manufacturing schedule of the company. The more preplanning performed the less disruption of the schedules and daily routine of employees. Management involvement at this time for scheduling assistance and project planning is essential. Actions required to install monitoring and recording equipment may have to be scheduled during line, business or free downtime, end of shifts, after hours or weekends if very extensive work is required for installation.

Also, plan ahead for installed asset usage if the monitoring equipment must be permanently installed. This will ensure other data gathering and monitoring equipment is ordered and available for normal business use. Also, if the monitoring equipment is placed in unfriendly or non-secure areas, it is protected and not likely to be damaged during regular working hours.

When the black belt and champion are in agreement the plan should be reduced to basic objectives time lines for completion of tasks and personnel assigned to complete the work. Then the Six Sigma team organization personnel informed of the tasks in their area to be monitored for improvement. This can lead to important information being generated later by the team when the program has begun. The black belt must also insist, and ensure, management does not micro-manage the program. Only the black belt has total control of a program, not management. Programs fail when management begin to micro-manage any program no matter for 'what' reason. Management hire personnel to be capable to complete and do the job they are hired to do. By continually interfering into a program will cause chaos, misdirection of personnel and a total lack of interest and directing team members. It is the black belt and Six Sigma teams that create improvements not management.

THE ANALYSIS OF THE ORGANIZATION AND COST OF QUALITY

The management person selected as the champion, take their direction from the CEO/President and his staff. The program selected is based on bottom line cost improvement. The finance accounting department can supply very important profit and loss information to narrow down the areas to be or need improvement.

In this analysis the "cost of quality" can be separated out so as to better develop areas requiring improvement. Cost of quality is typically broken down into three areas.

COST OF QUALITY PREVENTION AREAS

Preventative – quality analysis, planning and prevention
Appraisal – measurement to assure conformance to requirements
Failure – cost of defects not meeting requirements and customer service and repair

The effort (cost) must be spent in prevention at the 75% level with 20% appraisal and 5% failure analysis. When the appraisal and failure numbers begin to increase from these values, problems in business and manufacturing increase to proportions where the company must reevaluate their methods of doing business. Too many companies do not know their "True Cost of Quality." Most companies considered 'failure' or the creation of scrap their actual cost of quality and fail to put assets in the other two quality action areas, Preventative and Appraisal.

Cost of quality also includes rework, which may or may not be a measured item. If each are reworked or scrapped at these stations, the cost in time and materials is often not separated or charged as being a quality problem. Just the cost of doing business are considered a typical and acceptable business loss.

To keep quality costs in the correct proportions, program analysis of how projects are handled from sales to the start of manufacture must be evaluated. This will ensure all steps necessary for the highest state of their basic quality per operation (ISO9001) in operation of department functions are addressed.

COST OF QUALITY FOR SIX SIGMA

When a company establishes their quality base line for Six Sigma quality improvements, 95% do not know their actual COQ (cost of quality). Defects and their respective severity (number of) can be identified using a Parato chart. Each type and quantity of business or manufacturing quality defect is identified. A defect rate is calculated for the problem, usually associated with a single workstation for simplicity even if the same type of defect should occur at other workstations. Keeping the work stations separated will facilitate the problem solvers with fewer variables to consider in determining a solution to the problem. Also, the solution may be applicable for all stations down the line or at least a starting point to consider in their solution.

The question that must be answered is to know exactly what is considered and counted as a defect at a workstation. It is either a business or manufacturing problem not previously considered or a procedure not written to handle the particular problem situation when it occurred.

What often needs to be answered is what is considered a defect or problem in the operation. Is a defect considered when a product must be reworked, scrapped, or not available to ship? Has the problem always existed but never fully addressed due to some set of situations that now are more important or a problem newly discovered during the manufacturing operation? These questions need to be answered, as it is typical of a company having difficulty determining what a defect is, and where the defect is counted. Typically most companies do not count all of their known defects before the product is ready to ship.

Product defects can originate from any company operation. Problems can originate at the quoting and contract level, business operations and practices, design, engineering, and manufacturing that go into creating a product or service. What is considered a product defect, and where they are discovered, and accurately counted is very important to effectively implement a preventative solution to a problem. Once these business, manufacturing, or product problems are identified a preventative workable solution is implemented. This may involve the questionable parts being sorted for type of defect and then further inspected for scrap or rework. Then after repair or rework, returned back into the manufacturing system.

Business problems can be classified and broken down into several categories. Problems can occur at the order entry station, specification and design gathering of data, plus the dissemination of information within the company. This can occur as wrong information entered into the order and manufacturing system. These errors may not be recognized as an error until the order has been shipped and arrives at the customer. This may cause the customer problems if the product must be used by them immediately in their products or shipped to their customer. When the customer informs the supplier of a problem, it is often too late to immediately affect a solution. This leads to lost confidence in the supplier and is difficult to make up. This is why many companies have frequent meetings to discuss the product requirements to ensure the customer and the manufacturing operations are always in agreement.

The key to success is a coordinated method of communicating information between supplier and customer. It is too late if the error is not discovered until final inspection and time has expired for making a replacement.

Manufacturing and product problems are typically found at the plant and identified before the product is despatched. In some cases the customer finds the problem and a warranty problem created. A lost part is a COQ (cost of quality) as the product could not be shipped and sold for a profit This cost is often absorbed as a cost of manufacture and never identified as a cost of quality, which it really is in the operations cycle of costs.

All companies need to establish their own internal COQ program. It is discouraging to know how many companies do not have a cost of quality program. These company's track defects, scrap, and lost units but often do not relate these losses to actual loss of existing or future business. Much of their time is spent attempting to determine the actual cause of their problem defects and losses. The problem often encountered is that the problem solving team members do not know enough of the operation and never determine the "actual root cause" of the problem prior to the problems discovery point. Here is where the Manufacturing Process Control Procedures and FMEA's can be of monumental support in addressing the problems.

Be aware that these were developed using fishbone charting to be sure all variables were determined when the charts and procedures were developed. All of these operations must be completed before or during the company implementation of their quality improvement program.

The Six Sigma quality program was based initially on achieving major monetary savings. To accomplish this goal, the prior manufacturing and quality operations and procedures must be implemented. Monetary and manufacturing excellence is based on reward for continued quality improvement plus other additional benefits such as improved output with defect reduction, and meeting customer delivery dates.

Establishing the COQ in a company begins with teams from management, business, finance, quality, manufacturing, sales, etc. being formed and discussing their operations and how they interact and respond with their current customer base. Also, what and where do they want to move to in the near future to keep their business operations successful and build their business base?

Many companies have a difficult time in identifying their true quality functions in their operations and who should charge time to these cost centers. Operations must be identified, severity of loss, or problems weighed, and cost of manufacture established for these and all other operations.

Initially have your personnel report their time and material resources spent on corrective, preventative, or warranty action issues to determine and establish the areas actual COQ. Many accounting and manufacturing personnel's time will initially be charged to these work centers in calculating the cost of quality for these operations. These will be a one time charge or COQ against the organization and not considered further in COQ.

The quality cost ratio for program management to effectively show an improvement in a companies quality system should be in a time ratio of 70–20–10-percentage for prevention, corrective, and external/warranty actions. The corrective action heading can also be further broken down into appraisal and internal failure categories.

PREVENTION VERSUS CORRECTION

Prevention is the major key of quality. Six Sigma is the procedure for using the key to unlock the door of preventative actions to reap the advantages of benefits once completed. All defects and problems will never totally be found as more variable changes will occur, often unbeknown to the operators during manufacture, but the key is to minimize the variables variance and keep the system under control in "Real Time." This is accomplished by developing procedures for in-plant operations and inspection points to handle these unexpected situations when they occur.

Correction is a part of the key that must be implemented using the Six Sigma quality procedures. Correction without implementing prevention will never be successful in eliminating a companies defects and scrap. Always ensure the corrective action goes the full limit to ensure the correction solves the root cause of the problem and does not in itself create another problem further down the operation.

APPRAISAL COST

Appraisal costs are described as the effort expended in inspecting a product process during manufacture, to prevent defects.

Internal failure describes a product or process that will not pass final inspection and is usually scrapped or reworked. Does this mean if the part can be reworked it is not considered as scrapped? Or is it counted as a defect, which it was, at the station it was found to not meet requirements.

How charges are determined, the rate charged and the time and parts accounted for is very important. Is it counted as a loss in sales and subtracted at the sell price to the profit line? These are decisions that management must make when determining their true defect cost and COQ bottom line. Each company establishes their own rules as to how quality costs are determined. My recommendations are, counting any defect in the operation even if it is reworked and saved. If this is not done, management will never know the true capability of their suppliers and their own internal business and manufacturing operations.

One quality-consulting author recommended putting responsibility for internal and external failure cost on the manufacturing department. This ensures responsibility is assigned to manufacturing for the defect.

This would assign the quality accounting dollars charged are reassigned to the quality department for defect prevention at the point of discovery, which will be directed to the point of origin in the manufacturing operation for prevention.

The cost of quality is assigned to the department where the defect originated and was recorded as a loss or rework cost. Owning the responsibility of how, and where, the defect occurred is good as far as the assigned cost to the company, as this is where the costs originated and must be assigned. But, likewise the cost of corrective and preventive action to ensure the problem is solved is shared by both manufacturing and quality to solve and then eliminate the problem in the future. Each department will have to spend money to find and implement a fix. This is in the area of correction and prevention. Assigning cost of responsibility is key to determining where the costs originated and where quality dollars are spent for corrective or preventive actions. The company must know and accurately determine where the costs of quality originate to accurately assign the cost of quality to the total cost of the product.

This places the origin of the problem where it belongs so the quality department, with the assistance of the affected, problem originating department personnel, can spend their budget dollars to permanently fix the problem. Either way, it is a cost of quality improvement, no matter where the problem originated and is finally resolved and prevented. Quality must team with manufacturing to assist in preventing problems by being very proactive versus reactive, as that is too costly to the company.

Each department in the company should develop a continuous improvement plan. The cost for this will be charged as either corrective or preventative quality costs. The only difference is how these costs are identified, reported, and assigned. It is a CPQ (cost of product quality) deficit being turned into an asset. Therefore, remove the blame and focus on the cure with the dollars being spent for the most advantageous actions by the department making or implementing the improvements.

CALCULATING COST OF PRODUCT QUALITY

A list of twelve possible categories used for determining your internal CPQ are listed for consideration. If you can find more specific or related categories to identify your cost of quality items, add them to this list. Then add up each specific area/department costs and this total sum is divided by the companies annual sales revenue to obtain the CPQ as a percentage of revenue.

This example is related to the plastic injection molding business but any industry can add, and change, the variables in the example for their own internal evaluation.

Waste

Your company purchases raw materials, in this instance, plastic resin pellets in bags or boxes. Poor control of product in the warehouse results in damage. A hole in the package and material is lost, swept up and thrown away. No one reports the loss of this material to production or purchasing. When the job was originally quoted material for the parts plus a standard material loss factor was used for calculating the amount of raw material required for the program. This lost material cost was calculated in the piece part price as a resin loss factor for normal machine purge and product startup material loss. Therefore, should a major problem occur and excessive number of parts is scrapped, more material will have to be ordered to cover this part loss. This can severely affect the profitability of the program especially if losses are substantial and material costs are high.

Establish a policy in the warehouse to always report daily any loss on a 'Scrap/Material Loss' report to manufacturing. This will assist in maintaining an accurate accounting of product and material can be established and maintained for all customer programs.

Scrap

Startup production parts not meeting specifications are scrapped. This can involve a considerable amount of time and large number of parts for the molding system to reach equilibrium. This is to ensure acceptable product, product meeting the customers and quality assurance requirements, is made and product saved for the customer.

If it was determined that runner and out of specification parts cannot be used as regrind, this is another material loss for this job. In most instances, unless the scrapped parts are badly contaminated, the plastic material can be saved, and made into regrind that may be allowed to be used for other less critical parts that allow regrind. Never use regrind unless it is allowed in the contract as it will lower the physical properties and affect the color of products.

All regrind containers must be clearly marked as "regrind" and accurately identified as to type and lot of material it came from, and if it is to be used again. Always store the material in moisture proof and sealed bags, or a container, to ensure it is never contaminated with other materials in the plant. If regrind is allowed, the program cost is the time required for remolding more new replacement parts to meet the customers shipping requirements. If not allowed and you save it for a less critical job allowing regrind, cost of grinding and warehouse storage space must be considered as a cost of business for the current job or charged to the new program the material is used.

Keep accurate records of regrind use as a percentage allowed for mixing with virgin resin. Scrap regrind after the 5th pass as materials physical properties could be too degraded to be used for parts. Be very careful if color is critical for the product. The majority of pigments are now organic and over each successive pass of material through the manufacturing cycle, the pigments are degraded. This affects the color when regrind is used in even very low concentrations. Attempts to mix color concentrate with virgin and regrind can be done, but is very difficult to always obtain the correct product color as the regrind pigment is continually degraded on each successive pass through the molding machine.

Rework

Some parts or assemblies can be reworked if a component fails. Rework depends on several factors: 1. Can the product be corrected by rework? 2. Is the time and expense justified for the recovered units sell price? This latter question is not often considered when there is a push to meet a customers ship date. The accounting department can very quickly determine if the cost or rework is justified versus making a new part from the beginning. It also depends on the type of failure and where it occurred in the manufacturing or assembly process.

Repair

To repair a molded, blown, extruded, or other type of manufactured plastic product is often impossible or at the most cosmetically difficult to impossible. A color, surface blemish, or crack is not possible to repair. The plastic product or part must be replaced in the assembly or scrapped. Most plastic parts are not repairable by the supplier unless they are a sub assembly to the product. If a reoccurring problem part is discovered, it is usually corrected by redesign of the part or remanufactured to meet the customers requirements in the product.

Repair in the plastics industry usually refers to the equipment of manufacture, molding machine, molds, dies, assembly and decoration equipment or sub assemblies in the product. Consider this cost factor as one of the following: repair of a mold that was damaged; check ring replacement, grinder cutters sharpened, etc. based on your management's decisions. These can be counted against the job as they directly affected the manufacturing operation as a cost and down time.

Concessions

Parts not meeting customer requirements may be accepted only if the supplier grant a concession or obtains a price reduction from the supplier. This is often termed a deviation if the defect is deemed acceptable and not affecting the form, fit or function of the product. This concession or deviation can only be made by the customer and should only be acceptable if communicated in a written form and acknowledged by each party. These should only be for a minor problem allowing the use of parts not meeting contract product quality terms and not affecting the product function or safety of the customer. The customer may allow the supplier a deviation for the customer to accept the product.

Reinspection

If required to qualify additional manufactured products or requalify a prior manufactured product, the administrative and direct labor costs are both considered. This could also include storage costs if the product had been manufactured earlier and being questionable, put in storage for consideration at a later date as to the product meeting the requirements of the customer.

Warranty

If the supplier offers a warranty to their customers, a percentage of funds must be set aside for replacement of any products returned. If the company reserved 2% of product cost for warranty and had 4% returns for that year, the extra 2% is deducted from profits. This is one of the highest costs of quality to avoid.

Replacement

Charged as direct cost of material and time plus loss of profit. Must be based on sales price to customer. This is a warranty or lost shipment quality cost.

Additional Overhead Cost

Time required to order more material and perform problem solving analysis, inspection, and reengineering time when required.

Claims Adjustment

Any time spent making accounting changes or adjustments for returned product and concessions.

Packaging and Shipping for Returned Products

Replacing defective products to customers, time, material, packing, and shipping costs.

Goodwill

The goodwill of a customer is a difficult item to establish a cost figure. Loss of good-will can cost your company future reorders, a customer, and/or a referral to a potential new customer. Can be equated to rise and fall of their products market share as a percentage of unrealized profits.

Concessions

The primary action required is to ensure all four costs of quality; detection, investigation, solutions, and implementation are identified and reported accurately. Accounting can assist by providing the most intelligent and descriptive time and material identification for tracking inputs to identify specific quality costs of your products. A list of quality change cost categories is determined and this information is distributed to department supervisors. These supervisors will be trained in knowing when, and to what charge code their personnel should charge time and material to these cost codes. This will assist in ensuring the true identified costs from each area of the company, is reported accurately, as your system was established to report quality and defect costs.

Each company will track their quality costs differently. An example follows for such a company tracking their cost of product quality (CPQ). Personnel who have tracked their quality costs for over 17 years the same way at each plant, used this method within multiple plants.

This company tracks CPQ for almost all employee job descriptions and assign cost-of-quality activities to them for internal and external failure and appraisal cost. They believe almost every employees job falls into one of these categories.

Prevention Cost

Prevention cost are recorded separately and their thrust is to ensure enough time and money are spent on prevention and time not taken away from prevention to reduce the other quality cost areas. They categorize their prevention cost in some of the following areas:

- *machine operators salary*
  Time equivalent for prevention as cleaning tool surfaces, recording process variables, and charting, part inspection, using go/no go gages, minor preventative maintenance on their machine or auxiliary equipment used for the job.
- *team activities*
  Meetings and time to prevent reoccurring problems during working or after work time.

- *internal and external training of personnel*
  Plastic courses at the local technical school for technical jobs related to know and do a better job.
- *College and GED*
  Any and all advanced training for a college degree or attaining a higher knowledge level
- *travel and lodging expenses*
  Any and all expenses related for off site education and training seminars.
- *All supervisor salaries*
  Management assumes if they are doing their job correctly, they are preventing problems.

Experience has shown prevention results in lower scrap and failure rates. Reducing scrap reduces appraisal costs, because less looking is required to find problems. When scrap is reduced it is less likely a mistake will be shipped to the customer. This affects internal and external costs, less new or required rework, and lower customer complaints and returns. They discovered that for each 1% of scrap reduction, their external defects are reduced by almost 5 times that amount. This is a very good return on investment and reduction of CPQ.

Once this program is setup and running, monitor and audit the initial changes to ensure the data reflects the true identified problem area with the supervisors and their personnel. Accurate reporting of data is very important to ensure management that their true cost of quality is being documented.

QUALITY RETURN ON INVESTMENT

Now the calculation of the quality ROI (return on investment) can be determined for the company and each department if they are setup as a profit center in the accounting system. Use the following equation for this calculation of quality ROI.

$$\text{ROI} = [(y - x)/x]\,(100) = \%\ \text{ROI}$$

x = the costs documented for your quality program over a set period of time

y = monetary benefit you received in return, profits realized from your quality program

An example of this, based on spending \$150,000.00 to implement your quality program in the first year and an additional \$60,000.00 for support in the second year. Considering a two-year time period for the quality changes to take effect for the initial analysis the cost of implementing quality are;

x = \$210.000.00.

Tracking increased sales, less defects in manufacture and customer returns, all positive changes as a result of the quality program, yields a realized increase in profitability for the quality system of \$450.000.00 or y equals;

y = \$450,000.00.

The ROI for quality is:

$$\text{ROI\%} = [(Y - X)/X]\ (100)$$

$$\text{ROI} = [(450{,}000 - 210{,}000)/210{,}000]\ (100) = 114.3\%$$

This calculated ROI means in two years time you gained back your original investment plus an additional 14%. You also realized a gain in profits of the same amount that is a very good investment in your quality system.

This information, as reported by the major corporations, who accurately track their Six Sigma improvements have realized profits and savings into the billions of dollars since conception, and have forecast similar saving results for the near future.

Management must focus on the success of quality programs they can implement with the right personnel, training, and management support within their companies. There is a good payback in a capable and improving company business and manufacturing quality system, Six Sigma or not. It is important to keep quality alive and well in your company.

The quality function to use along with anticipated results are shown in Table 1.

DEPLOY SIX SIGMA BY FUNCTIONAL AREA, PRODUCT, PROCESS AND DEPARTMENT

The use of all quality methods will assist the company to ensure the customers requirements for a product or service are available, meets their specifications, and delivered on time. Six Sigma is a technique combining the necessary quality techniques into useable, profit oriented tools employees use for a win-win combination for supplier and customer.

Table 1.

Quality Method	Action or results of quality method used by company departments
QFD	sales reviews, know customer requirements and expectations used by supplier to ensure they meet customer expectations
QFD, Checklist	contract review to meet customer requirements and specifications
QFD, FMEA	engineering reviews to ensure all departments, design, manufacture, material, fishbone diagram product support inputs are understood and on time to meet customer expectations. Evaluate product and supplier risk during manufacture. Maps out all variables for each material and operation
QFD, procedures, checklists	purchasing – material completely identified (specifications complete, accurate, and in-depth to meet manufacturing requirements) as specified by the products design, quality and production control
QFD, check lists	finance – a profit can be made with the product and kept within costs. Accounting method to accurately track the true cost of manufacturing and quality in each department
C&A, FMEA, QFD, DOE, SPC, work instructions	quality – input into all areas of controlling the quality of services, material, business monitoring, inspection to support manufacture of the product and do business
FMEA, SPC, check lists, maintenance work instructions, process flow diagrams, and control plans	ability to support manufacturing in the capability to make product repeatability to meet the customers requirements and specifications for total satisfaction. Inspection points at critical and necessary hand-off points for the next operation
fishbone, FMEA, CpK, SPC, JIT	capable materials, equipment personnel and systems to make a quality product in a repeatable operation. Define all manufacturing variables for each operation from design to shipping for the product
QFD, fishbone, FMEA, SPC, check lists	pack and ship-ensure product is protected and meets customers delivery requirements plus customer service and repair – ensures order information is correct and any customer support is available and able to meet the customer requirements

When Six Sigma is initiated in a company, the decision of the selected champion as to where to begin is a major decision process. They will receive input from all corners of the company, CEO on down. The champion must filter out where the most important needs and savings can be realized to direct the black belt team to begin.

PROGRAM MANAGEMENT REPRESENTATIVES

It is also critical within a company to have a program manager for each program or set of programs within the corporation. They will be the focal point for all information regarding their customers programs. This focal point is used by the black belt to ensure any improvements are not in conflict with the customers requirements, and ensures changes for quality input are conveyed to their customers quality department and used to increase the business profitability.

PARETO CHARTING

The use of Pareto charting is an analytical method of counting and charting the severity and frequency of defect or problems occurrences of various possible business, product, and quality concerns. A method of categorizing, by frequency of occurrences, these concerns to enable quality control and business priorities when more than one concern is present. The implied Pareto principle states, a small number of concerns is typically responsible for most quality problems. The data is presented in bar graphs of all concerns identified and ordered from the highest to lowest number of occurrences. The ranking of severity is a secondary action conducted by personnel who have the information to rank the seriousness of each problem to the others within the company. What also must be considered is how these problem areas impact on the cost of quality further into the manufacturing operations as the cause of their failure. Failure to identify this type of problem will have no effect on the down stream problem area if not recognized at the beginning of the Six Sigma program.

Upper management must have this problem frequency and severity data to wisely select the most important program to begin the Six Sigma corrective action that impacts their customers and their company's bottom line profit and creditability in the market place.

Depending on the results of the Pareto analysis, a question that is frequently asked is; "Is the customer accepting, and happy with, the product and delivery?" What is often overlooked in meeting customer "upper management concerns" is what is it costing their company to meet these requirements? Where are the costs of quality the most? Is it in prevention, correction, or failure analysis? Also, are they actually meeting the customer requirements, and if not, why not? Are they close to loosing a key customer if they ship questionable product and do not respond to their corrective action requests.

The Pareto analysis, when performed and presented correctly, will show management the true cost of their endeavors to meet the customer's requirements. An example of this chart is shown in Figure 1, a Parato Chart of Defects.

If a company has to produce on average 20 to 30% more product, in the hope of meeting their customers delivery requirements due to the high loss in their manufacturing operation, they have a very serious quality problem. This is a very poor use of their manufacturing and quality assets but this happens in the majority of both large and small companies. This can cause delays, waste, and a constant problem of schedule delays with frequent rescheduling of orders to meet their key customers delivery requirements. This resulted in smaller customers having their orders further delayed on their suppliers manufacturing schedule.

When a company practices this form of 'bad business practices' such as quality concerns were not addressed and management was unable to make enough product to meet their customer's delivery requirements their business always suffers and customers are lost. Needless to say, frequent

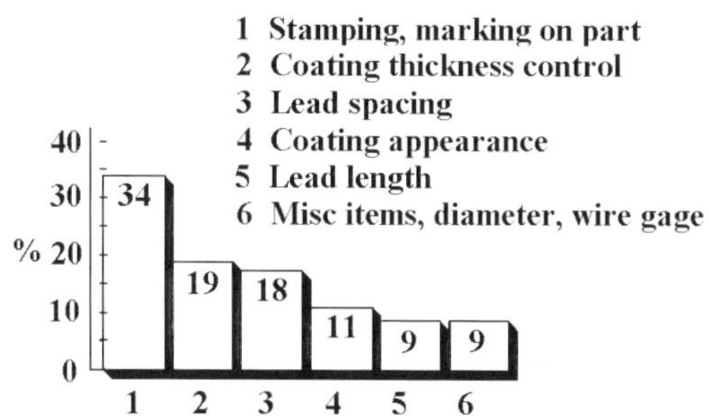

Figure 1. Parato charting for defects.

questionable product quality concerns occurred and were not addressed correctly or waived, by the question, "Has the customer complained?"

When the answer was no, the order was shipped and often without qualities approval. This situation continued with customer returns mounting and the customers quality concerns answered by unacceptable corrective action reports sent to upper management on the outcome of their internal MRB's (material review board). As a result, quality did not improve and shipping schedules for other key customers kept falling behind which eventually resulted in a lost customer. Their customer finally found an alternate, more quality conscious and reliable source, who when a quality problem occurred, produced and implemented a proactive corrective action that solved their problem.

Their suppliers continued follow up, assisted them with solutions to prevent and ensure this and other internal quality problems were eliminated to avoid future customer complaints. This grew their customer base at the expense of the original supplier.

Using the Pareto chart, the problems, often many small ones and often not directly in the area of concern, must be investigated as the probable contributing factor to ongoing quality problems. Remember, just plotting problems without solving them is a wasted effort. Manufacturing personnel want to see solutions as no one wants to produce an unacceptable product.

Customer complaints of late delivery must flow back through the organization to find the root cause of the original problem of why the product was not delivered on time. The problem is late delivery: the cause occurred way before this and must be addressed, and solved, to eliminate it. The process of analysis and elimination using the fishbone or cause and effect analysis to determine the root cause is often used. Along the trail back many other factors will be discovered that were caused by the problem and can then be corrected based on their impact on the late delivery problem.

Therefore, where to begin is decided on where the greatest need and monetary return for the company and customer satisfaction resides. This is usually looking at the company's top 20% customers and solving their problems, that in return solves their other 80% customer problems. Each company must make their own decision on what program to begin. As seen, most programs will impact improvements in other areas that will be addressed in the initial or follow on programs. Improving one area will have a cascading effect on other areas. In fact, solving root cause problems can eliminate other problems in the business and manufacturing chain. They can also show areas where additional improvements, even minor, can improve process and product quality leading to greater customer satisfaction. So do not look at just the head or point of origin of a problem, search for the base root cause.

STANDARD OF FLEXIBLE IMPLEMENTATION DESIGN ACROSS VARIED BUSINESS UNITS

To not dilute the effects and results, the implementation process should proceed only at full time commitment for the Six Sigma team members. This is not an auxiliary program but a full time job. Any areas not initially affecting the successful conclusion of the program must be documented for future Six Sigma program consideration. Stopping to fix all problems discovered, not-affecting the initial problem area, will not be beneficial at this time. The team must focus on the solution of the initial program not dilute their efforts as new opportunities are discovered for later solutions.

The decision to focus on area, product, process, or department is selected as already discussed. Do not deviate from the primary Six Sigma program goals and objectives.

ISO9000 IMPROVEMENTS BASED ON SIX SIGMA RESULTS

Remember, if already ISO9000 certified, your quality system was audited as being in compliance with the standard. The ISO9000 quality requirements are shown in Figure 2, in triangle form. The top of the triangle is the quality

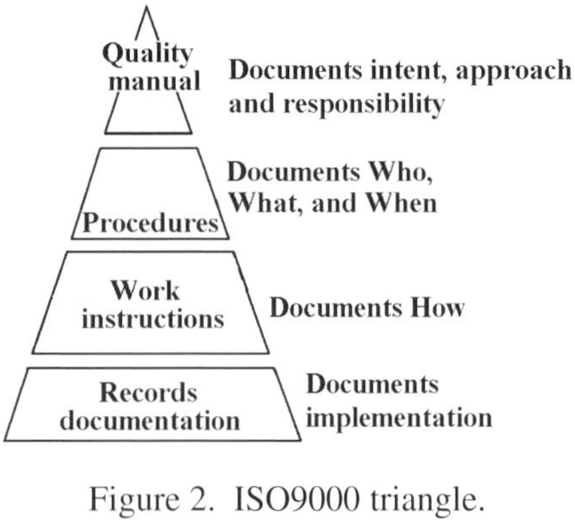

Figure 2. ISO9000 triangle.

control manual listing the requirements and how the company will respond to them. The lower sections, procedures, work instructions, and documentation list the remainder of the requirements.

QS-9000, the automotive side of ISO9000 is composed of the same requirements as ISO9000 along with additional requirements of subscribing companies. These may be company-specific, division-specific, commodity-specific, and/or part-specific requirements in addition to QS-9000. It is important to remember that it does not mean your quality system is the best it can be if you are certified as meeting the requirements.

Being certified only means that your quality system as you have written it, and as audited, meets the auditor's interpretation of the ISO requirements. Any quality system can meet ISO but only the best system will meet your customers requirements all of the time. Improvements are now necessary to continue your quality upgrades. During the course of doing business better and new methods of doing business will become apparent.

The decision of company and department managers then requires their decision to make the change now, wait for completion of the Six Sigma program, or try to incorporate the changes immediately. In any situation, the champion and black belt team leader must be brought into the decision process. They along with the department manager can decide if the change would benefit the success of the program or wait until the program is completed.

Only the team will be able to evaluate the effects of a new change as beneficial or imposing more unknown variables that could mask or create new problems in the program. This is similar to evaluating a process. Change and evaluate one key variable at a time to see the effects on the process. Inject too many variables for evaluation at one time and the effects of each on the process are often not easy to determine. Stay with the programs plan and schedule and only consider new changes if their anticipated effects can result in a positive result on the program.

BALANCE THE RATE OF DEPLOYMENT FOR MAXIMUM EFFECTIVENESS

One method employed for rapid deployment of change is Kaizen, which is one of many quality techniques. For the record, Kaizen is not the method used for Six Sigma Programs. What then is the Kaizen method of business and process improvement? Developed to obtain positive results in five days time, with minimal in-plant cost, not including consulting costs, and with scheduled successful completion of all action points within 30 days after completion of the initial Kaizen.

KAIZEN

Kaizen employed in its true form, is allowing ten or more completely knowledgeable, business and quality strangers, forming among themselves and your personnel teams, coming into your company for a five day intense blitz of carte blanche authority to make process decisions, equipment line changes and reassignment of employees when and where required. An important item is the changes occur so fast, upper management is not aware of the changes until after they have happened. Also, no capital cost is involved, only the immediate creativity of the teams relying on using on-hand resources, with improvement decisions made in committee and acted on immediately on the factory floor.

They then monitor their actions, gather information on their results and report the improvements obtained in a final meeting of manufacturing and teams. Results can be impressive with poorly laid out manufacturing lines restructured for improved flow, better utilization of assets and personnel and realized savings with improved output and quality. These are not change for change sake but actual realignment to improve flow and work methods and product through put. It is often associated with a fast move to lean manufacturing ideals.

In very complex systems Six Sigma is a more in-depth analysis of a problem to develop a lasting solution or process improvements. Some say, Kaizen is only the most obvious changes that can be made with minimal cost and time. This is true, savings are attainable, cost can be reduced, and productivity increased as reported by companies doing Kaizen programs. But, complex programs reducing defects from 20,000 per million to 3.4 per million require more time and in-depth analysis than allowed in the 5 day blitz.

A typical weeks schedule, simplistically stated is:

Day 1: Employees involved assigned to consultant teams and trained in basic Kaizen techniques.

Day 2: Teams collect data from "Real Time" processes and make fishbone charts of processes for determining variables of manufacturing processes.
Day 3: Teams determine what fixes are required and "do it". Fixtures made, work place realigned, tools made more accessible, machines moved, instructions posted in work areas.
Day 4: Teams gather new data and measure effects of change
Day 5: Team leaders and members present results to management using charts and new layouts showing short term and anticipated long-term results.

Kaizen does work but with the short time spent at a company only a few highly visible fixes are applied to assist in process improvement. Processes critical to a company's product in many cases do not permit these short-term adjustments to gage the entire process over a long time period. Therefore, excellent for quick fixes and material and time, motion reductions.

I visited one Japanese owned company that performed a daily Kaizen at a specified time each day. They reported on problems, established teams to fix the problem, set a completion schedule, and expected results within the time period with the assistance of management, when required. This appeared to me to be only a company method to address officially, with management's involvement, corrective actions that were not always allotted enough time or personnel to completely address the true cause of the problems. Personnel still had to perform their daily duties while solving the identified problems. As a result professional moral suffered, problems of the same type kept reoccurring, management was never satisfied with results and customer loyalty was wavering due to chronic reoccurrence of the same or similar problem.

Performing a Kaizen at a company can make improvements in work simplification, staging resources, parts, tools, and fixtures where they are needed thus saving time, motion, and worker duplication of actions.

During a Kaizen teams schedule their work to interface with employees on all shifts to learn, modify, and improve their jobs. If redundant actions require elimination of an employee, the most knowledgeable worker is placed on a new company resource team to assist, after the Kaizen, in writing new procedures, work instructions, and seeking out new methods to continually improve quality and flow of work through the plant. No employee is ever eliminated, only reassigned to an equal or more responsible position. The true intent is to improve processes not eliminate personnel as many company reorganization plans have done in the past.

After each change is made an analysis of results, process improvement, time, and cost are analyzed and documented. Teams meet frequently with employees to make the changes and evaluate their results.

Participants in a Kaizen can come from a multitude of industries, sponsored by the consulting firm initiating the program. In this manner a vast amount of knowledge, from all sources, business to manufacturing, are descending on your operation to look, decide, do, and evaluate changes in your operations.

Will or can there be opposition, yes, of course. But, when changes show a positive output, it is up to management/supervision to ensure the changes after implementation, are used as developed. This is one reason employee buy-in is necessary as required by management so that after the change, employees do not go back to their old, more familiar, ways of doing their job.

Internal Kaizens can be conducted by company employees after seeing the results of their first, and after receiving additional training in objectives, problem solving, decision making, and implementation of the changes and monitoring the results. Similar results are attainable with Six Sigma only requiring more time and less personnel. Results can and will be longer lasting.

SIX SIGMA PROGRAMS

Six Sigma is more process, operation, and statistical driven using reliable data collected over longer periods of time to show if greater control is possible to reduce defects and improve process and output. After improvements are made they are then monitored to show compliance to the changes and how successful were the reduction in defects and problems.

Six Sigma is deployed in an organized plan. Objectives and goal driven, using any and all quality techniques to arrive at a successful reduction of defects into the Six Sigma range, 3.4 defects per million.

A milestone chart is developed with checkpoints along the time line to encourage and map completion of tasks. The check points are decision times to ensure task data has been selected for review and decisions can be made on how to continue the progress on yet to be completed items. These can be purchasing of equipment and systems to control, monitor, and update equipment and systems to reach the stated goal.

During this period many tasks may have to be completed to continue forward in the outlined program. The key is to have the black belt keep the team and their improvement progress moving to the goal while ensuring all tasks are completed to continue the program. These are not quick fix items or realignments of operations but an in depth analysis of the operation to ensure it meets the goal of cost reduction and continued quality and output improvement in the company operation systems.

SIX SIGMA OPPORTUNITIES FOR PROBLEM PREVENTION

ISO9000 and QS-9000 both when referring to corrective action say they are efforts taken to eliminate the recurrence of a problem, or by definition of a problem that has happened before and was never completely solved. Corrective action taken more than one time is time wasted as the problem was never correctly fixed. We daily see this occurring in business and industry with many reasons given as to why the absolute solution was never implemented. Time, effort, understanding the problem, lack of training, procedures incorrect, or needing revision, etc. are some the many reasons. The primary reason besides lack of time is the true solution is not known or able to be determined and keeps reoccurring. Management is also at fault as they do not specify that these problems must be solved or their growth like a weed will strangle the lawn.

This implies that whenever a problem occurs it must always be solved immediately. True but in real life seldom happens. Time is precious as is manpower and assets are not always available to accomplish the goal of prevention of all problems.

To assist a series of ideas and a part of the Six Sigma Team's learning curve must deal with the analysis of problems and how to solve them as they work on improving the business and manufacturing systems of their companies. Follow the directions given for problem solving and use the quality methods to their best output of data necessary to solve the problem. Look for opportunities for the prevention of problems and use the following as idea initiators when any task is performed. When a solution to a problem appears in your mind, write it down, discuss it with your supervisor and others to get their input and most of all do not let it fall dormant. Keep up the action to solve the problem. Be proactive and fill out a problem solution form and submit it to your quality or Six Sigma team collection problem box of consideration.

Opportunities for problem prevention can be discovered by performing the following operations during your work day:

1. Review systematically CpK values for your key manufacturing processes and business operations. Analyze them against historically established levels to see if trends could initiate investigations of causes before nonconforming product are made or services performed.
2. Measure and monitor key equipment variables such as run out, wear, vibration, and oil consumption and temperature against acceptable thresholds and equipment manufacturers specifications. As results approach the specification limits, but before they are exceeded adjust, repair, replace, and always use scheduled maintenance programs before the equipment fails or produces nonconforming product. Insist that maintenance is performed as scheduled and per the equipment suppliers recommended plan or as required per your operating requirements. Do not wait for an item to fail before servicing the item. Also, if matching items as bearing and gears working in conjunction with each other wear, always replace them as operating units. New parts working with old parts, will wear out sooner requiring replacement of the entire gear assembly that results in a more expensive down time and cost.
3. Depending on your current method of manufacture, Lean, Just-in-time, or per factory order, institute a minimum/maximum ordering system for raw materials and purchased components. Base this on established consumption rates and order completion times avoiding problems associated with depletions, out-of-date, or wrong materials ordered and not discovered until the order was to be produced.
4. Conduct annual reviews of all non-conforming material reports to identify trends, reoccurring patterns, top ten defect codes, and cause codes. Then assign problem solving teams to find solutions and implement them at their lowest level in the company.
5. Review delivery times for all significant raw materials and purchased component parts. Negative trends may trigger involvement with the suppliers or subcontractors, even though no late deliveries have yet occurred.

6. Establish and conduct trend analyses of important customer requirements to determine shifts in areas as volumes and tolerances that may result in adjustments in labor and acquisition of new equipment. Ensure there are sufficient assets on hand to accomplish the job without delays, breakdowns, repairs, and replacement of critical equipment and assets. These analyses may utilize data collected during quality planning exercises.
7. Critique all inspections and variable inspection data to identify undesirable trends and adjust manufacturing processes. This is performed with a check list or inspection setup procedure so no nonconforming product is produced and must be rejected.

These are some of the items that must be performed to ensure problem prevention is in place and working. Aggressively search out problem prevention ideas. The more that are implemented and in place the fewer number of problems will occur. Do not hesitate to follow the solution to a problem into another department where the problem may be created, exist but not yet a problem, or is known but does not cause them a problem, they work around it and pass it along to the next department. The best and most satisfying approach is to be proactive to identify true preventive action and ensure it is implemented, monitored for effectiveness and is working.

Some ideas or techniques for investigative and analytical investigation of preventive action are:

1. Decide by department, group, process, product, and business level what program to begin. In most cases the Champion and their staff will make this decision. As the program matures there will be an abundance of programs that can be worked on and solutions developed to prevent business and manufacturing problems.
2. Fully understand the activity or administrative activity as well as operational department being affected and analyzed.
3. Evaluate all inputs to the process or activity thoroughly, and not just the most obvious. The use of the fishbone diagram is one of the better methods to use to develop all of the variables of an activity or manufacturing process.

4. Consider all potential problems and causes of failure. Discuss with the operators the problems they have daily and what areas they believe would assist in improving the product or process. The floor personnel and daily workers are your best source of information in Real Time fashion.
5. Identify those resources, such as quality records and reports that will show the effects of outside influences. Also, ensure that all machine variables are documented so an analysis of their influence can be analyzed if determined relevant the problem.
6. Decide which of those resources should be regularly analyzed. Select the tool or tools best suited for the plan of attack.
7. Determine who is best suited to perform an analysis of the data. Often the closer the person is to the problem the more biased they become or are committed to owning the operation and do not want others to interfere with the operation. Also, is the correct data being collected to really analyze the problem. Here the team approach is best used to see the whole situation with unbiased eyes.
8. Settle on what are the variables to look for in the review. Describe the problem in its most basic and simplistic terms.

To assist in understanding the nature of true preventive action it will be helpful to list some of the basic characteristics or attributes of activity leading to the identification of preventive actions:

- Periodic
- Planned
- Investigative
- Systematic
- Analytical
- Deliberate
- Proactive
- Predictive (what can or will happen if conditions are allowed to continue as is)
- Not responsive or reactive
- Involving knowledgeable personnel
- Utilizing quality data, records, reports, information, and more

- Based on understanding of the quality system, processes, equipment and such
- Requiring management support and assets

REENGINEERING EXISTING INFRASTRUCTURE TO FACILITATE SIX SIGMA

Reengineering manufacturing or reorganizing business operations may be necessary if the problems indicate this is necessary. This often requires rewriting new procedures and work instructions for the affected departments.

Carefully evaluate any reengineering need to be sure a procedural change or work instruction is not being followed is the reason for the occurring problem. But, if findings indicate the procedure is lacking the control or communication required to eliminate a problem or improve a process, then reengineering is warranted. Before implementing any reengineering change perform the quality tasks of evaluation as if the change was already in place.

Reevaluate the customer and departments QFD matrix to ensure all requirements are met by the change. Perform a new fishbone or modify the old one to ensure no variable step is missing in the business or engineering analysis. Perform a new FMEA analysis and review the process flow chart and control plan to see if the problem area is noted, what steps are implemented to safeguard the problem not reoccurring or what new, if any, variable may now result from the reengineering change.

Also, are new or updated check lists, monitoring, and work instructions, project cost, and training required to facilitate the change being implemented, to work as designed and produce the desired results in product or service quality.

When all the data, changes, and questions are answered, only then can the decision to reengineer the system be determined. Any reengineering must show a positive profit payback to the company in savings, product reliability, and quality coupled with increased customer satisfaction.

It is important to only make the changes as required to improve the processes. Consult with business and line personnel on the intended changes and obtain their input and their agreement the changes will improve the process, their job capability, and eliminate future problems. Some people resist change and if their agreement, ideas, and buy in are obtained early in the program, the reengineering process can implemented with total support of all personnel in the department.

HOW TO BALANCE THE RATE OF DEPLOYMENT FOR OPTIMUM EFFECTIVENESS

Rate of deployment involves many considerations. Is the fix or improvement a business or manufacturing system: Each one will require an expenditure of time and possible new or modifications of equipment plus new or rewritten procedures and work instruction changes and personnel training to complete the program. The rate of change depends on the complexity of the change, personnel acceptance, and implementation of equipment or systems for the change to be employed, plus training in its correct use.

EXAMPLE OF CHANGE CORRECTLY IMPLEMENTED

The customer service representative (CSR) must obtain customer information accurately. This involves order processing, checking inventory, and manufacturing schedules to accurately supply their customers with ship dates of product. This is easy if your information processing system is kept up to date and always operating in "Real Time" input of plant inventory and shipping information. Also, if the information needed to input and process this information is user friendly and supportive of the customer service representative requirements and companies business operations.

Order entry errors can be very costly to the supplier and customer if the product is not made and delivered on time, the correct order quantity, and meeting the customers product specifications.

Order entry, customer service, is the front line customer contact. Having managed a customer service department for a major plastic resin supplier, I am very aware of the requirements for getting the order correct, follow up with the customer, and watching to ensure production has the product made, packaged correctly, and in inventory ready to ship when required by the customer. Order entry information is used to assist the plant in scheduling their production runs and the quantity of material to be produced and package type to meet customer requirements.

It has been proven using a split computer screen information display, customer information and available, real time, product inventory very user friendly and customer oriented. It also has production and projected manufacturing quantity and dates for the customer service operator's immediate use. The CS representatives also need to have all the required fields of data required to complete an order then update production requirements on a series of successive order entry and material availability software files.

This allowed the CSR to take, verify, and fill an order while being able to tell the customer when the order would be ready to ship. Also, if the CSR entered a wrong product code or selected a wrong material by product number and package type, if the two did not match, a code would alert the CSR to the problem and a speedy correction implemented in "Real Time" with the customer still on the phone. The use of e-mail is also speeding up information transfer. Orders can be sent in and automatically acknowledged by the mail program. Verification is made simple where before it required a separate transaction the CSR can now with a few key strokes complete the order and all parties know the whole story of the order.

The specific customer order and information screen also had information on past orders so repeat and blanket order releases could be updated with the customers purchasing as added information discussed during the taking of an order. The CSR knew more about the customers business than often the purchasing agent and they were very appreciative of the CSR's excellent follow up and assistance. Therefore, the order entry system was virtually error free except for quantity, package type, and delivery dates. This can now be addressed by e-mail order confirmation after the order is entered, automatically acknowledged and confirmed to the customer.

To implement this system a change was required in order entry, the plants MIS (manager of information systems) had to be involved to initiate a change. This presented a problem as other plants used the same order entry system with each plant requiring a slight modification to their order entry system as the production and recording of inventory at each plant varied for finished goods. One change would affect all plants order entry system.

As a result an even minor change would have to be queried to all other plant CS order entry systems to see what impact it would have on their procedures. This could take days and weeks and then time to change code and trial the system. It would also impact production input and the main computer network site for each plant. A simple change would have to be system and cost justified before being accepted for the entire order entry system. As a result, the supervisor of the CSR department and information system managers for each plant inputted their recommendations, needs, and effects the change would make with each plants and operating divisions systems. Their input and discussion lead to a successful program with each division obtaining, not exactly what they wanted, but close enough to make the new program successful.

Any change of this magnitude must be thoroughly thought out, mapped (fishbone) and analyzed to ensure the change does not cause problems elsewhere. Rearranging a plants work flow, combining equipment into manufacturing cells adding quick tool change improvements and a vast array of other improvements take time, cost money, and must show a positive savings in time, personnel, and product output with improved quality to the company before being implemented.

After implementation of a Six Sigma program, many other smaller savings, work improvements, and ways of conducting better business practices will be discovered. If time allows, during team down-time waiting on completion of changes required for the major program, these smaller projects can be investigated. Then when time allows, with the approval of your champion and upper management they can be started. A good way to use your team talent to keep their skills honed during a slow period of the major program.

Rate of deployment depends on the program, required assets to be gathered, and implementing the changes before the team can begin monitoring and gathering data on the new process to verify results leading to Six Sigma compliance.

REFERENCES

1. Taylor, C. M., "Preventive vs. Corrective Action: The Horse, the Barn Door, and the Apple." Quality Progress March 1998: 66–71.
2. Schonberger. "Work Improvement Programs."

Chapter 7

Six Sigma Improvements in Business and Manufacturing

OBTAIN IMPROVEMENTS IN BUSINESS AND MANUFACTURING, BENCH MARKING AND INDICATORS

Improvements can always be made in any business or manufacturing operation. The decision required is, "Can the Operation be improved?" If the intent is to reduce cost in achieving additional improvements, labor, (personnel reductions), procedures (work instructions and operations) and time then a Six Sigma program may be required.

In some situations a "lean" manufacturing approach may be the key. Use it initially to evaluate redundant operations, misuse of personnel assets, problem backup areas, quality situations, communication errors, etc. Do a fishbone to ensure all variables are considered in the evaluation!

A source of improvement for business and manufacturing can be the addition or upgrading of computer software programs improving operations and to assist in reducing the workload. These systems use "intelligent" data entry and recovery screens which slave to other programs inputting and outputting data that other operations have access to in "Real time" operations. Before a major change is made in an existing company software program, make sure there is system compatibility within the company's hardware and software components and systems. Too often changes have been made and then some departments make it known that their software or hardware systems cannot interface with the new system. Always, do a compatibility analysis of the hardware and software in and to be added to the system. Also, talk with your suppliers to make sure the system is really what it is said to be and acceptable for your organization.

SOFTWARE SYSTEMS FOR SIX SIGMA QUALITY

Six Sigma Software

Companies can obtain quality software with subjects from 'auditing to quality training'. There are more than 34 subjects available. There are over 370 software suppliers for these 34 plus quality topics. Not all software suppliers provide all of them, but many provide similar software programs only in a different format or style of output.

The problem a software customer has is deciding which software programs are best suited for their specific monitoring and reporting operation. This is a very difficult question to answer since all software programs under one heading attempt to provide the best and complete services to assist their customers. If one program works well, they will try to sell their other programs based on initial results and customer satisfaction.

But, their potential customers, without actually trying the program has no idea how user friendly it will be for them. Also, is the program user friendly within their existing software programs. Can they interface and use the same storage unit in each type of program is a very important basic question to consider. Having evaluated several software SPC programs and not being a computer expert, look for the best user friendly system, with a good text and help section to answer, in your language, any questions or problem that happen when trying out the program. Also look for data collection software that will collect and store data in files such as Microsoft Excel or Access software programmable files or other accumulation, sorting, or storage system. The data must be useable with the software program you propose to use.

Some companies provide software that can be down loaded from the Internet for a set number of days for evaluation. Here the evaluation of the help system is really put to the test as no manual comes with the down load. If "Help" is not really user friendly, then neither will be the program in my experience. Based on the number of companies looking for good quality and test personnel on the Internet, requiring software quality analysis and test engineers, the market is booming. Also, as the computer industry continues to grow, the demand for exceptional personnel will never diminish.

An analysis of software to meet Six Sigma needs was published in *Control Engineering*, January 1999, page 66, by Dr. Terry Heng, Motorola's

VP of Corporate Software Development stating in my words, development of reliable software is possible, "By following standards, tools, and techniques promoted by the Software Engineering Institute (SEI) of Carnegie Mellon University, Pittsburgh, PA".

Dr. Heng states, fast track software will reach about the 3-sigma level of confidence, add one more man-year of continued effort and a 4-sigma quality level is attained. With two additional man-years of work, the quality level will achieve or close on Six Sigma.

SEI ranks software as follows:

SEI level 3 equals about 5.8 sigma
SEI level 4 equals about 6.0 sigma
SEI level 5 equals about 6.3 sigma

They rate NASA's space shuttle software at level 5. Motorola produces software at SEI level 4 with some elements at level 5.

Regression testing is used to debug software and some programs with over 10 million lines of code take a long time and at great expense. As debugging tools and methods get better and faster the current level for debugging tools will exceed SEI level 3 or 5.8 sigma very soon. The company's only answer is to use the information provided. Ask your intended supplier if they use the SEI level qualification scale, what it is for their programs and then sample the product to see if it meets your requirements and cost.

SIX SIGMA (BUSINESS) ORDER ENTRY SOFTWARE REQUIREMENTS

Customer service order entry computer system and data input screens must be user friendly allowing all departments' easy access to company information. Personnel must often in their daily work query the computer data system using screen names or identification specific to their use and requirements. It is important to keep the system both secure but also accessible for all personnel in the company to use and obtain information on their operations.

For example, in order entry, each customer will have an identification name, assigned customer number, ship to addresses (if multiple locations) with often different delivery dates, and variances in the product ordered. A suppliers product is identified by supplier trade name, number, or an agency identification code. All personnel supporting the customer may request customer and product information for their specific job interaction areas and it must be accessible without (often spelling specific or abbreviated) difficulty.

MANUFACTURING DATA ENTRY

Often a need to know query into the company information system is to restrictive. The inquirer needs to know a specific customer name, number, or identification code before the information is accessible. Having a general index listing of specific customer information for making a selection to identify a specific customer to enter the system will save time, expedite the flow of needed information for decision-making and greatly aid in finding customer information to complete an order.

Production needs a "Real Time" management schedule of raw material flow to the manufacturing floor. Personnel need to know when product is received for package type, and where stored in their inventory, and ready for delivery to the manufacturing floor. Systems for inputting completion of manufacturing order information or steps of manufacture for a customers order are invaluable for tracking product through the production cycle. This will aid production in knowing in "Real Time", especially for mass produced similar items used for multiple customers, how and when to reschedule their machines for change over to produce other products required for other customers.

BAR CODE TRACKING

Bar coding is an asset necessary in the manufacturing operation. Business operations and product manufacturing requires tracking in "Real Time" all of their operations. Product in process and completed goods placed into inventory need to be identified as to where their location is in the manufacturing process. The sooner they are identified on the plant floor and scanned into the computer system will it aid production control in locating and knowing where and in what stage of completion are products and processes. Bar coding on work orders and product labels work boxes of material in process, when scanned after specific operations, can easily be identified and located.

Product progress in manufacture for count, packed box weight, and location in the system when manufacturing steps are completed, inspected, and packed is important for "Real Time" material and product management. This gives the company the ability to know immediately where they are in the production cycle. Then, if a problem develops or a customer changes an order, they can react immediately as the situation requires as the information is always current in their system. Plant personnel also know where each order is in the plant and its stage of completion for each customer.

Successfully monitoring raw materials, work in process, and finished inventor can assist companies in maintaining a lean manufacturing system and a cost conservative organization. When their suppliers implement this system, a steady stream of quality, on-time materials will flow into the plant and onto the manufacturing floor. This reduces inventory, frees up additional space for profit making operations, and reduces the work load of the employees. With more companies requiring JIT (just-in-time) deliveries this is one method it can be accomplished. Six Sigma programs when implemented by customer and supplier to meet their information and material flow can save each considerable cost and improve profits for both including the customer. Performing a Kaizen may also be considered and can be successful when well planned to streamline manufacture. Be sure before doing a Kaizen that the system will actually be improved not just the assembly line straightened out.

BENCHMARKING

Establishing and using benchmarks in a Six Sigma or any quality program is required to know the state of the process or operation at time zero and how much of an improvement has been achieved both during, and at the conclusion, of a program. The use of the milestone display chart will assist in keeping a program on schedule plus visually show plant personnel their involvement in the improvement process.

Most Six Sigma improvement programs require interactions with other departments. These visible charts can then be an advantage to alert and ensure materials, equipment, and services are ready and available when required to continue the improvement process. Then when the services are in place, the benchmark reference is used as a guide to see if the goal was accomplished. Benchmarks can be used as targets of accomplished throughout the program. Goals can be evaluated during the program with work plans raised or lowered depending on the results achieved to date.

Visual indicators show the results of the improvement program. They are tracked after a benchmark starting point is determined to see if the programs goals are being achieved. Indicators in a process will show visibly which way the process is moving, is it stable, and in control. These indicators can be plotted on a control chart to visible show program progress and may guide the team to consider and evaluate other key variables if the trend is in the wrong direction. These indicators must be constantly evaluated for relevance in actual predicting the outcome of the program. When a variable change does not have an effect on an indicator, it may decrease in importance in the evaluation process. Use the indicators to keep the process improvement on course for a successful conclusion.

In business improvements, indicators may not be as obvious. Asking question of personnel using the improved process may be a good indicator. Responses such as, "why wasn't this done before", "it really saves me several redundant steps I had to do before", etc. and also "negative feedback". Comments such as, "I find this to difficult" or "the information needed is never available or on time", etc., "someone else does the same entry in another department", etc, can be a positive influence when explored for improvement.

A good Six Sigma team black belt will listen to all comments and feed back and present this information to the team. They can then discuss what positive changes can be made to eliminate the negative responses. One of the most important resources a Six Sigma team member has is, "listening to their fellow employees and operators". No one knows it all, but the workers who perform the task know more about how they perform their jobs and what could possibly be done to improve the process.

SIX SIGMA IMPACT ON IN PLACE QUALITY SYSTEMS

Six Sigma program results are best realized at companies certified to a national standard as ISO9000 and QS-9000. Any improved quality system in conjunction with ISO is an advantage as is QS-9000, automotive quality certification program and other individual ISO programs for specific organizations and businesses. Some of these other ISO programs are: AS9000, ISO14000 and others as they are developed specifically for their industry.

The better the companies quality program, the easier the application of Six Sigma methods for business and manufacturing process improvements. With the new ISO9000-2000, 34 new add on quality requirements, implementation will be even more enhanced and improved. More items are now required to be in place per the standard, not, "if the company wants to make them a requirement in their quality system." The reasons for this are the company has already made a commitment to providing a quality product and service to their customers and the sections were developed to assist in documenting their actions to meet this goal.

Being ISO9000 certified does not imply process control is in effect at the company. It does imply the companies quality system is in compliance, as the company quality system was developed to satisfy the ISO requirements but not always to meet the requirements of their customers.

The level of quality output is only as good as the quality system implemented by the company. Each company, to the same ISO standard item number, can meet it in the way they set it up in their operating quality system. As long as their system meets the items requirements it can be certified as in compliance. After certification it is expected that the company continue to remain at the same level of quality or better yet improved from what they were when audited for certification. This does not always happen and the supplier and their customer are the losers. Management must continually insist they never lower quality and continue to achieve improvements in their business and quality systems.

APPLYING SIX SIGMA TOOLS FOR CONTINUED IMPROVEMENT

All of the Six Sigma quality tools and metrics (statistical process control, charting, and analysis) can be added into the companies departments, services, and manufacturing areas. The old axiom KISS (keep it simple stupid) can be expanded to KISS OFF (keep it scientific, simple, ordered, fundamental, and fun). Quality can be fun when you are working to prevent future problems even in organizations that resist change. It is important to remember what the letters of the axiom stand for as keep it.

Scientific means, using the quality control metrics that are necessary to monitor the process.

Simple, use Standard and easily understood metrics to gather the desired data with process control charts with calculated UCL and LCL with a minimum of five data points at set intervals. These points then reduced to a single plotting point will usually be representative of recording the control of an ongoing process within reasonable control. But, as the process approaches Six Sigma, individual readings will become more important when taken at prescribed, timed internals.

Ordered, means the data was gathered in a method easy to obtain, record, and use. The data can be electronically obtained over a predetermined time or cycle period that will show the repeatability of the process or system under study.

Fundamental, means relating to the quality or manufacturing process or program under analysis. Gather only the data that is required to show the process control points and the variables that control the operation.

Fun, means just that. The job must be made to be enjoyable and all personnel who have a stake in the operation able to impart their own abilities into the program. They own the process and program and want to make it a success. Make it a win-win program for the operators, employees, management and corporate management. Keep away from threats of what will happen if it doesn't work. Be positive so creative and open thinking will always prevail. The carrot not the club as too many management personnel from the old school like to use.

These data points are usually gathered as individual readings, summed, averaged, and plotted as a process point on the control chart. Other metric data is also generated in the software program, standard deviation, range, etc., which is used to show the variance of control around the mean within the process.

PRODUCT CAPABILITY

The use of additional metrics CP, CpK, CR., etc. are used to show the capability or repeatability of the process. The mark of a true quality, process control person, is knowing what metrics to use, when, and how to interpret the collected data. Without this expertise and knowledge to interpret and see trends of normalcy in the data, it can be a hindrance when too many changes are made in the control of the process. When to leave the process alone comes from training and experience and will be imparted over time and training to the team by the black belt. This will be discussed in greater detail later in this chapter.

All processes have a degree of variability based on the process, material, equipment, wear, support services, maintenance, operator training, and plant environmental conditions, to name a few areas where process variability can enter and affect the process. Understanding how these variables affect the process when they change, how change can occur over time by outside influences and how to react when these occur so to not over control or adjust the process to maintain "Real Time" control of the process.

Ordered means gathering the data required cycle-to-cycle, or at other predetermined internals. Then reducing it quickly to useable information. The "Real Time" mode of data usage means just that. Once the cycle is over, your measurement internal for the process, the data is immediately collected, calculations performed, reported and distributed to interested personnel. The data made available for decisions on the cycles performance and trend can now be acted on and the process adjusted if necessary.

Measuring a process in a high volume and fast moving operations need the data as soon as it is available. Having to wait any appreciable amount of time, without seeing an out of control problem occurring can easily produce hundreds of rejects in a few minutes. Depending on the process, it can even break equipment and damage operations further into the process if on a progressive advancing manufacturing line. In these cases machine safeguards were installed to shut the equipment down before any serious damage could occur if an "out of control" product, process, or material variable occurred. You must consider in the setup of a process, safeguards for equipment to protect the operation and personnel

Fundamental implies what is basic to accurately measure the process variables. The closer the process gets to Six Sigma, the more difficult it is to measure process variations. This will be discussed in greater detail. Therefore, the measuring and recording equipment must be more accurate, in calibration, and sensitive to very minor changes in all the process variables. This includes any supplier in-house receiving inspection tests plus acceptance criteria for the manufacturing process. FMEA's are an excellent analysis tool along with fishbone diagrams for mapping what variables and processes may be affected by out of control variables.

FUN AND ENJOYMENT IMPLEMENTING SIX SIGMA

Quality can be made enjoyable for all personnel involved in manufacturing a quality product. Keep improvements as a challenge to the team, the goal attainable, and production aware of the efforts the team is making to make their job better, easier, and more rewarding to them and their customers. When the fun goes out of the job, it is time to consider a new profession and leave. Our human thought processes work best on positive inputs. Negativism should never be permitted to enter the Six Sigma teams thought process. The team must also be allowed to make the decisions deemed necessary to reach their goal, their way. Micro-management is an evil to be avoided by upper management. They select the Six Sigma programs not rule them.

TESTING

To test processes nearing or in Six Sigma compliance requires a more accurate approach using methods with high accuracy and repeatability to measure any slight variance in a process. This is illustrated in Figure 1, for test parameters of the process. The following metric example will illustrate this technique.

METRIC EXAMPLE

As the process nears Six Sigma control, the process may drift ± 1.5 sigma from the targeted process mean value as calculated for the control limits.

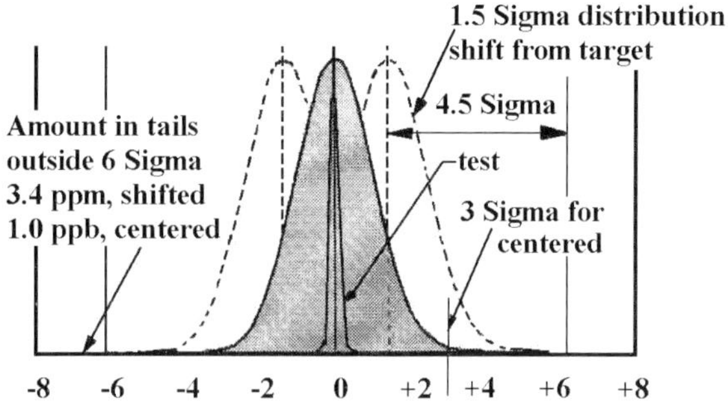

Figure 1. Six Sigma processes.
(Adapted from reference [2])

When this occurs, the tails of the bell curve normal distribution will be at the extreme end of one side of the ±3-sigma normal process limits. This means for Six Sigma control, only 1.0 part per billion is outside Six Sigma for a process centered on the mean. For a drift of 1.5 sigma from the mean for Six Sigma control, 3.4 parts per million will be outside the Six Sigma limit.

If the process drifts, 1.5 sigma from the mean the process control charts monitoring the process will note the shift. It is now up to the six Sigma team to determine the cause or reasons for the shift. The same tools of analysis, DOE, FMEA, fishbone, CpK software, and reviewing documentation of raw material tests, machine maintenance and wear factors and other documented information is used to bring the process into or reaching Six Sigma control.

Problem solving skills are now employed to review the data, collect new data, and even perform high/low value analysis use DOE if the process still cannot reach or maintain continuous Six Sigma process control. In all situations try to use the most logical problem solving solution and analysis. This may be a review of the FMEA, fishbone of the process, or problem solution logbook for the machine and process. Then if the problem still seems unsolvable, the DOE should be considered.

The solution may be harder to find since the drift caused by a variable slightly out of control may be very minor. But, it may be enough to affect the process and it may not occur continuously for a trend to develop and identify it as the primary root cause of the problem.

Insure all tests, process, maintenance, and other data collected during the analysis are in your systems computer database. This will save the team precious time when they analyze the data from all the sources in the process as a group.

AN EXAMPLE OF PROBLEM SOLVING

A manufacturer of plastic covered dinnerware storage containers over a succession of days producing the same part in polyethylene noticed the following occurrences. When a lot of material change occurred for several machines running polyethylene, defects were immediately produced for all these products. Adjustments in the process variables could not maintain good parts on successive cycles. The process was holding very tight prior to the last material change.

To attempt a solution to this problem the team looked for anomalies in the data that could not produce parts to specifications. Data was collected from the process in "Real Time" indicated why the process shifted and corrections in the process changed. Also, data was collected to monitor if a material or environmental change was noted in the data. Data collected was lot data now being used, have plant/equipment changes occurred, or any other change noted or a new operator making a variable adjustment. All personnel were questioned if any variable control to the process was changed and if all the data was documented. This is when the team's problem analysis training paid off, seeking the elusive variable change or changes.

Discussions with their material supplier and reviewing incoming lot test data found the root cause was found to be slight shifts in the molecular weight distribution for incoming material. The slight variance was noted from incoming material tests as shown in Figure 2, for incoming tests of materials as noted on the curves. Retests of material in the system for molecular weight variance showed a major shift from the target mean for material tested now in the processes material feed system.

An overlay made from the molecular weight curve for an acceptable lot of material, indicated a major shift from the molecular weight curve for material tested. Additional discussions with the supplier indicated the material certification for the last lot shipped was typical lot molecular

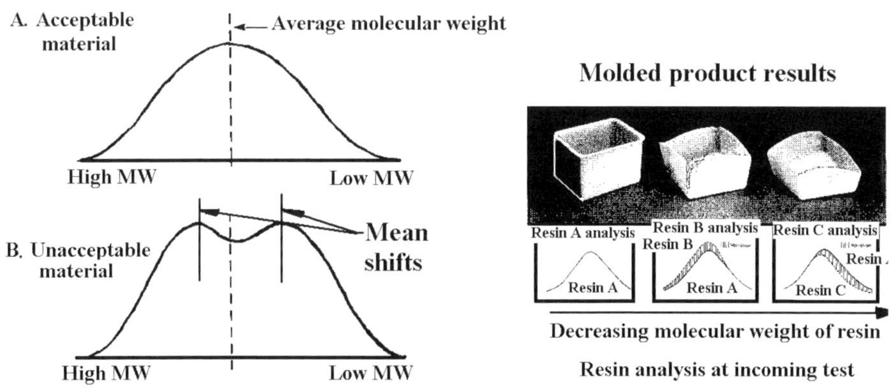

Figure 2. Incoming material testing. Effects of molecular shifts on product quality for low density polyethylene resin variances.
(Courtesy of Millipore Corp. All rights reserved)

weight data, not representative of the material lot shipped. This was not, as requested, specific lot data for molecular weight distribution for each and every lot shipped from the material supplier. The problem turned out to be two fold, incoming inspection, accepted this as specific lot data in error, and did not test material samples to verify the data was in specification.

This resulted in the material in one storage silo, consisting of 80,000 pounds was very suspect. This material was shipped to the customer in separate hopper trucks consisting of 40,000 pounds per load. A test of material in a second silo met processing and material requirements. The material feed was rerouted to the good material silo for feeding the machines and after extensive purging the process went back on line producing good products. A call was made to the supplier to discuss the problem and to seek corrective action and resolution.

PROCESS-CAPABILITY

Before determining if your company and all your processes need to be in Six Sigma Process Capability it is important to discuss what you can get for your efforts and evaluating your processes for how you get them to Six Sigma. What is a capable process? Some quality assurance experts define a capable process as one having and maintaining a CpK index of at least 1.33. This equates to a maximum defect rate of 63 ppm while others say a maximum of 3.4 ppm is the true capability process meaning of Six Sigma control.

The difference between your company's quality processing goal of a CpK of 1.33 or a CpK of 1.5 (Six Sigma) is only an additional 60 defects. If your production run is large, more than 20,000 parts, you can anticipate random defects for a process of CpK just meeting 1.33. When your production runs are greater than 200,000 and above, even if your process is within Six Sigma or a CpK of 1.5, you will still experience random defects due to random variability of the material, equipment, and processing variables. Therefore, for small runs that many companies experience in their daily manufacturing, it is more important to know the methodology of attaining an acceptable process capability of CpK 1.33 or a CpK of 1.5 than debating over which CpK value should be selected for your companies manufacturing goal.

No matter what run size and CpK goal, the manufacturing department along with quality assurance will be working together to improving

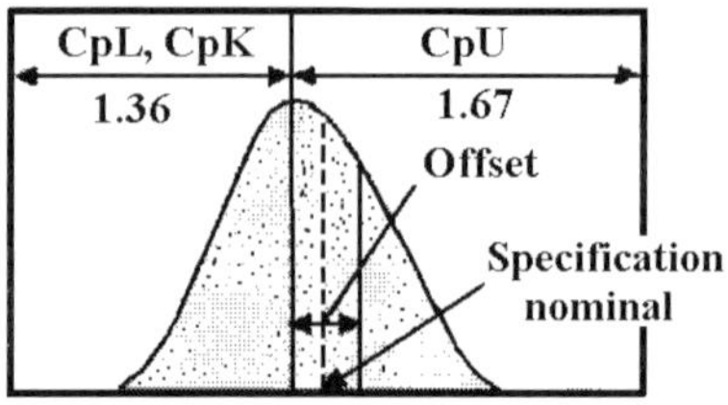

Figure 3. In process-capability study of cycle adjustments the Gaussian curve indicates a CpK of 1.36.
(Adapted from reference [3])

processes. These will be processes that are substantially below your selected goal or daily, ongoing processes that have had their CpK dip below your goal, due to a variable in the process going out of control.

An example of process capability can best explain this concept. A molding shop producing a product historically for years has a specified minimum CpK on the operation of 1.33 that is attainable and repeatable each time the job is run on the same machine. This is shown in Figure 3, for a process-capability study of a process running at a speed of 60 seconds overall cycle time. The process and cycle, now running at an overall mold open to mold open of 60 seconds, is set to produce the required number of parts in a specified time period. This was consistent with the machinery demands and utilization of the current customer base. The customer was very satisfied with the quality of the products shipped weekly.

Manufacturing engineering and quality assurance performed a process capability sturdy on this job and others so they would know if the process and other variables were as capable as possible after having the machine overhauled. The floor supervisor and manufacturing engineer were satisfied with the machines cycle and operation. The following data on pin length for a key customer product characteristic was evaluated on parts from a four cavity tool. A part from each cavity was initially evaluated to show the balanced runner mold was capable of producing repeatable parts cycle to cycle from each cavity. After this was confirmed, only one part from a specific cavity was measured for the process study. After 30 cycles the data points were accumulated and used for the analysis.

Data:

- Specification = 1.250 in. ± 0.005 in.
- Sigma = 0.0011 in.

• $\mu = 1.2495$

The sigma (process standard deviation) and μ (process mean value) were based on measurements taken from four individual cavities. Five parts were collected from each cavity for five cycles. When these were confirmed to be within specification, the single cavity was selected, cavity one, and a total of 30 parts were collected per hour and measured. These parts were considered representative of the production process.

Process-capability formulas were used to determine the process had a CpK = 1.36. As a result of the study the manufacturing process was deemed satisfactory and monitored as was typical during the manufacturing cycle. The process was considered to be capable and was in statistical control. This information was confirmed by the customer since product defects were not reported.

The customer requested additional product and a new manufacturing engineer was assigned the task of meeting this demand from sales. The initial process decision made was to reduce the cycle time by 50%, or 30 seconds, to meet the new order volume. The shop floor supervisor believed this was too great a decrease even though the material reacted, parts were ejected looking good, successfully. No visual change except part weight was lower by three tenths of a gram, which was being used as the process stability indicator.

On the first sample taken after the increased cycle time was set, the R chart indicated that the process was out of control, as confirmed by the change, lower part weight. The cycle time of 30 seconds was increased 50% more to 45 seconds and on the third sampling group the process yielded an R-chart reject. This is shown in Figure 4, on the range control chart.

With time running out and material being scrapped the engineer with help from quality assurance ran a DOE (design of experiments) to determine the

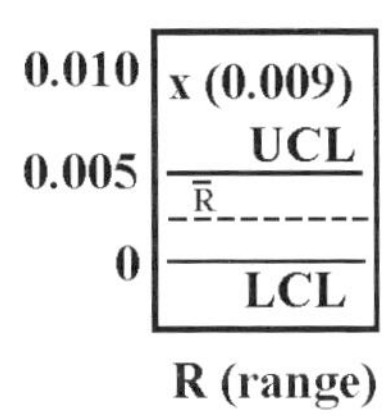

Figure 4. Cycle time change sends process out of control. (Adapted from reference [3])

critical factors in the process. Not surprisingly, cycle time was found to be the most important factor along with screw speed that affected melt temperature and setup time in the mold cavity. These were the critical variables and it was decided that mold cavity temperature would not be adjusted to assist in reducing the cycle because of the critical pin length.

Since there had not been any problem running at a 60 second cycle, the manufacturing engineer wanted to characterize the process at the 45 second cycle to see if the R-chart rejection was a random event and the cycle not yet in equilibrium. Quality assurance agreed to this request with the restriction that all product would be inspected 100% until the cycle stabilized and proved satisfactory for repeatability of product manufacture. Samples were taken as before until sufficient data points were obtained.

Analysis showed the mean had increased to 0.2497, and the standard deviation had increased to 0.0022. After calculating the process capabilities, the manufacturing engineer calculated a Cp of only 0.81 and realized that even with perfect centering, the process could not produce at a sufficient quality level to meet the company and customer requirement of CpK of 1.33. As a result the cycle was reset to 60 seconds, equilibrium again obtained, and the CpK again recalculated. The control charts again resembled those of the initial process and the manufacturing process extended into the second shift to meet the customers quantity requirements.

Process-capability is a very valuable tool and goes along with the "do it right the first time" manufacturing philosophy. When correctly used it will aid in the selection of equipment, materials, speeds, and other variables that can affect on going product quality. The process-capability formulas and language for calculating the CpK information is listed.

$Cp = USL - LSL/6$ sigma

Cp = Inherent process-capability index

USL = Upper specification limit

LSL = Lower specification limit

Sigma = Process standard deviation (obtained from a representative, random sample of at least 20 parts).

$CpL = \mu - LSL/3$ sigma

CpL = Lower process-capability index

μ = Process mean (obtained from a representative, random sample of at least 20 parts).

CpU = USL − μ/3 sigma

CpU = Upper process-capability index

CpK = min {CpU, CpL}

CpK = Process-capability index

Notes:

If μ is at the nominal dimension, Cp = CpK

CpK is always equal to or less than Cp.

PROCESS-CAPABILITY INDEX

The PCI (process-capability index) (PCI = upper specification minus lower specification limit divided by Six Sigma product standard deviation) was calculated and new data collected. The PCI index calculated during the process out of control went from 0.5 up to 1.33, a process in control, but not yet within Six Sigma values. The incoming test laboratory was then required to draw material samples for molecular weight testing every two hours, from the primary feed system (at the machine feed hopper) to verify molecular weight distribution. As a result, any minor variations in material would be adjusted at the machine with screw RPM's (revolutions per minute) speed adjustment. The speed was monitored to know what revolutions were required to keep the process in control even if molecular weight varied ever so slight. This control required on-time testing and feed back to the setup/process technician who would make adjustments in screw speed when required. The quality department, on their speed control data recorder, verified the change.

After this procedure was implemented and shown to work, PCI values were calculated showing even better improvement with a PCI of 2.0, being attained showing the process was in Six Sigma capability. The testing capability index (TCI = upper specification limit minus lower specification limit divided by six sigma test standard deviation) now remained in control.

$$TCI = (USL - LSL)/6\bar{\sigma}$$

With the process in Six Sigma control, production wanted continual feedback to confirm nothing changed in the process. Team members now required more precise and more frequent testing to detect any slight variable change in their processes Six Sigma operation.

This raised a new question and decision point. How precise does a test system have to be when manufacturing reaches Six Sigma? The decision can affect capital expenditures, planning in data collection, and feedback to the manufacturing floor in "Real Time." The decision required evaluating the cost of a precise and unbiased test system.

If the investment is low for achieving the control, for long running, high end user product quality, then make the investment. But, if the price and technology prohibit achieving the needed precision, then more information is needed to assist the company in making the right testing and capital expenditure decisions.

Testing the process involves answering questions that are dependent on the level of capability of the process. When approaching Six Sigma manufacturing capability, process variable testing requirements involve more demanding and sophisticated, exact, levels of control and testing.

There are three types of testing which must be considered for adoption to accurately track and control the process. Also, remember each customer have their own level of acceptable product quality they will use in selecting a supplier of their products. They will want what they contracted for and customer needs will vary from precise medical and electronic products to less precise products as plastic trash containers and throwaway drinking cups.

TYPES OF TESTING, APPRAISAL, CONFIRMATION AND CHARACTERIZATION

Appraisal Testing

Appraisal testing determines how well a product meets the customer's needs for typically form, fit, and function. This is the method used for testing when manufacturing-process variables are unknown and must be established to make acceptable products. When the products characteristics satisfy the customer's expectations and specifications it is deemed satisfactory. It is then released to the marketplace regardless of the control requirements of the manufacturing process used to produce the product.

As the company manufacturing process moves to Six Sigma to reduce defects, improve output, and reduce cost, appraisal testing should be replaced with confirmation and characterizing testing. These forms of testing move from, "Is the product conforming to specifications?" to "Is the manufacturing process causing the product to change?"

CONFIRMATION TESTING

This testing method determines if there are changes in the process. It is used when manufacturing variables are known and the process is in statistical process control. This is when "Real Time" testing of product is critical to output quality. Product dimensional data and statistics are gathered as soon as possible so that process variable adjustments can be made in "Real Time" control.

Each process will have an effect on the product. Verify when the product will reach equilibrium or stability before taking a measurement. Since all plastic products require a specific amount of time to equalize to their final dimensions, it was a habit to "prejudge", the amount of post mold shrinkage. This is where the final dimensions would stabilizer after the part was just removed from the tool in a hot and unstable (still shrinking) product condition. A method for determining, hot as molded dimensions versus cold after typically a 24 hour cool down period, were developed to achieve an estimated material shrinkage rate to meet final part stabilized dimensions.

Parts can be measured just out of the tool by using go/no go gages or fixtures based on hot dimensions that on cool down are known to achieve specific dimensions after a predetermined period of time and will meet the customers requirements. These dimensions can be used to estimate variable adjustments in "Real Time". This requires more preplanning and may or may not be required depending on the customers specifications. I offer this as consideration when initially adjusting the manufacturing process to get it into acceptable product control for defect reduction at the start of manufacture. By using the products hot dimensions a skilled technician can adjust the process in "Real Time" when they know what the products final dimensions will be after cooled down.

Then as the manufacturing process improves and remains in process control, product testing and measurement becomes less critical. Control of the process variables ensures the products variables remain in control and lot-to-lot repeatability and characteristics have consistency. Monitoring and controlling manufacturing variables in "Real Time" will eventually lead to confirming product variables are achieved. This testing along with process data will reaffirm nothing unforeseen has occurred in the manufacturing process.

PROCESS TEST CONTROL (DESTRUCTION TEST)

A destructive test can be run that will illustrate if the process cycle variables are in control. The test is run on a critical part section, such as a known high stress point. This can be the falling dart drop weight test. This test can be run immediately after a part is ejected hot from the tool or after it has cooled to room temperature. A test that can be used especially where melt fronts meet, using an impact force at a critical section, can be used to determined if the part survives, nothing has changed in the processes variables. A process change this test can detect that may not be recognized by the data output is, for example, a blocked mold vent or cooled melt fronts meeting causing a cold weld line, melt temperature too high causing a breakdown in the resins physical properties, toughness, heater band burn out or out of control (too hot), injection speed too fast, shear heating the resin beyond it's process range, loss of mold temperature control, or a valving problem in the injection molding machine control system. Any one of these process changes can cause product quality problems. The drop weight test can detect these changes when established at the start of production.

To use drop weight or impact testing for analysis, begin by selecting a tool product cavity that exhibits a marginal response to pass/fail results in dimensions or toughness, if this is a requirement for the product in service. Select a section on the part that is critical to its use and where a process change would show a lack of toughness in the product. Then, when the part passes, you know all parts from the cycle are good. Always test the part from the same part cavity on an established interval as every fifty cycles. Similar test can be conducted for other materials and products. The type of test is dependent on the product, material, and end use requirements of the product.

CHARACTERIZATION TESTING

This testing is used to describe the effects of process changes on a process or product. This is the most precise form of testing which measures the smallest changes in a control variable.

Characterization test examines a manufacturing process by characterizing the effects on a product for a known set of process variables, and any changes occurring in the process. This testing method assumes something in the process has changed. This requires sufficient testing to characterize the process variables that changed the product and to determine it's near term effects on the product.

Characterization testing will require more stringent variable analysis using quality methodology to collect the more subtle variable changes to attain a higher degree of process precision. This will involve more sample measurements using more specific and greater accuracy measuring instruments. Always use highly trained inspectors and technicians to ensure the data collected is as accurate as possible. Use the most specific sampling methodologies to obtain the highest accuracy in your data as the key to success.

In a mature manufacturing process-Six Sigma- testing has two requirements; confirmation and characterization.

CONFIRMATION

To examine for a process change through product examination and characterization is qualifying the effect of a process change on the product.

Using TCI (testing capability index) confirmed a Six Sigma process was reached with a value of 2.0. But, the TCI metric is not adequate to quantify the effectiveness of the test system. Therefore, a more adequate test method is needed to evaluate and maintain higher process precision control.

VARIANCE RATIO

The variance ratio, a characterization test, is capable of qualifying test system effectiveness as manufacturing processes improve to reach six Sigma capability.

Test variance can slant and hide subtle process changes. It is important to know and compare the testing variance to the observed variance and not to tie them to specifications.

The CVR (characterization variance ratio, (Cv)) compares the variance of the observed product test results to the variance of the test system.

$Cv = s^2 \text{ observed}/s^2 \text{ test}$

The variance of the test system is calculated from the test control products. This is illustrated by the use of the right triangle for characterization variance in Figure 5. The observed variation is related to the test system variation component and the product variation component:

$C^2 = A^2 + B^2$

or observed variance equals test variance plus product variance.

In our example, molecular weight distribution was determined through continuous testing the PCI (product capability index) increased from 1.0 to 1.33 and then with increased supplier and incoming periodic testing versus their regular, one time test, the PCI rose to 2.0. During this period the TCI,

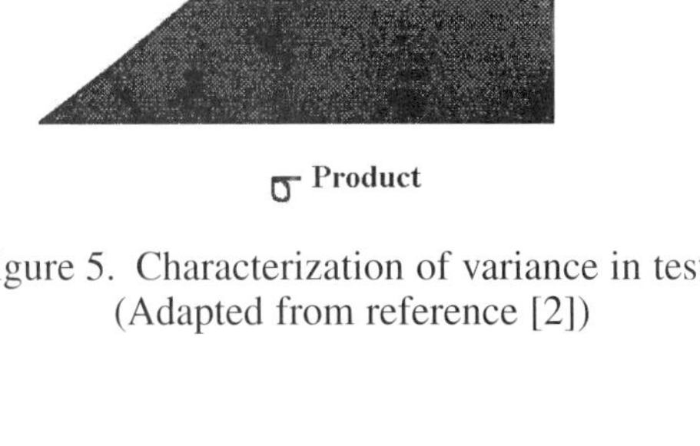

Figure 5. Characterization of variance in tests. (Adapted from reference [2])

(test capability index) remained at 4.0 with a standard deviation equal to 0.833 with the material specification remaining unchanged.

At this time a team member recommended a change in how the output curve data was plotted and used to compare to the ideal or suppliers specification. A finer plotting pen point was used for the output plot of periodic tests.

This resulted in an overlay transparency being used, overlaying, using the same scale the supplier supplied, chart as a direct test comparison to the shipped material. This allowed a more precise visual chart to evaluate any shift from the specification. This establishes a shared reference chart both used to determine a reference when a shift in molecular weight was large enough to require a process variable adjustment. Before this comparison chart was made, too many machine adjustments were made which caused more problems with the process than were solved. We also had our differential scanning calorimeter calibrated at the same company as our resin supplier's calorimeter. This helped to ensure greater confidence in our test comparison and reduced the risk of making a poor process adjustment when necessary.

A table was constructed to map the process improvements and serve as a guide to test variance capability. This data is shown in Table 1. The data

Table 1. Process Improvement.

Item	Original	Continuous Improvement	Breakthrough
PCI	1.0	1.33	2.0
TCI	4.0	4.0	4.0
Observed standard deviation (sigma)	3.333	2.500	1.678
Observed variance ($sigma^2$)	11.1088	6.2500	2.7778
Test system variance ($sigma^2$)	0.6944	0.6944	0.6944
Cv	16.0	9.0	4.0

(Adapted from reference [2]).

reflected as improvements in manufacturing increased the CVI (characterization variance index) was decreasing because the periodic testing of resin remained stable at a TCI (test capability index) of 4.0.

It was noted a TCI of 4.0 did not reflect the changing balance between testing and observed variance. As a result, the Cv showed that the test system was not keeping up to the manufacturing process improvements.

Since Cv is only a percentage of the test system variance compared to the total observed variance as process improvement increases, product variance decreases. Therefore, a point is reached when the test standard deviation dominates the standard deviation of product measurement.

When this occurs, it becomes increasingly difficult to detect any further improvements in the manufacturing process. Simply stated "Improvements must be made in the measurement process of the process." Donald J. Wheeler first stated this in "Evaluating the Measurement Process" (SPC Press, Knoxville, TN, 1989).

Wheeler states for example, "a Cv of 10 equates to 10% of the observed variance being attributed to the variance of the test system". Therefore, "a Cv of 5 equates to 20% of the observed variance being attributed to the variance of the test system".

As a result Wheeler suggested a more accurate and complex metric called the DR (discrimination ratio)

$$DR = *0* \qquad \{[2\ s^2\ \text{observed}/s^2\ \text{test}] - 1\}.$$

He stated, "For simple measurements, it might be well to work on the measurement process when the DR falls below four or so". A DR of four is equivalent to a Cv ratio of 8.5.

This is shown in Table 2, with examples of Cv and DR" with examples of Cv and DR ratios and their resulting interpretation.

THE RISK OF MAKING A BAD DECISION

In all process improvements, the probability of making a bad decision, denoted as the Risk Factor, in the process must be minimized. When moving a process from three to Six Sigma control, you want the lowest probability of making a bad decision as even a minor change can have major affects on the control of the process.

Therefore, a TCI, (test capability index) value that reduces this bad decision probability is recommended for use in evaluating control of the

Table 2. Interpreting Cv and DR.

Cv	Wheeler's DR	Total Observed Variance	Test System Variance	Percent Test Variance to Observed Variance
6.24	6.24	10	0.5	5
4.35	4.35	10	1.0	10
4.0	4.0	10	1.2	12
3.0	3.00	10	2.0	20
2.64	2.64	10	5.0	50

(Adapted from reference [2]).

process. This is shown in Figure 6, as the process approaches Six Sigma control, the bad decision probability can be reduced. Selecting a TCI of value 2, as shown, results in the Risk Factor being almost zero.

When a test result is developed, a statistical inference is made on a population of test samples from which the statistic is calculated. As an example, if the outside diameter of a hypodermic syringe tip, where the

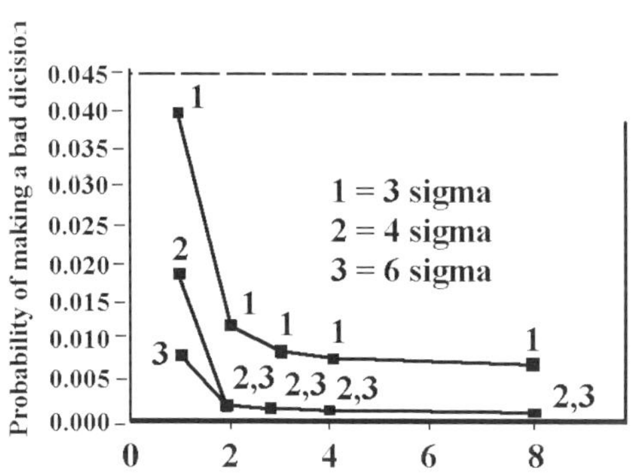

Figure 6. Decisions versus TCI at various sigma levels.
(Adapted from reference [2])

needle is attached using a twisting press fit with an outside diameter specification of 0.077 ± 0.001″ and it actually comes out 0.075 ± 0.002″, how can we quantify the error of making statements of meeting the specification based on the hypodermic population given measurement sample statistics?

The answer, that can always be debated is, an OD (outside diameter) tip tolerance of 0.001″ and product statistics show an average of 0.002; after post mold shrinkage, how can you quantify the error of making statements of syringe OD in the population versus sample measurement statistics?

Referring to Figure 6, TCI values are plotted from 1 to 10 against the probability of making a bad decision given a test value at 3, 4, and 6 sigma manufacturing process capabilities. The graph shows as manufacturing capability goes from 3 to 6 sigma, the probability of making a bad decision decreases as the system is in, and maintained in better process control. This is reinforced when TCI values are two and greater. With TCI values greater than two, the probability of each capability becomes asymptotic, an almost horizontal straight line on the base line scale. With four sigma and higher process control, there are almost no bad decision probabilities which straight lines out at a TCI of 3 for 4-sigma manufacturing. As a result a TCI value of 2 should be adequate for confirmation testing of the process.

A bad decision is the summation of two probabilities. Telling the customer the average syringe OD is not 0.001: (when it really is) versus telling him it is, 0.001 (when it really is not). These two probabilities when added is the probability of making a bad decision. They are categorized as producer and consumer risk.

The decision of what specific data collection information the customer receives depends on several factors:

1. Plastics have specific post mold shrinkage dependent on section thickness, gate location, and mold cavity temperature plus others.
2. Tool and material temperature can affect part dimensions and post mold shrinkage and final product dimensions.
3. Shrinkage values are dependent on how long after manufacture the samples were measured, plastics stabilize at different rates.
4. Who performed the measurements, with what instrument, and was it calibrated.
5. Were the specifications for measurement specified in the contract, when parts are to be measured, and where on the part for only critical dimensions.

6. Did the customer agree on an average of the measurements or must all recordings be within the specifications.
7. How critical are the deviations to the customer, either ±.
8. How large a population was measured to arrive at the end use measurement sent to the customer, AQL specification limit specified and what is the average inside diameter of the needle boss required for a twist press fit on the tip of the syringe?

Items 3 to 8 should have been addressed and answered before the program began by engineering, sales, tooling, production, and quality assurance. Education of the customer in material, tooling, and manufacturing variables is necessary to determine if the tolerances specified are critical to the customers product. In this example, the needle attachment is a press and twist fit to the syringe tip. Is the seal and attachment of syringe and needle adversely affected by this variance? It must be determined early in the program as it is very critical for dispensing the correct dosage every time to a patient. After the fact is too costly for each partner as time, tool changes, process adjustments, and liability are all factors to consider. In good business negotiations, answers using the contract check list and other check lists must be obtained to meet the customers specification and reduce Risk of Business and product variances.

TRACKING MANUFACTURING USING EXISTING PROCESS INDICATORS

When manufacturing nears or is in Six Sigma control, are current process indicators sufficient? The answer is yes; otherwise you would not have achieved this tight process and product control.

New and/or tighter test verification methods should now be used to monitor and maintain this control. Also, an item not always recognized, is factoring in a wear rate factor for the process. Maintaining tooling, cooling and machine maintenance is very important, as simple as replacing filters for air, water, and hydraulic systems on a more frequent internal. Wear in the tooling and processing equipment will also affect the control parameters of the process. Gate and check ring wear, contamination of the machines hydraulic oil cooling radiator, and tool cooling channels and piping can cause, often unrecognizable changes in the system, which will affect the process variables of your manufacturing system. This will take longer to occur but once wear of equipment and temperature control of the operation is lost, so will Six Sigma control of the process and product.

Monitor these systems and keep them clean and replace worn parts before they cause processing problems. When a tool is pulled, bring it back to as close to new condition as possibe. Tool maintenance is less costly than tool down time for repair, lost production, missed shipments, and lost customers can result.

These simple preventative operations are indicators of performance capability and if not maintained in good condition can quickly affect the process. These are cost factors that must be considered in the cost of quality problem prevention. When carried out before manufacturing begins it will result in keeping the process in Total Quality (Six Sigma) process control

Also, consider using only the best machine controls and sensors to ensure the output data is controlled and accurate as possible. These were identified by Denes B. Hunkar as his "Gold Standard" process control output recorders he used for complete and repeatable accuracy in his work on CpK and the repeatability of the manufacturing cycle for injection molding. If unsure of the controls sensitivity check with your suppliers who can come in and monitor the sensors output to known standards. Not all machine suppliers install the best controls on their machines, as this is often the buyer's decision at time of purchase. Verify the degree of control your machine and equipment can achieve before attempting a program for process improvement. Also, ensure the repeatability of operations by conducting CP and CpK analysis.

VERIFICATION OF SIX SIGMA MANUFACTURING CAPABILITY

Begin by qualifying all your machine and process data variable outputs. These are used to measure product and test variance so the rewards of the improved process will be verified with increased output, quality, and lower cost. Problems will never cease but be assured if you create the best manufacturing and monitoring system to reduce risk in the process, the rewards to the company will be very evident.

Calibration and verification of new and existing equipment and systems before their use in a process must now be a requirement for typical manufacturing procedure. Training of personnel will continually reinforce these "must do" items. The cost in process and quality loss is to high not to maintain the best systems possible with continued work to keep and improve all systems, business and manufacturing in your company.

The black belts job will never be completed. After the Six Sigma methods and processes are embedded in the company culture, they must continue training new and existing personnel in these quality methods and philosophy. This will ensure it never becomes routine but a continued movement to improve all company operations. The black belt then continues to be the leader for continued process improvements for new products and services. Their results will culminate in continued profitability for the company and growth in their business. This is proven daily by working with your suppliers as shown by General Electric and other major companies that results in cost savings, having a better-qualified supplier, and growth for each other's business and customer base.

The last key reward to be gained from Six Sigma is the company infrastructure always striving for improved quality. This task will never end. It is a challenge for the future to always be, "the best you can be".

WHAT REALLY IS SIX SIGMA AND IMPROVED PROCESS CONTROL

Six Sigma methodology and practices are: increasing business and manufacturing operations to their highest state of operation by reducing personnel and manufacturing problems to their lowest recordable level. This is accomplished by using the existing quality tools; statistics and metrics to measurer and analyzes their operations to reduce and eliminate errors resulting in increased costs to the company in lost business and customers.

Companies must build their business around their customer's needs and requirements. This includes any external activity requirements as well as their own internal-process improvement programs of operation. A three sigma manufacturing operation is costly with 67,000 defects per 1,000,000 parts produced. Reducing this to four sigma, 6,220 per million defects is now fairly typical for manufacturing but still to high a number as shown in Table 3, yields results in 3.4 defects per million, a difficult but obtainable goal.

It takes a dedicated non-micro management team using a black belt improvement team to achieve this result by employing a strong and resolute

Table 3. Significance of Sigma on Results.

Sigma numbers	Defects per million
1.5 σ	500,000
2.0 σ	308,300
2.5 σ	158,650
3.0 σ	67,000
3.5 σ	22,700
4.0 σ	6,220
4.5 σ	1,350
5.0 σ	233
5.5 σ	32
6.0 σ	3.4

Six sigma is better than twice as good as three sigma, it is close to 20,000 times better.
(Adapted from reference [4]).

muscular approach to business improvements. The company "culture" must be managed to be supportive in the drive to quality improvement as illustrated in Figure 7.

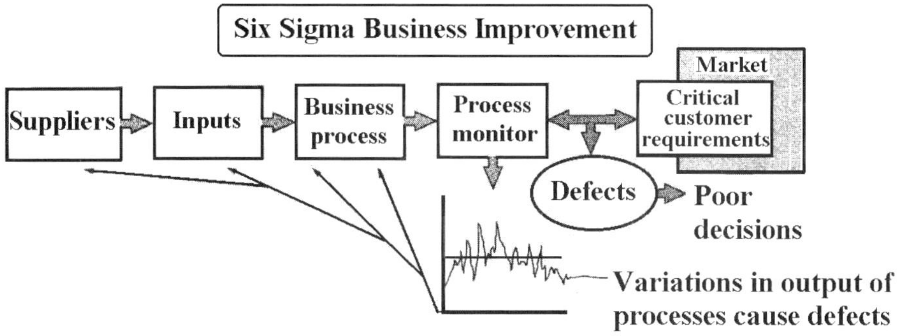

Figure 7. Six Sigma business improvements root-cause analysis of defects. (Adapted from reference [1])

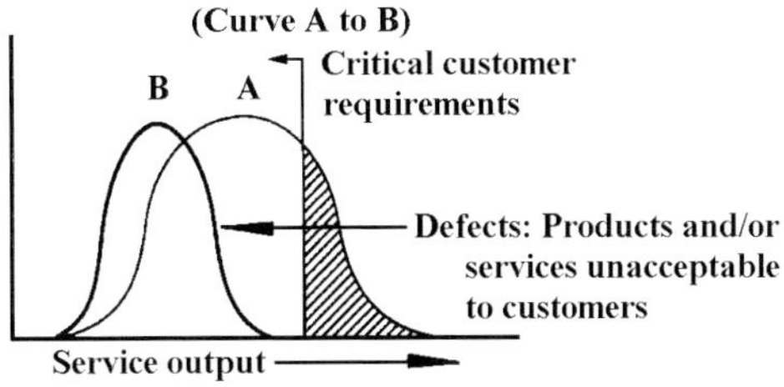

Figure 8. Six Sigma can reduce variation and move product and/or service outputs permanently inside customer requirements.
(Adapted from reference [1])

SIX SIGMA METHODOLOGY

The goal of Six Sigma is to align a companies business operations strategically to its marketplace (customers) and deliver real goods and service improvements (cost savings for each). The objective improvements must move the business and goods attributes within the customer's specifications by reducing defects as shown in Figure 8.

How this is done for any operation is shown in Figure 9. General Electric, Plastics Division, developed their plan of Six Sigma operations that have proven successful in their Six Sigma improvement programs. The quality metrics were put in place for analysis and information gathering, the only difference is what they call them. At General Electric they use CTQ (critical to quality) versus the ASQ (American Society for Quality), standard quality methods using QFD, (quality function development) and many others, FMEA, DOE, CP, CpK, fishbone, process control charting, etc. which have been in use for many years.

The intent of the program is to gather the customer and company critical improvement information. This information is then measured, analyzed, and evaluated so good business decisions can be made to improve the operation by reducing problems and defects.

Throughout each step of the Six Sigma process, tollgates, and evaluation points are established on time line charts. These show where check and balance points are needed to be established to ensure only quantified results were achieved before proceeding onto the next step of the improvement program. This assists in ensuring only the best practices are used consistently and all variables have been considered and if they are capable

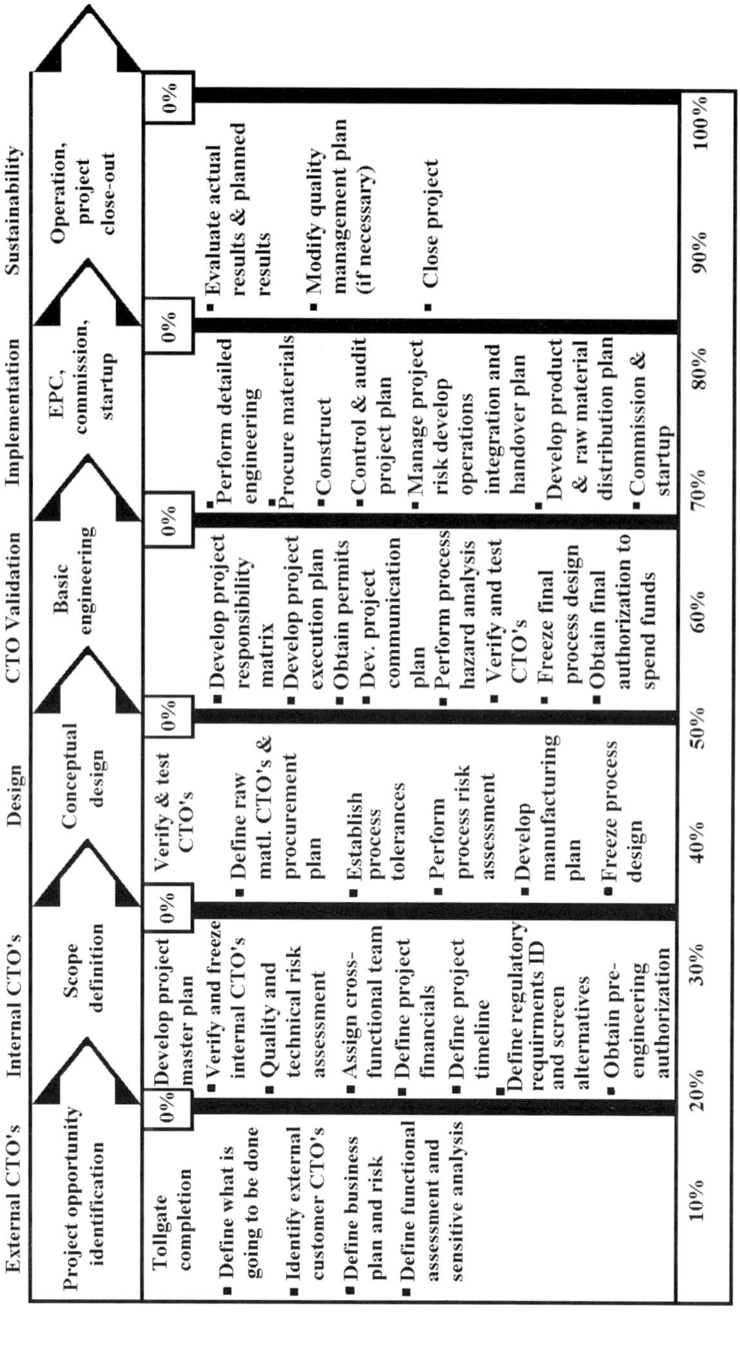

Figure 9. Designing for Six Sigma capability.
(Adapted from reference [4])

of being maintained in control. This should include both internal and supplier services and products.

IMPLEMENTING THE SIX SIGMA PROCESS

If a supplier of raw materials, internal or external, cannot guarantee the quality required, this would be a decision point in the process asking the question, "how can it be accomplished before proceeding?" For example, as in any process if the design is questionable, the material varies, the tooling is not adequate for control of dimensions, then how can the best-controlled operation ever be able to achieve acceptable products, cycle-to-cycle? The answer is; "it will be impossible", as too many external, uncontrolled variables will control the products quality, not the manufacturing process. This occurs too often because the organization was not willing or have the management guidance and procedures in-place to prevent problems from happening. If this procedure were in place at the start of a program, the tollgate would have stopped further progress possibly at the initial product design review for the product or process. Without the tollgate this would have resulted in only three sigma or less quality being possible for the process.

This problem was experienced when consulting with an OEM for automotive headlamp bezels and lenses. The suppliers manufacturing system was found to be out of control. It was discovered when evaluating their manufacturing capability for CpK on two of their eighteen injection-molding machines, we were only able to obtain a CpK value above 0.03, for one machine, and essentially zero, 0.00 for the second. They were continuously manufacturing products for their customer but their quality and manufacturing process control was extremely out of control. This company was ISO9001 certified, but their quality procedures and work instructions were lacking in maintaining the quality control required for their customers products.

This implied they were not following or using procedures and work instructions that were capable of producing a high repeatable quality product, cycle-to-cycle. This was a management problem that permeated through the entire company resulting in very high defect rates. This resulted in having to produce more parts to just meet their customer specifications using 100% inspection. As a result deliveries consistently fell behind, secondary operations such as plating, produced unrecoverable products, resulting in scrapped products and as a result continued to lose money. To solve their problem they implemented new management with strong belief in Six Sigma process control and the ability to implement this quality system.

IMPLEMENTING THE SIX SIGMA IMPROVEMENT PROGRAM

Procedure 1. Committed Management Leadership

Leaders for implementing Six Sigma need the ability to make knowledgeable and tough decisions affecting the future success of their business. This is defined as "edge" in a book *The Leadership Engine*, written by Noel Ticky and Eli Cohen.[1] Two leaders of their industry possessing this "edge" were Jack Welch of General Electric Company and Larry Bossidy of Honeywell.

Leadership is not only delegating but also delegating responsibility to other committed, responsible, and knowledgeable leaders that have the "edge", to lead the Six Sigma program forward. This includes driving the Six Sigma projects and teaching their other managers these same hands-on methods. This leadership stairway is illustrated in Figure 10, from leadership commitment to incentives for all.

Procedure 2. Integrating Using Existing Initiatives, Business Strategy and Key Performance Measures

Management integrates Six Sigma methodology into all facets of the company using Six Sigma methods to amplify in-place methods, business strategies, and performance metrics. Honeywell (ex-Allied Signal) implemented these outside of manufacture into business/office operations. It has improved the product development process by moving new products faster to market. General Electric spearheaded Six Sigma into their globalization companies at GE Capital Services, and other GE financial and service operations. It is lead by a senior business leader, their top champion, to solve critical business problems, known to exist, to achieve their financial targets.

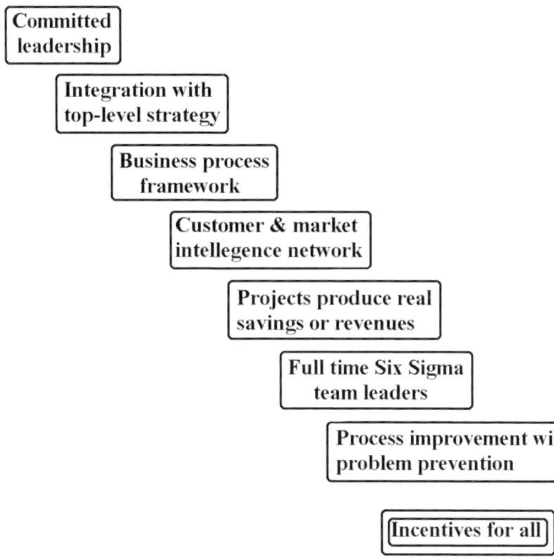

Figure 10. Six Sigma implementation stairs for success. (Adapted from reference [1])

GE ties bonus enumeration to Six Sigma success with bottom line financial improvements for their top business leaders. General Electric is dedicated to having all of their employees obtain a minimum of green belt capability at all of their operations, both business, manufacturing, and service related.

Integrating Six Sigma improvements strategies at the business-unit level while complementing the company's long-term goals is key for success. This is illustrated in Figure 11, and is usually lead by a senior management team focusing on driving home its importance to all their employees.

Procedure 3. Framework for Process Thinking

Before any quantitative analysis gathering can begin, you must identify what systems are in place and being used to assist you in meeting your customer requirements. Basically, establish a base line for improvements. To do this you must rigorously map the existing business processes to develop and analyze quantitatively your performance. Using QFD many customer and supplier questions on system requirements can be defined. The system

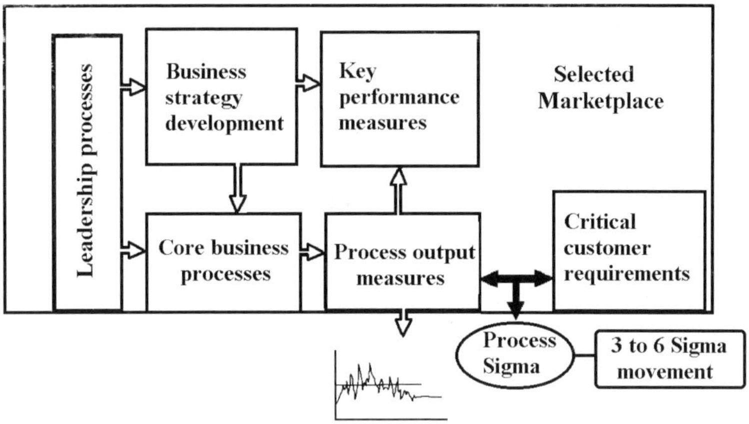

Figure 11. Integrating Six Sigma with business strategy.
(Adapted from reference [1])

must know what the customer requires for their use in confirming the quality required is attainable from your company.

Six Sigma is closing the knowledge and information gap between CCR's (critical customer requirements) and your business and manufacturing capability that is termed "process sigma." The distance between your capability and CCR's is used to prioritize your Six Sigma efforts. The narrower the gap, the closer you are to meeting consistently your CCR's.

Therefore, when your champion initially selects a program, discuss it with them, to ensure the program is not an isolated project with low customer importance. Be sure a program can become a part of a structured framework of improvements, as these will realize a faster rate of total success. These programs are more focused and efficient because improvement in one will translate through all other programs to begin later. They focus attention directly on product output and market demand instead of relying on an individual's intuition. Black belts are not just followers but guides to assist in the most return for the Six Sigma's team efforts.

Procedure 4. Disciplined Gathering of Customer and Market Intelligence

Six Sigma will assist the company in staying in direct contact with their customers needs, existing levels of customer satisfaction, and gain them loyalty in being a preferred supplier of their products. Interaction of supplier and customer key contacts is reinforced with disciplined inter-company communication at all levels. Relationships develop which are key to learning customer satisfaction at all levels. Customer input can assist in knowing at-up-to-the-minute where the market is going.

An example is the yearly build up of tools for sale at Christmas. One supplier to Sears began the build up in earnest in mid-summer continuing to the holiday. Then excess employees were laid off until the build up rush began again. The supplier often was not able to meet all orders, even as extended working hours. Therefore, a solution was reached to allow the supplier to build parts all year long, keep a level work force, and inventory at customer expense, to meet the Christmas rush. It worked and a win-win developed for each with inventory costs minor when compared to profit realized with increased sale volume with products available in the stores for sale. If insufficient warehouse space was lacking, renting semi-trailers was their answer for storage. Product was monitored to insure no environmental damage occurred during the time from manufacturing to shipping to the customer's stores or shipping point.

Often some CCR information may not always be given, not as an unintended secret, but often not realized at the time the QFD is conducted. New requirements can occur at anytime and hopefully are immediately transmitted to the supplier. They should be transmitted as an ECO (engineering change order) using an ECR (engineering change request) form to be acknowledged with any updated drawings and instructions. This is used to ensure the supplier has complete understanding of the request. Responsibility for any WIP (work in process) must be determined and if the change can be made for completed units. If not, renegotiations must be accomplished before a status is achieved for these goods. A request for a price reduction can also be transmitted based on competition threats to you or your customer.

This requires a closed loop intelligence gathering and information transfer process to be in place to gather customer and market data. This data must then be translated into hard measurements that can be analyzed regularly and compared to business output processes and transmitted to the supplier. This is shown in Figure 12, as defining customer and market requirements.

Maintaining a close-loop, information gathering and feed back system, on both the customer, competition, and the market will enhance the company in

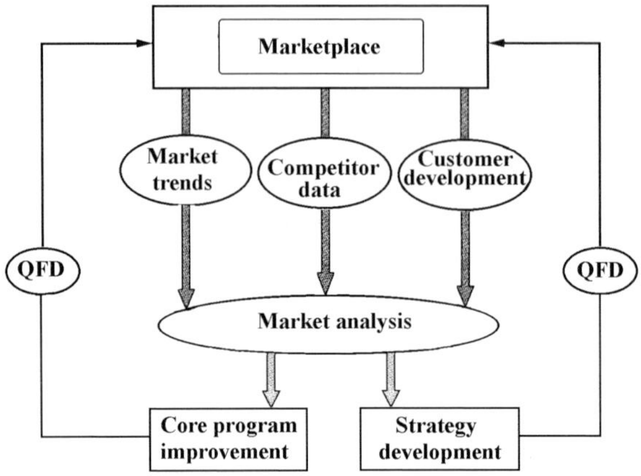

Figure 12. Defining customer and marketing requirements.
(Adapted from reference [1])

planning expansions, financial assets requirements, changes in customer needs and buying habits, and anticipating trends that can affect their bottom line.

This becomes very evident when custom molders provide products to the major automotive companies and consumer point of sale as; Sears, Wal-Mart, K-Mart and others because their schedules can change weekly, and missing a delivery can spell disaster for a small to medium size company. Many suppliers use this intelligence to plan and set their manufacturing schedules, a month or more in advance of even receiving a purchase order when indicators show a trend developing. They can then query their contacts at the customer to receive as up-to-date information as possible to firm up their production schedules. Your best information can often come from a supervisory person in the customers production department. Then validation at the upper management or sales level can confirm its accuracy. Just do not over use or abuse this personal contact if proven very reliable.

Procedure 5. Projects Must Produce Real Savings and Revenues

Returning a positive cash flow from a Six Sigma program is the goal. In the past other quality improvements programs lacked a true monitoring of revenue even when successful. As a result many earlier quality programs such as, TQM, zero defects, quality circles, etc., were not continued as results on the bottom line, which mattered to management, was not always available or documented. Too often the only cost of quality information was the salary numbers for the department, rework, scrap, returns and warranty replacement items. As a result many were not continued as results on the bottom line were not accurately tracked, documented and reported. Reducing cost of quality requires defined programs with personnel assigned tasks, completion data, monitoring with results reported weekly to management for analysis support when required. Six Sigma projects new to company management may insist on short-term payback that is in conflict to Procedure 3, long term with payback improving a chain of identified, successive projects.

This is a problem for many companies tracking their cost of quality even before starting Six Sigma programs. The tracking of their Six Sigma personnel is attainable. But, documenting the true costs of quality without great consideration of what data is to be actually collected is very difficult. The costs are usually broken down in Prevention, Appraisal, and Warranty. The responsibility is on the quality personnel to work with their controller to develop charge codes for describing the type of operation performed and accurately documenting these costs. This is very important so that charges for time and material will be correctly identified for the category of quality prevention work they are performing.

Establishing a cost of quality baseline begins before any program is started so results can be monitored. Quality problems begin with poor production forecast planning and with ordering the wrong material that delays the manufacture of a program. Receiving the wrong certification or inspection information is a quality problem and efforts must be made to ensure all related quality cost areas are identified and charged so an accurate baseline and future quality costs are developed for the company's true cost of quality program.

SIX SIGMA PROGRAM DEVELOPMENT

Short-term Six Sigma projects are usually selected to improve process efficiency, (lean manufacturing methods), reduce rework cost, fix identified problems at the site of origination, increase capacity (pull versus push work flow) to achieve lower operating costs are often easily identified. Successful completion of defect reduction, solutions to on-going customer service and support problems, or meeting on time delivery are all possible if wisely planned in short time lines. But, the underlying problem root cause may still require identification and the problem solved which creates a 'cart before the horse' syndrome on occasions. Any identified solution to a problem even in short programs, reduces both supplier and customer risk and are keeping the customer better satisfied. Six Sigma success breeds success, and increases customer satisfaction and generates new sales in the long term.

But, remember once these changes are implemented after ensuring they do not cause a new problem further into the manufacturing operation, go back, identify and eliminate the root cause of the problem then implement methods to ensure what is in place, will remain in control with additional Six Sigma improvements in the system.

SIX SIGMA PAYS ITS OWN WAY

The goal of a Six Sigma program is it pays its own costs in savings, if not immediately, at least from the second year of being implemented. The initial Six Sigma program savings potential requirement was set a $170,000.00. Realistic for major companies but often to high for some medium to small companies. Too often the accounting and finance departments have no idea, due to poor tracking of cost methods, where these savings can be realized.

Therefore, establishing a Six Sigma program budget with anticipated payback is required to find financially realistic programs for improvement. Each company will establish their own financial baseline for beginning a Six Sigma program. General Electric reported in 1996 as suffering a small net loss in their first year but subsequently attained savings of more than $750 million through 1998. Projected addition monetary saving through fiscal year 1999 were estimated to be $1.5 billion. Mr. Welch reported projected addition global savings of over $8 to $13 billion a year by eliminating inefficiencies and lost productivity. Allied Signal (now Honeywell Corporation) documented savings of $1.5 billion since 1991 with their Six Sigma programs.

Procedure 6. Full Time Commitment to Six Sigma

Six Sigma lead personnel, known as Black Belts, running projects at a company full time, will lead an intense application of personnel trained in the methods and metrics to successfully complete a program. At major companies, rigorous training in quality tools, root-cause analysis and statistics occur over a four month period with one week each month devoted to their Black Belt specific training off site. Using, JIT training methods personnel after each training session go back to their companies and apply the training to selected Six Sigma projects. The result of this training is a cadre of change leaders who can map, measure, analyze, and develop improvements in almost any business or manufacturing operations.

Large companies utilize their black belts to manage or assist in numerous projects at the same time, which may be 3 to 5 improvement teams. These teams are training progressive, certified green belts who will assist the team members when the black belt is assisting other teams in their selected programs.

Selection of team members should be a cross-section of the employees in a company. The number of members selected should fit the need of the improvement projects the company implements singularly or several concurrently. This also depends on personnel needs in day-to-day operations and how much change the company can absorb. Six Sigma teams are dedicated full time to projects unless delays, not of their origin or waiting on parts, may occur.

Team members may come from all levels in the company as long, when evaluated; they possess the knowledge and learning ability to be a black belt or team member. Often a technician can with their extensive knowledge and technical capabilities, be trained and become a better black belt leader than others due to their inter personal skills with plant personnel.

Procedure 7. Reward the Achievers

Companies who reward success at all levels of their company based on achievements, are also rewarded by employees who strive to achieve the goals of the company. Six Sigma offers these recognition challenges to all who are selected for the program.

Some companies have implemented new recognition programs for Six Sigma professionals and executives who have managed six sigma successful programs in their departments. General Electric's Jack Welch required advancement to higher levels in the company, executives on down and master black belts a bonus tied to being Six Sigma trained. This is the reverse of having managers meeting shipping and sales volume monetary goals at the expense of sacrificing quality.

PROCEDURES FOR IMPLEMENTING SIX SIGMA

Some companies may not be certified to ISO/QS9000 even tier one automotive suppliers due to internal company reasons as, they see no benefits. They may believe others bought their certification or their quality is good enough and accepted by our customers and the extra cost is not justified. These answers came from a tier one suppler who was obtaining currently a 50-ppm defect rate, about 5.6 sigma. What was not known was their cost for this level of quality. This company was performing daily Kaizens with teams selected to solve the problem identified. This was the cause and affects solution process and can be very expensive if the root cause of the problem is not identified and fixed. This can cause a misuse of good quality personnel being overworked, not actually preventing problems, and loosing sight of the company quality goal. As a result of this the company had a higher than usual turnover rate of quality personnel.

STEP ONE: ASSESSMENT OF THE COMPANY ORGANIZATION

A review with department managers using the QFD analysis method is necessary to determine what they and key staff members know about their customer's requirements and the company's competition. This is shown in Figure 13, to ensure all business and manufacturing personnel know the same information. A good method to begin this process is questioning each business unit individually, documenting their answers, and finally having a structured summary meeting with all department key personnel to access each department's answers. This will provide a common reference base for determining the relationship of operating units to the customer and competition and how the departments are involved in the daily operation of their department business.

Before the meeting, ensure each department's responses are circulated to the other key unit managers to compare data and develop how their inputs can improve information flow at the summary meeting. Remember each operating unit will view the customer in their area of operation, as will their competition. They will also discuss how effectively they interact with their own in-house departments to meet the customer's requests and requirements. What is also needed to be discussed are the supplier assessments of how the customer's personnel interact within their company and with the supplier's personnel.

A list of specific business, operations, and process questions are sent to each department head for answering in their initial meeting. They should also develop a list of questions, with their own answers, following the QFD matrix format for their departmental answers, if not already completed.

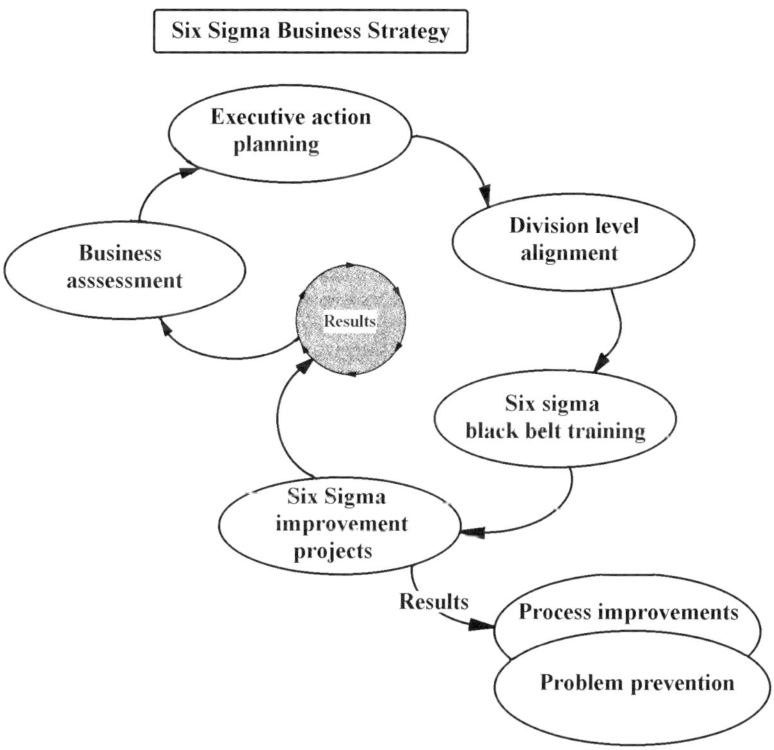

Figure 13. Implementing Six Sigma business improvement strategy. (Adapted from reference [1])

If a black belt is already on site, they can assist the unit managers in developing and understanding the operation of the QFD if they are not familiar with the process. If not, then the quality manager will assume this responsibility. It is very important these initial meetings be format structured, an agenda, location, and meeting time with the anticipated length of the meetings time, to obtain the answers to the questions. Keep out personal feelings, hearsay, and guesses. If the answer is not known, say it, and then get the answer. Ambiguity must be eliminated, only true facts and input. So state this in your agenda for these meetings. Also, request their answers have backup data if collaboration is necessary!

In the meetings agenda, set a realistic time period for each unit to respond, especially the first few units scheduled. Their responses will inform upper management just how informed each unit is to the customers requirements. This will assist them in the initial selection of Six Sigma projects that need implementing to improve business, quality, and customer relations. It will also tell upper management how well managed each unit is in the company structure in relating to the customers satisfaction and risk reduction.

A series of internal company questions to be answered are:

1. Have you performed a QFD with your customer?
2. Are business processes mapped, QFD, FMEA, fishbone, C&A, SPC, CP, CpK? etc.
3. What process are currently in effect, ISO, QS, other quality metrics and are they being followed and audited? (always ensure the data is presented)
4. Are they sufficient to maintain control or do processes need improvement?
5. Attach a separate list with why needed. Justify each for customer potential, time, labor, and cost savings.
6. Who do you communicate with at the customer, their position, and assistance provided you?
7. What is known to be important to the customer at your business unit level requiring product and service requirements? (present data from your and the customers QFD)
8. Is your department capable of meeting the requirements of the customer on a continuing basis?

9. If not, what areas need improvement, list each and any deficiencies needing improvement?
10. Are they measured and how? (show data on the results)
11. What level of control is achieved and does it meet the customer's requirements?
12. Level of customer complaints and quantity problems within last six month, Parato charts. (show the data and results of quality compliance to prevent future problems)
13. Does your department have the assets, equipment, and trained personnel to meet customer's requirements? (provide maintenance, calibration, CP, CpK, and data plus supporting information)
14. If not, on a separate report attached, what is required in your department to meet these requirements? Be specific to detail the needs.
15. What is needed from your support department to assist you in meeting these requirements? Be specific and have you discussed them with the department head for their assistance?
16. Have you requested support, if so what, where, and the response from the support unit.
17. Are any of these needs now under way or completed and the effect on your operations?
18. Is business, program management, or process unit responsibility assigned, what and to whom?
19. Do you know our company/unit ranking with our customers: Competition and their ranking? (sales input, quality assurance data)
20. What is known about our customers in other market segments we now supply? (sales and marketing)

Competition Information Requiring Answers

1. Who are our competitors?
   (a) How close do we compete with them?
   (b) What product/service areas are we competitive in/not? Be specific!
   (c) Do we know competitor ranking with our customer?
   (d) If so, where do they rank? If not known, find out!
   (e) What do they provide or supply we cannot or are lacking at this time?
   (f) Are we price competitive?
   (g) Is design working on new products or programs to meet their current and future needs?

2. What competitive information do you need to become a better supplier to our customers?
3. What is the market direction from your unit's information?
4. Are new breakthroughs possible and if so what are they? Be specific!
5. How does your department handle competitive information and communicate it within the company?
   (a) Received on a regular basis, your source, and how reliable?
6. Can it be improved and how?

Additional questions can be submitted as long as company and department specific for the analysis. The time required for the initial meeting for discussion and feed back of information should be no more than two hours. Submitting the unit's answers to the management analysis team a day, minimum, prior to the scheduled meeting can assist in a more complete review of the answers and keep the meeting on schedule with better output and results. Any management questions can then be formulated prior to the review to make the information exchange as current and informative for evaluation. Anticipate management questions based on the information to be supplied and obtain these answers before any meeting.

Based on the department units review the management team selects the companies champion for assisting the black belt, Six Sigma team in their programs.

It should be obvious after the reviews who would be selected to be the champion. This person is a believer in quality improvement, already an identified leader, and with position and professional ability to provide assistance for the Six Sigma program team members.

STEP TWO: EXECUTIVE ACTION PLANNING WORKSHOP

The CEO/president with support of the newly trained champion will convene training and company vision meeting to decide what Six Sigma programs should first be developed in the company. Prior to this meeting each member is directed to review the results of the company department reviews and select a program that merits Six Sigma development. Backup must be provided for their decision, customer satisfaction rewards, ability to meet their requirements, anticipated rewards in cost savings, increased profits, labor savings, quality improvements, problem elimination, and any other justifiable benefits to be used to select this program and an alternate. Only two programs from each executive manager are required.

COUPLE THE SIX SIGMA PROGRAM TO THE COMPANY VISION STATEMENT

The company vision statement must be concise, apply directly to quality and customer product improvement and easily understood by all levels of personnel in the company.

It focuses on their ISO/QS-9000 quality program tying in Six Sigma as the primary tool to improve the company's success to their customers and markets. Plus use common language to develop a culture of continuous improvement by integrating other business, manufacturing or service initiatives into the company structure.

Six Sigma eligibility programs are reviewed during the program selection meeting. Input from each team is requested with company benefits summarized and then the group selects the primary program. Selection is based on either anticipated short-term success results to firmly embed the program in the company structure or a more long-term program, possibly affecting multi-departments.

A company wide communication statement is then issued outlining the program, benefits and rewards anticipated. Also, employee assistance is spelled out when requested by the Six Sigma system to empower their personnel and implementation team members.

An orientation plan is released, submitted in each employee pay envelope outlining what Six Sigma is, who can participate, and requesting their support in the selected program(s).

If the black belt team leader is not yet selected, the executive group then decides who within the organization will be selected as the program leader. The ideal black belt selected should have a background in problem solving, know quality methods within their range of current responsibilities, self-starting, and a fast learner with the ability to teach others what they are taught. Good communication skills and a skilled listener are also other attributes needed. A person with an established creditability within the organization who has already made, recognizable improvements, knows the operation of the company departmental functioning and able to operate without direct supervision as team leader and able to communicate on all employee levels in the company is an additional required attribute.

An accomplishment variable detector and analyzer, self starter, and goal oriented person is more desirable than your brightest and smartest engineer as the black belt will lead a team of individuals with the technical skills required for the program. Other knowledgeable personnel in the company can be called on to assist the team when their specific knowledge and skills needed. The black belt need not have all the desired skills, as other members of the team are there to assist and complete the program successfully.

The final step of the executive team is to agree that the financial assets and personnel will be available, without question, to ensure the Six Sigma program implementation has the assets and drive required to be successful.

STEP THREE: GATHERING INFORMATION

The operating unit or department selected for the Six Sigma program must now complete any remaining information gathering not yet completed as outlined in the company and customer department list of questions. All other departments must complete their list of information development questions.

At this time the chosen black belt, if untrained, begins their four month, one week a month dedicated Six Sigma training, at a training source selected for their knowledge, ability and success of training black belt leaders. Depending on training company selected there may be a wait until the next training session begins. During this lull in startup activity, the executive group has several options.

1. Bring in an outside source (consultant) to begin quality methods (metrics) training in functions the champion was introduced to in their training. These can be QFD, FMEA, C&A, fishbone analysis, DOE, CpK, SPC, and process control planning and charting plus methods plus methods for problem solving.
2. Use existing in-house personnel, quality assurance manager, statistician, process control methods, etc. using training videos and tapes and in-house workshops on topics in item 1. The Internet is also a valuable source for training aids which have software programs which can be used to train new personnel and refresh team members in how to perform specific tasks not frequently used. Also, the department for the program

can meet with the team members and discuss how they operate and where they believe help is required. What is not recommended is doing nothing!

With the lead Six Sigma program selected, the team members are chosen for team membership abilities plus the experience and knowledge they bring to the team in knowing how the company operations function.

Their training begins using method 1 or 2, learning how to solve problems, team building skills, metrics, and how the team will function, individually or as units within the team structure.

Data currently available on customer and company requirements plus the ability of the company and departments to meet these requirements is analyzed. Then personnel are selected to perform tasks to gather additional specific information and metrics for determining solutions, improvements, and cost savings for the selected programs. Work flow and operations management are also discussed with employees performing these tasks to analyze how they can be improved and what is involved to more correctly and efficiently complete the tasks. Interaction with other departments is also evaluated and how they communicate information and work flow with each other.

After the first black belt training week, the preliminary efforts of the program begin. These are assessing the areas determined with the variables that control or guide the process. These processes are determined by analyzing the customer and competitive information, QFD, fishbone variables diagram, process control plans, and FMEA for where current or potential problems can result and when problem and defects and errors are documented using C & A (cause and affect) analysis reasons for their occurrences.

Information the team finds lacking is obtained by asking the department why this occurs, what is required by the customer to improve and add value to the product and what the company must do to meet these requirements.

Processes are analyzed for repeatable operations, CPK, in their current setup and what variables, from the fishbone, will affect improvement changes for the process or product with the base line information on the problem or process is obtained that may require several weeks or more.

During this time period a program schedule is developed based on known data with assignments, action plans, and time lines developed plus milestone points set based on Figure 9, for the team to follow. Keep to the plan, complete required work objectives and use this knowledge and information when ready to move onto the next improvement step and not before! Plan each action step as complete as the initial data allows but also build in flexibility should the path indicate other variables to pursue in the improvement or solution.

STEP FOUR: JUST-IN-TIME TRAINING

There is no falling behind in the schedule if all members complete their specified tasks on schedule. Obstacles will occur; information late in being obtained or must be developed. Other uncontrolled reasons may occur to cause the information gathering process to fall behind their black belts training schedule. The black belt must, and will, make allowances and modify the schedule for his teams training when necessary to ensure all the information is available when the next step in the program requires it to be used. There are prescribed steps to be completed and information gathering is one of the more important operations to have accurately completed. This initial information is used to establish the process or programs base line for improvement.

Black belt and team training will include the following areas:

1. Defining and planning project goals.
2. Establishing the critical path or program time line schedule with milestones of process/program evaluation and analysis points for review by the team to evaluate their progress.
3. Mapping out the process, fishbone, FMEA, procedures, process control plans, work instructions, check lists, etc.
4. Measurements, metrics, control charting, types and results anticipated and how to use the information for improvements, when and how much to change the process, etc.
5. Analysis of existing business and manufacturing processes, QFD, control plans, FMEA's and process procedures.

Team members learn how to measure and rank customer and in-house factors upon which the customer bases his buying decisions. They also learn the concept of statistical thinking and analysis of data plus how to perform regression analysis, design of experiments, conduct sampling, and calculate business and manufacturing process capability, CpK and Six Sigma process variability and control information.

Also, as the process reaches Six Sigma control, to analyze the test variance versus product variances to determine what variables, if a change occurs, are visible for analysis based on just test variance capability factors.

As the first three months and one week end, the black belt is considered fully trained in Six Sigma processes and methodology. During this time period the black belt team leader changed from student to teacher (four times) while leading the team. They also become the facilitator of training knowledge by leading the team in applying their new knowledge of Six Sigma principles directly to their companies business and manufacturing problem and improvements. The pressure on the black belt and team may be great to show results but management must not miss the true intention of the program. The programs intention is to produce an improvement in the way the company does business, measure it, and show it can and will stay in Six Sigma control after completion. These changes must also improve the company's profitability which is ultimate goal. Measurement of cost and savings must accompany the solution with assistance from the controller's personnel.

At the beginning of any program, management will always want a rapid return on their investment. Even when everything goes right, projects of data gathering completed, analyzed, and improvements implemented take time to show results. When the assets required are not available within or the service work required not available when requested, including parts and equipment for replacement or repair, then additional time must be requested by the Six Sigma team. During these unscheduled program hold periods, regular business operations must continue.

The schedule will be extended or altered to meet these changes in the program. This is not always a disadvantage as it provides the team time to perform other reviews such as gathering additional statistical analysis data that may uncover other areas in the program that influence the program variables that may assist in better, more reliable and specific methods of improving or performing the operation. There are always tasks the Six Sigma team can accomplish during these programs holding periods until the program can be resumed and completed.

This is often difficult for medium to small companies as key personnel are at a premium and most staffs are very lean with personnel managing more than one operation. Management must realize this and accepted the cost and time away from their normal daily duties. The Six Sigma teams are not broken up or their efforts deflected from their primary goal. If this is not adhered to, Six Sigma can at their company become just another quality improvement process that can be listed as, "It did not work for us"!

Since 1986 with Motorola, General Electric, Honeywell, and other large and small companies have obtained major business and process improvements and dramatic (multi-billion of dollar) savings, the Six Sigma quality improvement program does work when applied correctly and supported by management.

OPTIMIZATION THE KEY TO ACHIEVING SIX SIGMA CAPABILITY

Any business, manufacturing, or service operation can be improved. Each company has established a method best suited for their organization to perform their respective tasks. Some are better than others; some companies have more assets to implement quickly the new methods and technologies as they become available.

Others do not, will not, or cannot afford them. But, when shown a way to improve an existing operation, with minimal cost and higher payback, increased customer loyalty, and customer growth will conclude the potential savings must be realized.

The steps required to implement a Six Sigma program with black belt training is expensive. Training a black belt can vary from a few thousand dollars up to $40,000 by some training organizations. By exploring all available training organizations, good Six Sigma training can be very economical for the program savings to be attained when completed in your company. To do business costs money, the first item learned in Business 101. The key is obtaining the greatest return for each dollar spent. Here is how it is done with a trained Black belt.

REFERENCES

1. Blakeslee, J. A., Jr., "Implementing the Six Sigma Solution." Quality Progress July 1999: 77–85.

2. Edson, J. et. al. "Testing Needs Change for Six Sigma Process." Quality Magazine November 1999: http://qualitymag.com/articles/1999/nov99/1199f5.asp
3. Fine, E. S., "What is this Process-Capability Stuff Anyway?" Quality Magazine March 1997: http://qualitymag.com/articles/1997/mar97/0397ft.html
4. Harrold, D., "Designing for Six Sigma Capability." Control Engineering January 1999: 62–66.

Chapter 8

Six Sigma Keys to Success are Control, Capability and Repeatability

With the Six Sigma program underway it will take time to reach its final goal. Management may at any time change its goal as to how close to Six Sigma the operation will achieve. It is not always necessary to reach the actual Six Sigma quality level of 3.4 PPM if the savings to be realized are not great enough over the initial cost of the Six Sigma program. There may be other Six Sigma programs of equal value more critical than the current program to reach Six Sigma.

As programs are developed and completed there will always be more areas, processes, and business units that require improvement and prevention of defects and problems. Focus on these high visible programs and make them successful.

During all this work, never forget what was already accomplished. The completed programs must be audited, reviewed, and maintained in their completed status. To lose control of a completed program is a setback never to even be considered. Resources of personnel, time and assets are needed to ensure they are maintained in "Real Time" quality of operations. Training new and existing personnel must never cease. Monitoring and upgrading, working with suppliers, testing and verifying data is an on-going process never to be neglected. As leaders of the programs move and retire and leave the company, new dedicated and trained personnel, black and green belts, must replace them. In a number of cases I was told by company personnel that back in 1997, we were in manufacturing control and operating within four sigma limits. We are not now, as people left, programs changed, management objectives varied and action was directed in new directions or was forgotten. Procedures were never upgraded for existing process changes as new products were added. A sad state of affairs, but unfortunately true for many companies operation. Therefore, the company must, once in control, follow the Six Sigma keys for continued success.

THE THREE KEYS TO SIX SIGMA SUCCESS

The three main keys to success are control, capability, and repeatability and they have been around for decades. The only problem was how employees interpreted them in their organizations. Top management often only talked about wanting quality improvements in their operations. Some top management did not fully understand the improvement process or take the responsibility to ensure their company and operations had and maintained a good quality program for their operations.

Technology today can control, with repeatable and verifiable capability, almost all of their business, manufacturing, and service operations when the variables of the process are known, understood, and controlled. The emphasis is on management to ensure a reliable quality system is identified, implemented and keep it in control and constantly improving.

MAINTAINING PROCESS CONTROL

Any well designed process executed correctly and with accurate and repeatably controlled systems in place can attain and remain at a high repeatable performance level. Using existing quality metrics as control charts, plotting the operations major variables as temperature, pressure, time, speed, and a products dimensional and physical effects can identify when the process is in control and also predict when the process is becoming unstable.

A process control system is only as good as the equipment in the system and accuracy of control to monitor the process control variables required to maintain the process within specifications. Operational inputs and settings of process control when set correctly can control a process very accurately. But, if variables change, adjustments are required to bring an out-of control process back to a controlled process.

A process can be improved and then maintained in control when the significant variables influencing the process are:

1. Identified, major and minor variables.
2. Understood, for effects on the process.

3. Optimized, for repeatable and improved product output.
4. Maintained, at their optimum level and performance.

To this end, Six Sigma is based around these four variables.

MEASURE, ANALYZE, IMPROVE AND CONTROL THE PROCESS

Six Sigma identifies these as Measure, Analyze, Improve, and Control (MAIC). Black belts are trained in the use of the MAIC phases of improvement. To optimize a process using MAIC, several successive passes of variable adjustment are usually necessary. Each pass eliminates a key variable, or identifies a key variable contributing to low process control and output. There are several software analysis programs that rate a processes capability for repeatability. Their use will simplify analysis of the data but are not necessary for continual process improvement if not available for your specific operation. It just takes longer as calculation and analysis must be evaluated using other quality assurance methods.

These software programs, based on frequency as shown in Figure 1, for Pareto charting of out of tolerance variables. These variables are usually compared to a standard or a system variable known to contribute to control of the process often termed a "gold reference standard." It is also very

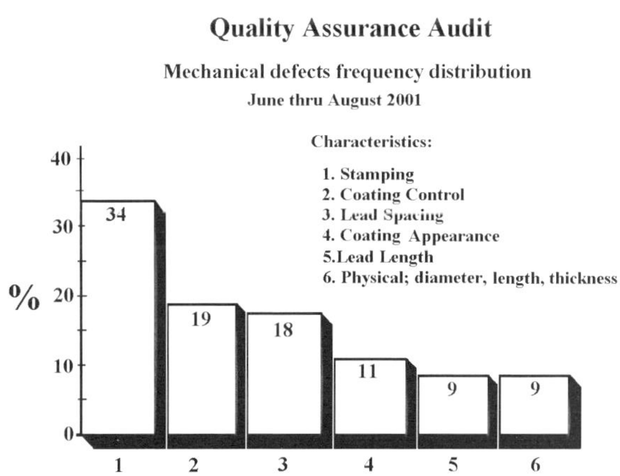

Figure 1. Parato charting of variables.

important that data gathered during the analysis is gathered using "gold standard" sensors. These are sensors with known deviation tolerances and calibrated to a high degree of accuracy before beginning the gathering of any data. Also, the person using the standards and gathering the data knows how to install the sensors, when they are reading correctly, and what to do should they indicate an out of control situation. Only then can data be gathered and recorded for the system under evaluation. With each successive cycle evaluation of the system, any out of tolerance variables (specifications) are identified, with the software adjusting or comparing the variables analysis to the next variables higher or lower degree of repeatable control.

This analysis operation depends on which operations of the process you are analyzing such as, improving or attempting to establish a base line of existing control of the process for future improvement. When all out of specification process variables are eliminated, the program and operation outputs a level of repeatable control the process can maintain as long as no further variables go out of control.

This requires the establishment of a process baseline of variability for the operation performed and equipment capable of repeatability.

EXAMPLE OF BASE LINE CAPABILITY

With the process base line established for new processes estimated for required product control, the process can be improved by working backwards correcting or adjusting the variables known out of control or specification as defined by the software program as it records the variables output during the operation. This adjustment, on only one variable at a time, is performed on the out of tolerance variable by correcting the identified variable from the prior capability level which recorded it as out of tolerance for the process. By working backward and adjusting and correcting the variable, the process can be improved and brought back into control. This is accomplished by adjustments, corrections, or repair of the operation by successively making improvements and solving variable problems until the operation is in the control required to product repeatable and quality products.

Some software programs identify possible causes (a trouble shooting guide in the program) of any out of tolerance variables so the correct process adjustment can be implemented. The data standard the software program uses, compares the current variables values to a known and verified standard capable of obtaining control of the process. These values are termed "gold standard" values or data used in the software program was developed over a series of controlled tests written around a set of established capability levels for each stage of process capability for the operation or process.

If material, plant systems, or the environment are affecting the process, it will act on one of the process variables but may not necessarily be identified or easily recognized as the contributing variable. A variables affect must be identified as contributing to the out of control situation for the process. Most equipment suppliers have equipment process limits established. It is now the plant engineers responsibility to ensure these machines and systems operate within these control parameters and if not, why?

During the program when the process still remains out of control use the fishbone, FMEA, C&A and other quality tools for evaluation and then run a DOE to identify the major contributing variables from all the identified process controlling variables in the entire operating system for the process.

Something as small as a machine hydraulic oil filter being dirty can cause either intermittent or continuous pressure problems depending on how contaminated is the hydraulic fluid filter. Very fine (micron sized particles) can eventually foul a control valve requiring a real serious analysis to determine the problem or it could result in a catastrophic failure of the pressure system if the valve is blocked open or closed.

Question? When was the last, or if ever, your machines hydraulic fluid system was cleaned or analyzed for foreign contamination or breakdown of the hydraulic fluids viscosity level?

Always consult with your equipment supplier for their recommended maintenance on your equipment.

As shown in Figure 2, there are four MAIC phases (measure, analyze, improve, and control) of process control augmentation with eight basic and possible more quality tools and metrics.

The Sigma Breakthrough Technologies recommended for use are listed and are not necessarily used in the order presented. The black belt must decide which technologies are best suited to develop the analysis for their program. Emphasis is placed in the Six Sigma programs to follow a proven road map or structured analysis to improve process quality. But management must remember that all managers are not the same, and they must rely on the experience, knowledge and training of their black belt to successfully manage and complete the program.

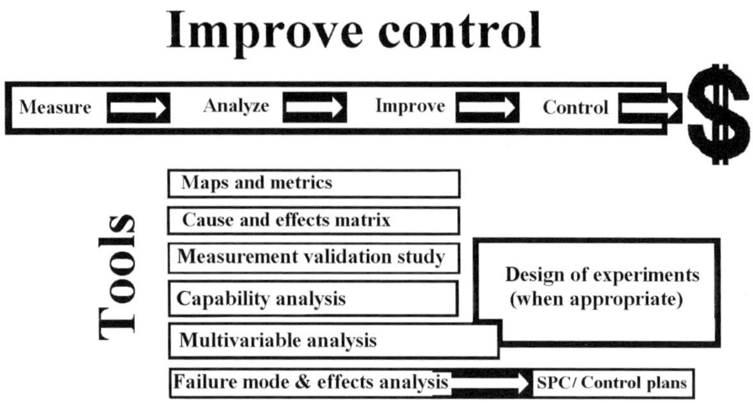

Figure 2. Four phases and eight key tools for Six Sigma. (Adapted from reference [7])

THE TWELVE STEP IMPROVEMENT PROCESS

A sequential set of twelve process and problem solving steps were developed to ensure a thorough examination of a problem or process is achieved. Before these steps can be applied the process variables must be mapped out. Several methods are available to use other than listing the obvious variables controlling the process, which are your starting point or base line for any analysis.

The twelve steps for process improvement:

1. Determine the critical process to quality variables.
2. Define the required process and product performance specifications.
3. Validate, measurement, and control system, equipment, method, and procedures.
4. Establish the current processes capability for repeatable product quality.
5. Determine process upper and lower control limits for output data.
6. Identify sources of variation, machine, material, mold, auxiliary and plant support systems.
7. Analyze and screen potential causes of variation to identify the key variables requiring tighter control.

8. Determine and test variable variances for their interaction on process and product.
9. Establish process requirements for each key process variable with control limits determined to control their variability.
10. Validate the process control measurement system to verify the controls are capable to produce repeatable operations within control limits and the product is within specification control limits.
11. Determine the process systems CpK (capability) to control the key variables.
12. Validate the variables measurement system, statistical process control to maintain the process and the product to stay in control and the worst case for process limits as shown in Figure 3.

In every process the major controlling variables also have controlling variables associated with them that also interact and influence the degree of control of the major variable and process. To determine the "degree" of control the variable is actually capable of maintaining on a process, these sub variables must also be considered. Sounds complicated, which it is, but with sufficient information and knowledge these sub-variables can be identified and then controlled and their variable effects on the major controlling variable minimized.

Using the fishbone diagram predominately and listing every item that could affect its control on the process can identify the sub-variable that may be affecting the major variable. The need to focus in greater detail will assist

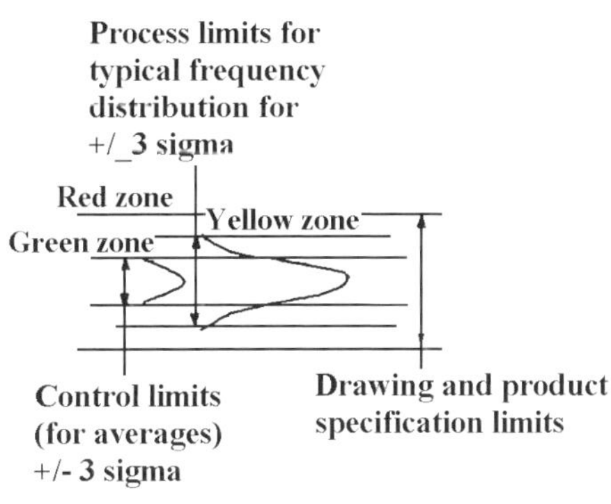

Figure 3. Manufacturing limits.

the Six Sigma team in discovering every conceivable item or factor that can have an effect on the control of the process. An example of this traceable detail is shown in Figure 4, considering only one area in this control of temperature, the injection molding machine heater bands.

As illustrated, a simple set of three zone heater bands have a multitude of sub-variables which affect it's control, reliability and operation to just input a set amount of controlled heat energy to the barrel to assist in melting the resin pellets. Any one of these variables can affect the control of heat input and affect melt temperature of the molten resin in the barrel of the machine.

All of these variables can be controlled, checked during the process improvement and verified as correct or within tolerance. It just takes the correct, calibrated, and heat sensor tipped measuring tool, the pyrometer. The team has to be aware of, and ensure, the variables remain in tolerance during the molding process. Then, as an example, if a melt temperature problem is detected it can be evaluated by the team using equipment and tools to discover where the problem originated. A simple pyrometer, in calibration can do this very quickly. Also, having an observant operator to ensure no external factors are affecting the bands heating output. These can include a fan blowing on the machines barrel, insulation barrel blanket not used, loose electrical connection, burned out heater band, or bad rheostat controller for a set of heater bands.

Six Sigma uses both SPC (statistical process control) and SQC (statistical quality control) tools to monitor, measure, and control a process for repeatable product output. Six Sigma methods add tools, discipline, and the knowledge to improve any process in any industry.

At the beginning of any new process the control limits must be established to maintain the product within the customers requirements and

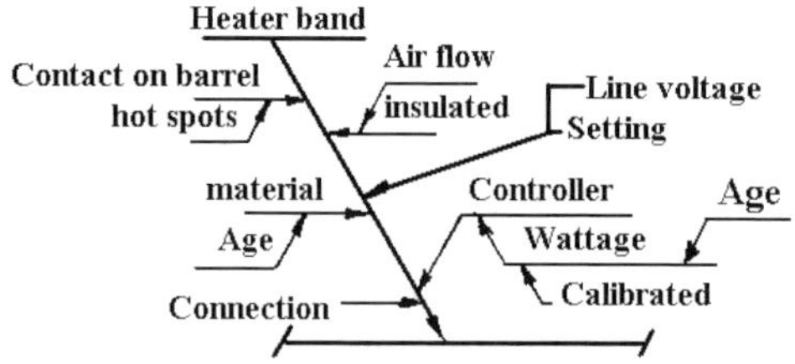

Figure 4. Ishakawa "Fish bone" diagram for heater band breakout.

specifications. Once the specification is established, the process is controlled within these limits. During the manufacturing process they are evaluated to determine if they are correct or tight enough for controlling and maintaining the tolerances required for the product. The method used to establish the process control limits for a new or existing process, for both upper (UCL) and lower (LCL) control limits, is easy as discussed in detail earlier in Chapter Five. Control limits must be recalculated when the products variables show a trend to go out of control or when tighter control of the process is desired.

Establishing the manufacturing process Steps 4 to 9 involves monitoring control chart methodology. Then operating the process and making adjustments to obtain as tight an initial control for the process is currently capable of maintaining, without additional process variable analysis.

VALIDATE THE MEASUREMENT SYSTEM FOR PROCESS CONTROL

Before determining the CP (capability process) of the new process, item 11, it is very important (part of the six sigma process) to validate the measurement system collecting the process output data. Excessive data variation in the measurement system can cloud important variable changes for improving the process and even make it impossible for steady state operation. These variations in the measurement system, often very minor, can make achieving process capability improvement difficult to impossible no matter how much effort is put into improving the process.

The capability of the measurement system must be assessed and periodically reevaluated using established statistical studies for accuracy, repeatability, reproducibility, stability, and linearity. This implies the measurement and output equipment is on a scheduled calibration program and if any instrument reading is questionable then the instrument must be recalibrated. Keeping a spare set of your machines calibrated gages, valves, and sensors is very important to be able to replace a questionable controlling or data collection device or instrument with a known calibrated one. Good measurement systems have the following characteristics:

1. Variations due to common known tolerance range values, not special causes.

2. Total variability of the measurement system must be more accurate than the sensors recording an output of the product specification limits.
3. Measuring device scale increments must be no greater than one-tenth of the smallest scale increment for either process variability or product specification limits.

As the process is being brought into control there may occur a product dimension that varies too much. This can happen even when the process appears to be in control. The process engineer must now determine what variable or variables are in the process, often several may have influence on the dimension for analysis purposes. A cause and affect analysis is first conducted to identify which process variable affected the results obtained in the product by adjusting its value. If it cannot be found or was only a random out of control reading, it could have been an anomaly in the equipment. Monitor the system to ensure it was random and not a trend leading to an out of control condition. Even a minor operation change in cycle time operation can product this condition. Ask the operator if a condition occurred they had to solve not noted in the process record to cause the questionable reading. Also, ask the operators if they observed any spikes of data variation that quickly is corrected on the next cycle or series of process cycles.

When all process variables are evaluated the next question is which variable to adjust in conjunction with all the others, either up or down, and specifically how much. To evaluate each in successive order would take too long and if more than one needs adjustment, which one. The process must determine which variable is or may contribute with another variable to be the controlling variable affecting the product dimension?

If the problem persists there are two courses of action to employ. The first is connecting up the capability analysis system discussed earlier and evaluating the entire process and using the trouble shooting guide in the program to identify the out of control variables. The second should this not solve the problem is to perform a DOE evaluation. The DOE can evaluate a multitude of variables in a fraction of the time required to evaluate each variable individually.

The DOE for process control evaluates all selected variables in either their maximum or minimum process value range at, essentially the same time, only in a randomized and ordered sequence of planned trials. This analysis takes much less time and through process numerical analysis and statistics, can pinpoint the potential controlling variables for adjusting the process to meet the product specification while having the process in control. This process for conduction a DOE analysis is easy and very time saving for determining the variable or set of variables that have the greatest controlling affect on the process under evaluation. An example of a DOE is presented in Appendix B. The example shows the metrics and methods to determine the variables affecting a process or product dimension created by the process variables.

This is how Six Sigma methods are used to control the process for a product to meet it's specifications and keep it during processing within statistical process control. Now, with process control established, the process capability must be analyzed and improved for repeatability, centered around the process mean, CP, and in Six Sigma control for continued repeatability, CpK, to meet the customers CTQ (critical to quality) requirements.

MEASUREMENT OF PROCESS PERFORMANCE

There are four metrics used to measure process performance and to show if the normal, initial Six Sigma distribution of values are centered around the mean of the processes upper and lower control limits.

These are CPP (potential process capability), CpK (process capability indexes which assesses current process capability), Cpm (analyzes the process CP which assumes the process mean, (μ) equals the process specification target (T) and that (T) is the midpoint of the specification tolerance. Finally, the St (instability index) is used to examine the process over extended time periods.

In this example, moving to Six Sigma process control assume the process is now in three Sigma control with variable analysis completed. At this time a controlled process is running but producing product at three-sigma quality or at too high a defect rate. This means all events (variables) falling beyond the three-sigma limits or in violation of the process operation rules have been identified. The process is now in statistical stability and the parameters identifying the distributions of deviations from expectations have become constants.

SIX SIGMA PROCESS CAPABILITY

All metrics and quality methods used to reach three sigma are not utilized to fine tune the process to Six Sigma control. This is done using the four statistical analysis methods and then eliminating any process bias in the process and finally knowing when to leave the process alone.

CP is the ratio of specification width over process spread

CP index = (USL − LSL/6 sigma)

where the specification tolerance (upper specification limit minus lower specification limit divided by 6 sigma, the standard deviation of the limits. The CP index assumes an ideal centering of the bell curve around the mean of the specification. This is the goal when either 3 or Six Sigma process control is achieved or reaches the design target for the process.

In the real world few processes are centered on the target, mean value. In fact the bell curve may be shifted off center or have more than one center point and still show an acceptable CP value. The initial objective of the process being in control as shown is attaining a CP value of 1.0 or 1.33, the value assumed to have the process centered around the mean, this is shown in Figure 5.

NON-NORMAL DISTRIBUTION CURVES

Curve 1 is the normal bell curve distribution with a specified sigma limit process.

Curve 2 is highly skewed to one control limit line.

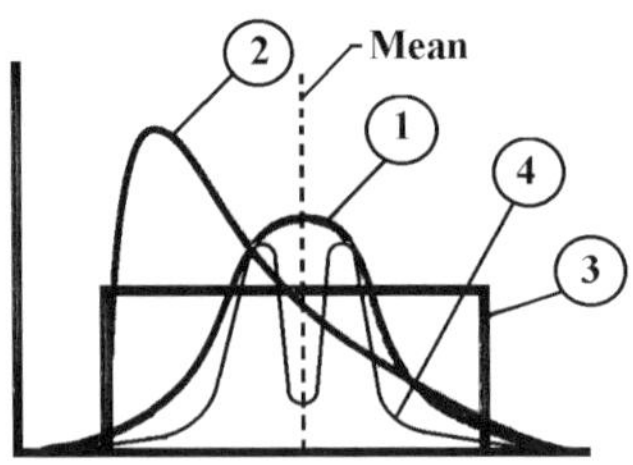

Figure 5. Non-normal distribution curve.
(Adapted from reference [1])

Curve 3 is a rectangular distribution.
Curve 4 has two bell shaped distributions of data centered but not necessarily centered around the mean.

But, centered around their own individual data distribution established means an off center or multi-centered process is penalized for shifting from where it is, to where it should be for good repeatable process control.

CpK is the measurement index used to measure real process capability with the off-target shift applied.

$$CpK = CP\ (1 - k)$$

The value (k) equals the specification target mean (T) minus the process mean (μ) divided by one-half of the specification width.

$$K = (T - \mu)/1/2\ (\text{specification width})$$

With CP used only as an indication of process capability, it is assumed the process mean (μ) equals the specification target (T) and (T) is the midpoint of the specification tolerance.

All curves are within CP limits but vary greatly from the ideal target (T). Therefore the need for CpK is mandatory to access exactly how the data is spread out in the process.

When the data is presented, curve 2 through 4 the process is not capable of Six Sigma control. The spread is too variable and inconsistent with the definition of 1.5 sigma maximum variance from the target mean (T). One or more control variables are out of control and must be corrected!

INSTABILITY INDEX

Assignable causes in the process:
The use of the instability index (St) will provide a method for extended time periods for analysis of data spreads not as severe as curves 2 to 4. The (St) index method determines whether assignable causes are still present in the process.

St = (number of out-of-control data points divided by the total number of data points) multiplied by 100.

The ultimate goal is an (St) of zero. Any number above zero (0) tells the engineer the percentage of data points remaining out of control for the process.

When, (St) = 0, the process control limits should be recalculated, based on the tightened limits, proceeding to Six Sigma control.

OBTAINING SIX SIGMA PROCESS CONTROL IN "REAL TIME"

The Six Sigma approach to quality management objectives is to attain as low as possible the number of defects per million, specifically a 3.4 PPM defect level. This is accomplished by controlling the inputs and process control variables of the process by allowing the process mean (μ) to only drift from the process target (T) a maximum of 1.5 sigma. This requires the process deviation to be small enough to meet the specification limits of:

$$T = \pm 4.5 \text{ sigma.}$$

When CP = 1, the USL and LSL values equal the control charts upper and lower control limits. For Six Sigma control a CP of 1.33 is common. This assumes the process mean (μ) equals the specification target (T) with (T) the midpoint of the specification tolerance or ($\mu = T$).

But, as discussed a non-uniform data curve, when using CpK, may result that erroneously says the process is in control but is actually not.

The indexes CpK and CPM recognize the (μ) may not be equal to (T). Therefore, the following indexes are modified to compensate for the normal variance.

$$\text{CpK} = \text{Min.}\ [(\text{USL} - \mu)/3\ \sigma] \text{ or } [(\mu - \text{LSL})/3\ \sigma]$$

$$\text{CPM} = (\text{USL} - \text{LSL})/6\ \sigma T$$

where: $\sigma T = [\sigma^2 + (\mu - T)^2]^{0.5}$

Then; both CpK and CPM converge on CP as (μ) approaches (T).

As discussed, process variability has two components, ($\mu - T$) and (σ^2). Any deviation from the target.

($\mu - T$), either fixed or variable, is not wanted. This is analogous to accuracy where bias and precision are intermixed.

Both capability indexes, CpK and CPM, acknowledge a fixed process bias of $(\mu - T)$.

Six Sigma allows this bias to vary, as it is a normal occurrence in a process. To have zero bias would mean every variable in the process is in exact control that would be every process and quality persons ideal situation.

To assist in reducing process bias several techniques have been used. Reducing the time internals between data points has resulted in marginal distribution data points with a constant mean and variance. This plot can show a stable process on a control chart making control of the process easier. But, sequentially plotted data may be structured, non-independent and auto-correlated. Variance is present but less visible than shown on a time estimated data plot. Future process trends will be easier to predict and can be used to forecast the next process data observation.

When the variance between forecast and target appears large, the obvious process is to use the forecast to adjust or correct the trend toward the target value. This use of data to adjust the process so the new target is closer to the process target, which results in the forecast deviation becoming nearly zero. When successful the deviation from target should more closely resemble those from an ideally stable time trace. Some process engineers even split the variance difference to judge if an adjustment will move the plot in the correct direction so as not to over correct or adjust the process to drastically for any one process adjustment.

This process of methodology assumes there is a controlling factor (X) that exists when changed at time

$$x(t) = X_t - X_{t-1}$$

can compensate for the observed deviation. As an aid when process adjustments are made the Box-Jenkins manual adjustment chart can be used, using the EWMA (exponential weighted moving average) chart extending the practical use of control charting of the process under change.

BOX-JENKINS AND EWMA CHARTS

The Box-Jenkins charts are used for making adjustments in a process. Control charts monitor the progress of the process. Box-Jenkins charts make the assumption the process is unstable and continually moving and it employees the EWMA as a forecast.

The EWMA at time (t) equals ($\hat{y}_t + 1 = \hat{y}_t + \lambda_{et}$)

Where:

$e_t = y_t - \hat{y}_t$

is the observed error at time (t), and (λ) is the EWMA parameter,

($0 < \lambda < 1$).

The EWMA parameter (λ) can be estimated from a historical time trace with the best choice for (λ) leaving the error (e_t) randomly independent. This means (λ) is simply chosen to be in the internal ($0.1 < \lambda < 0.4$).

This leaves the selection process fairly narrow with the Box-Jenkins methodology yet robust enough to any (λ) selection in this range.

Using the Box-Jenkins manual adjustment chart begins with the assumption the process is still not stable. It uses the EWMA as a forecast of variance and it's model is an integrated moving average (IMA) model of order 1, the IMA (0, 1, < 1) model.

It operates on the first variance (different readings) of the data trace, which means there is never a constant (μ) or process mean value. Therefore, the observed deviations from the process target (T) that remain are considered automatically correlated as there is no process mean.

The time trace (t) then has only a local level effect that can be estimated (due to the auto-correlation) and used as a forecast for process movement. Then the forecast values can be compared directly to the target (T) and any deviation then used to correct the process.

When the process and adjustment are reasonable (minimal) with the system approaching tighter control, the deviations remaining after each adjustment should approach ideal stability. This will result in random variable independence on the process, minimum constant variances, and result in an on-target mean value. This procedure is robust in obtaining tight process control and works very well even when its precise mathematical requirements are relaxed.

The use of control charting and the Box-Jenkins adjustment methodology support each other and provide the process and quality engineer the best opportunity for obtaining product yields, with minimum variance, on target.

PROCESS ADJUSTMENTS

If a decision is made to adjust the process after every observation,

set $(g\ x_t = -\lambda e_t)$.

The constant (g) is used to keep the measurement units aligned. When the process is adjusted after every observation, the forecast is always zero and the observed deviation from the Target (T) becomes the error (e_t).

The correction (x_t) is then simply proportional to (e_t).

A regular control chart is now used to monitor (e_t) as protection against unanticipated special cause disturbance in the process.

As the process moves closer to Six Sigma control, adjustments become less and less. The quality process engineer will establish a criteria or values as to when and how much an adjustment will be made. The decision to make an adjustment can only be made when the EWMA forecast exceeds these established values.

A control chart is developed plotting the observed process values and their EWMA values simultaneously. Then whenever the EWMA crosses it's adjustment boundaries;

$(g\ x_t = -\lambda\ e_t)$

an adjustment is made with (e_t) the deviation between EWMA and the process target.

The plotting of (y_t) the observed variance over time using a control chart provides special protection from unexpected special caused disturbances when showing process stability.

The time to use the Box-Jenkins Analysis for determining when, if ever, to make manual adjustments of a statistically stable process is a study of knowing when the process needs adjustment. As Six Sigma process control is reached, process control charts are used to monitor the process as tightly as possible.

This means all variables are within their specifications as are the process equipment and raw material including process control settings for temperature, pressure, time, feed rate, speed (RPM), etc. Also operator interactions (protocols) have been established, for repeatability (cycle-to-cycle) efficient and capable production with acceptable and in tolerance product being produced.

The process control charts are now used to monitor the process to identify opportunities for potential process improvements and to verify stable process performance has been attained.

The process and quality engineer must always remember, W. Edward Deming's, funnel experiments stated, "Adjusting a process once statistical stability exists, only serves to inflate the total variance", of the process under analysis.

DEFINING PROCESS STABILITY AND REPEATABILITY

Process control charts (Shewhart control charts) are used to monitor a process after the processes UCL and LCL (upper and lower control limits are calculated).

These visibly show the process engineer/operator what effects their control adjustments to the process have with individual observations, or patterns of observations falling within or beyond certain established limits. If outside of these limits, they launch a search (analysis) to determine the cause of these variances on the control of the total process.

When observations fall within the established limits, the process is not adjusted. It is left alone to shift within its normal variable range of values. When stability is attained, capability indexes are often constructed to measure the processes ability to meet the objectives of the manufacturers process.

Processes are not always in a non-stationary condition, meaning never stable due to the many external influences on the process. These contributing factors are age of the equipment, raw material lot variances, tool and machine wear often not noticeable, plant environment changing daily, heat exchanger temperature variance due to cooling water instability, catalyst and pigment decay due to over heating, use of regrind varying in quality and feed ratio and rate, operators changes and the process aging as it runs.

This creates a process level that continuously wanders; the process becomes non-stationary and requires continual actions for verification of input variables meeting specifications, attention to operating conditions, and adjustments for control.

Processes using automatic process controls can compensate, within their set limits, process drifts from the set target values, the mean. But, most processes still require the occasional operator adjustment. These adjustments can be determined by using the Box-Jenkins adjustment charts and tables. "Real Time" process data is also used to bridge the gap between, estimated manual adjustments and adjustments by automatic controls.

REAL TIME PROCESS STABILITY IS DEFINED

Measurements of (y_t) are made at time (t, $t - 1$, and $t - 2$), on continuously manufactured products.

With the target value equal to (T), assume the process mean,

$$[E(y_t)] = \eta$$

This implies (η) is a constant and equal to the target (T).

The process variance (Var) is then described as:

$$\mathrm{Var}(y_t) = \sigma^2$$

σ^2 is also a constant and a measure of common-cause, independent, and normal distributed process variable events.

These assumptions then define Deming's statistically stable processes. This information is described and used in the following example for an injection molding process or any process with a multitude of variables.

HOW TO USE THE BOX-JENKINS BOUNDED MANUAL ADJUSTMENT CHART

The time to use the Box-Jenkins Analysis for determining when, if ever, to make manual adjustments of a statistically stable process is a study of knowing when the process needs adjustment. As Six Sigma process control is reached, process control charts are used to monitor the process as tightly as possible. This means all variables are within their variable specifications including the process equipment, raw material, process control settings, (temperature, pressure, time, feed rate, speed (RPM) (revolutions per minute), etc.)

Plus operator interactions (protocols), following work instructions and/or process control procedures, have been established, repeatability (cycle-to-cycle) manufacturing equipment operation, and monitoring of the process in "Real Time" for efficient and capable production with acceptable product being produced.

REAL TIME PROCESS STABILITY

Process stability in real time is determined by taking measurements (y_t) made at time (t, t − 1, and t − 2), on continuously manufactured products.

With the target value equal to (T), assume the process mean,

$$E(y_t) = \eta$$

this implies (η) is a constant and equal to the target (T).

$$\eta = T$$

The process variance is then described as:

$$Var(y_t) = \sigma^2$$

σ^2 is also a constant and a measure of common-cause, independent, and normal distributed process variable events.

These assumptions then define Deming's statistically stable processes.

This information is described and used in the following example for an injection molding process or any process with a multiple of variables.

A process control chart was constructed from recording 100 readings of data, shown in Table 1.

Table 1. 100 Data Readings of Deviations from the Target: Z_t.

–0.22	–0.04	–0.03	–0.11	0.14	–0.02	0.25	0.26	0.23	0.09
0.06	0.10	0.19	–0.17	0.02	0.16	0.26	0.12	0.10	0.15
–0.04	0.11	0.21	–0.10	0.16	0.24	0.23	0.15	0.12	0.05
0.11	0.12	0.02	0.13	0.06	0.29	0.34	0.34	–0.02	0.13
–0.02	0.28	0.18	0.01	0.23	0.34	0.24	0.07	0.10	0.22
0.11	0.19	0.02	0.00	0.09	0.24	0.14	0.20	0.28	–0.09
0.12	0.19	0.00	0.10	0.05	0.31	0.41	0.16	0.12	0.03
0.14	0.08	–0.10	–0.17	0.23	0.26	0.36	0.07	0.08	0.12
0.04	0.17	–0.08	0.09	0.10	0.38	0.29	0.00	0.11	0.03
0.04	0.09	0.11	–0.04	0.14	0.29	0.13	0.08	0.04	0.12

(Adapted from reference [3]).

With (n) equaling 100 data readings of data deviations from the target mean (Z_t) for a precision multi-pin PC (printed circuit) card connector made with a glass-reinforced polycarbonate resin.

The first (n = 50) deviations (recorded in time (top-to-bottom) left to right) are considered historic data used to construct a process control chart to investigate the possibility of using a Box-Jenkins bounded manual adjustment chart as shown in Figure 6.

The remaining 50 data deviations were used to illustrate the benefits that could be realized using the adjustment process shown in Figure 7.

The development proceeded as follows:

For the first 50 data points (process observations) an estimate of the standard deviation (σ) is developed by using the average moving range statistic. This is the average absolute value of the differences,

$$(Z_t - Z_{t-1})/d_2 \text{ (with } d_2 = 1.128) \text{ then equals } \hat{\sigma} = 0.100$$

with d_2 equaling the number of data point process observations. This is shown in Figure 6, as the Shewhart individuals control chart.

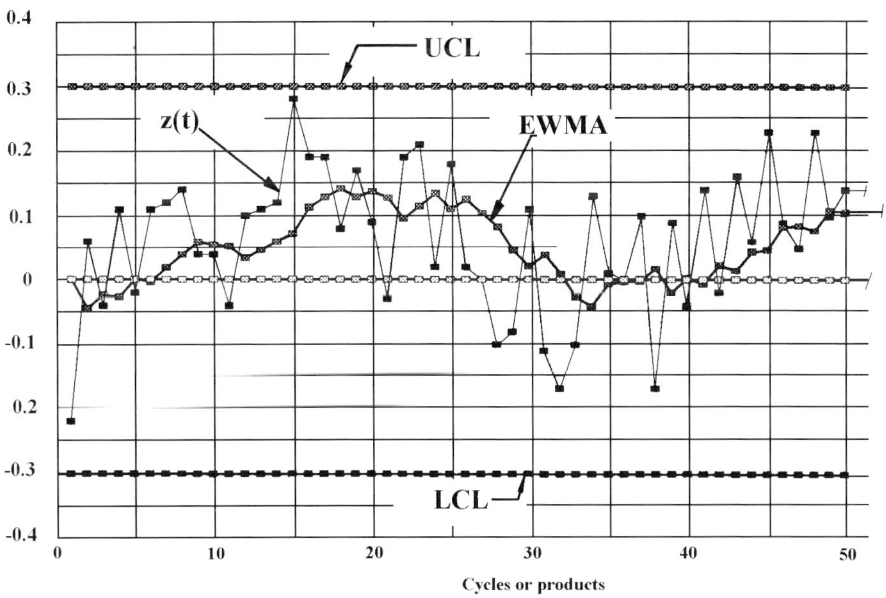

Figure 6. Shewhart individuals chart for Z(t) with EWMA.
(Adapted from reference [3])

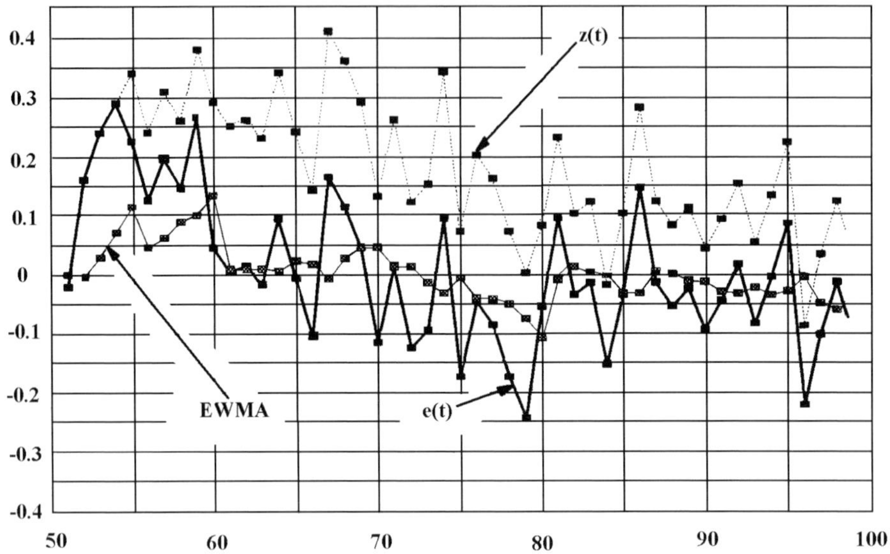

Figure 7. Bounded Box-Jenkins manual adjustment chart. (Adapted from reference [3])

The charting of the historic data (first 50 observations) reveals all values fall within three sigma limits. Also, 90% of the data falls within two sigma limits.

Even with this tight centering around the Target mean, the data still drifts and is not within Six Sigma limits.

Based on the drift visible from the process control chart Figure 6, the process is judged non-stationary. This can be verified by testing the hypothesis that the (Z_t) data were time independent by estimating their auto correlation coefficients or by constructing their variogram. These computations were omitted but do show the data to be non-independent.

With the process declared non-stationary. Using the EWMA (exponential weighted moving average) an approximate model for the first 50 data points can be generated to show the effects it could have had on the process for better control.

The EWMA model does not have to be the theoretical "Best" model to prove useful in controlling the process. As George Box has stated, "All models are wrong, some are more useful than others". This is similar to the original process control charting of the first 50 data points, which assumes the mean and variance (deviation from the target (T)) to be fixed and the errors independent.

To recognize the EWMA at time (t) from the standard deviation average (Z bar at time t) = ($\bar{Z}_t$), use the symbol (Z tilde at time t) = ($\tilde{Z}_t$) for the EWMA.

Likewise when computing ($\bar{Z}$) you must know the number (n) observations, a value must be selected for the EWMA parameter (λ) within the interval ($0 < \lambda < 1$) before calculating the EWMA or ($\tilde{Z}$).

The EWMA can be calculated from either of the two equivalent equations,

$$\tilde{Z}_t = \lambda\ Z_t + (1 - \lambda)\tilde{Z}_t - 1 = \lambda\ Z_t + \theta(\tilde{Z}_t - 1)$$

where

$$\theta = (1 - \lambda)$$

or

$$\tilde{Z}_t = \tilde{Z}_t - 1 + \lambda(Z_t - \tilde{Z}_{t-1})$$

The value of (λ) for the EWMA was selected as 0.20, a commonly selected value. This gives at each point in time a 20% weight to the current (Z_t) deviation from target value and an 80% weighted value to the prior history which is gradually discounted with each new data point.

USE OF THE BOX-JENKINS MANUAL ADJUSTMENT CHART

The use of the Box-Jenkins manual adjustment chart on the production floor is an excellent process control to be employed. Its use follows this implementation procedure.

1. Begin by co-plotting the EWMA ($\tilde{Z}_t$) along with the normal process deviations (Z_t). This is easily accomplished by co-plotting (Z_t) one anticipated value ahead. It is expressed as: "Tomorrow's predicted value equals today's predicted value plus lambda (λ) times today's error". This is expressed as:

   $$\check{Z}_{t+1} = \check{Z}_t + \lambda(Z_t - \check{Z}_t) = \check{Z}_t + \lambda\ (\text{error})$$

   Note that

   $$\check{Z}_{t+1} = \tilde{Z}_t = \text{EWMA (a forecast of } \check{Z}_{t+1})$$

or as the smoothed value ($\tilde{Z}_t$), of the time series at time (t).

Either symbol is appropriate for the EWMA.

This is shown in Figure 6 as the co-plot of the typical process control deviation (Z_t) and the EWMA ($\tilde{Z}_t = \check{Z}_{t+1}$).

The computation of these values is shown in Table 2.

Since based on prior readings, the first forecast, located in Cell 1B, equals zero, at time (t = 1). This is because there were no prior readings and thus the process is assumed to be on target.

No matter where you start plotting EWMA, the EWMA will over successive readings stabilize.

Since the value relies less and less, of importance on, the 80/20 relationships, prior to present values of deviation from the target, the older data has less significance on prior data readings.

With the use of EWMA, Figure 6, by adjusting the process to eliminate excess process motion variability about the Target (mean) it would be considerably reduced.

Table 2. Computing the EWMA.

Cell	Z_t A	EWMA B	Z_t-EWMA C
1	–0.22	(0.000)	–0.220
2	0.06	–0.044	0.104
3	–0.04	–0.023	–0.017
4	0.1	–0.027	0.137
5	–0.02	0.001	–0.021
–	–	–	–
–	–	–	–
–	–	–	–
48	0.23	0.078	0.152
49	0.10	0.108	–0.008
50	0.14	0.107	0.033

Cell 1C = 1A − 1B
Cell 2B = 1B + (0.2)*1C
Cell 2C = 2A − 2B
Cell 3B = 2B + (0.2)*2C

(Adapted from reference [3]).

The squared deviations around the Target yields,

$$\Sigma Z^2 t = 0.80$$

where the error sum of squares equals

$$\Sigma(Z_t - \check{Z}_t)^2 = \Sigma(Z_t - \tilde{Z}_{t-1})^2 = 0.59$$

This is a reduction of process movement of approximately 30% or stability with greater process control.

HOW AND WHEN TO ADJUST THE PROCESS

Does the process have to be adjusted after every observation? No, not every time! The bounded Box-Jenkins method is more practical and adjustments are not as frequently required. The task of multiple computations can be reduced to a minimum with only a negligible increase above every observation adjustment.

This slight increase in variable adjustment, vs. every observation, is negligible for the final adjustment of any systems deviations. It also provides more time for the process, variable dependent variables to act on the last process adjustment. This will also be verified by only a modest increase above it's minimum of the sum of squares of deviation about the target mean. Lets begin the adjust procedure.

The value (e_t) will be the observed deviation from the Target when using the Box-Jenkins adjustment procedure.

When no adjustment procedure is used the symbol

$$Z_t = y_t - T$$

is used only for deviations from the target value.

The setting at time (t) of a variable that will be used to adjust the process is denoted as (X_t).

X_t is the setting at time (t) of a variable that can be used to adjust the process.

This relationship is,

$$gX_t = (1.0)\ \eta t$$

with (g) units of (X) equal to one unit of the response (η), with the quantity (g) called the gain that is selected to be less than (λ) the EWMA value to be used in the process adjustment.

This permits the full influence of a change in the control variable,

$$x(t) = X_t - X_{t-1}$$

to be accomplished within the next time interval of process adjustment.

The required adjustment equation is,

$$X(t) = -\,[G/g]\ e_t$$

For our analysis for process adjustment (G) is set equal to the EWMA parameter (λ) with ($\lambda = 0.2$).

The gain (g) value is selected; selection is predicated on how much a variable change is to be approximated in the analysis, at ($g = 0.8$).

This results in an,

$$X(t) = -\,0.25\ e_t$$

To decrease the variance of (x(t)), the gain is often set to less than (λ). The process must be adjusted to compensate for the observed deviation from target, thus the ($-$) minus sign.

THE BOUNDED BOX-JENKINS MANUAL ADJUSTMENT CHART

The bounded Box-Jenkins chart is easy to use and graphical. It will assist an operator (if directed by their management) to make specific process adjustments within a pre-specified range. When an operator is told they can adjust specific variables within a set limit, when the EWMA forecast (Z_t) exceeds a pre-specified limit, ($\pm L$) the bounded Box-Jenkins chart makes their process changes very simple.

The limits for adjustment ($\pm L$) are established on the basis of engineering judgment, cost of control, being off target, customer requirements, and establishing a Six Sigma process control of the operation.

The example limit was established at ($L = \pm 0.10$).

This value is nearly equal to the process standard deviation of the earlier, first 50, observations data points, (Z_t) actual unadjusted deviations from the target mean.

Control (adjustment of the variable) is only taken when the absolute value of the forecast, the EWMA, ($\tilde{Z}_t$) exceeds the bounds ± 0.10.

When control is not required, the EWMA value is updated.

Referring back to Table 1, and Figure 7, the next 50 (now unobserved data points) deviations (Z_t) plotted as a faint line, the observed deviations using the Box Jenkins adjustment equations from the target (e(t)), and the EWMA ($\tilde{Z}_t$) are displayed within their respective bounds.

Next, Figure 8, shows a plot of the EWMA data points along with the approved process variable correction adjustments (X(t)), when the EWMA values crossed the (L = ± 0.10) limit line.

Whenever a correction (x(t)) is made the forecast is always reset to zero. The correction adjustment (s(t)) is obtained graphically from the nomograph, Box-Jenkins Bounded Adjustment Chart is shown in Figure 9.

The use of the nomograph is simple, press side, and easily understood by the operator to make the correct adjustment. The nomograph chart Figure 9, is used in the following manner.

1. The present EWMA (e(t)) forecast (in this example is 0.06), put a pin at (e(t) = 0.06) on the left (e(t)) data line.

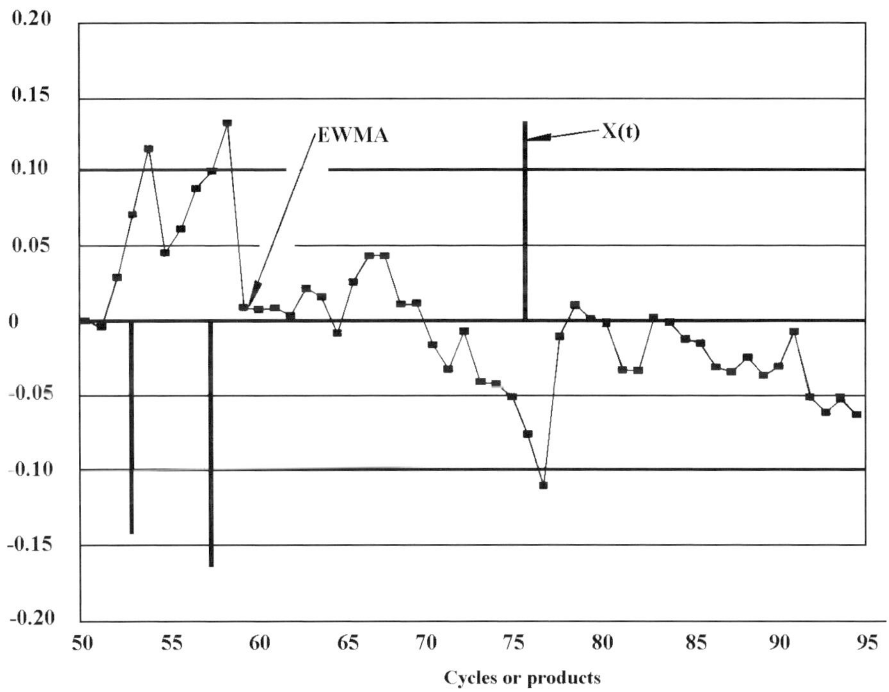

Figure 8. The EWMA and adjustments.
(Adapted from reference [3])

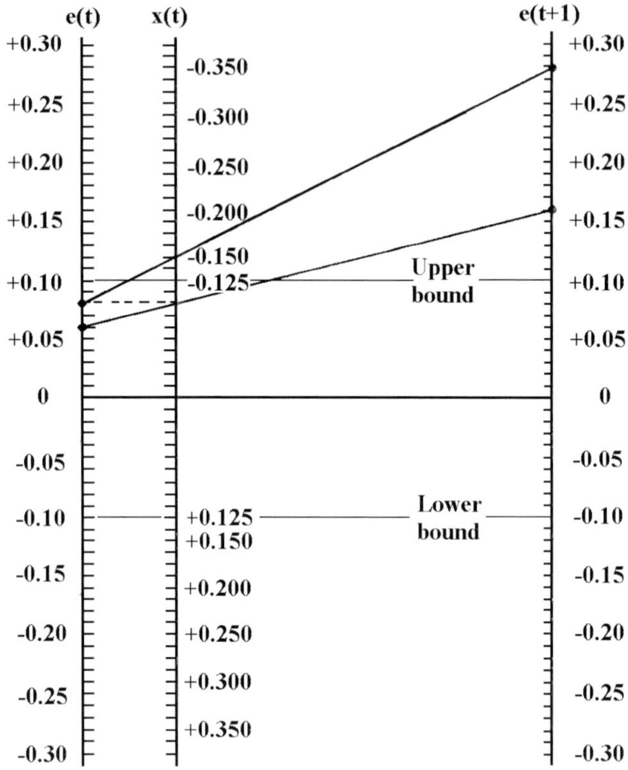

Figure 9. Nomograph, Box-Jenkins bounded adjustment chart. (Adapted from reference [3])

2. The next observation (based on a scheduled observation time line) deviation from the target is:

   $e(t+1)=0.16$, now put a pin on the right $(e(t+1))$ data line at 0.16.

3. The forecast EWMA (x(t)) can now be determined by using pins and string (or other methods such as drawing a straight line between the two points or pins) thus connecting the two points.

Where the string or drawn line cross the, x (t) line, located two tenths the distance from the left data line, e(t), the value $e(t)=+\ 0.08$, is read by drawing a horizontal line to the left data line e(t) is equal to $(+0.08)$.

Note
The scale on the x(t) data line does not agree with data lines e(t) and e(t + 1).

4. Nothing is adjusted now as the e(t) value is less than L = ± 0.10. The next starting point is with the (pin moved to, e(t) = 0.08). This will be the new starting point for the next deviation from the target observation reading.
5. Suppose the next e(t + 1) data point is,

   e(t + 1) = 0.28

   A new straight line is constructed connecting points e(t) and e(t + 1) = 0.08 to 0.28. This new line crosses the EWMA, x(t) line yielding an

   x(t) = + 0.125

   a value beyond the limits, the updated forecast line, above the L = ± 0.1 limit line, and intersects the line, x(t) = − 0.150.

This is the required variable correction to bring the process back to the Target within the next cycle or operation. (Note: when any temperature adjustment is made it will take a considerable amount of time for any adjustment to be observed, as the process must reach a state of equilibrium.)

In fact, the e(t + 1) value can even increase higher until heat loss or gain in the system reaches it's new equilibrium point based on the variable adjustment. Continue to plot observations but do not readjust the variable until the effect has stabilizes. Whenever possible, temperature adjustments should be one of the last variables to adjust due to the time delay effects on a change.

6. Once this adjustment is made, the pin or starting point is repositioned at zero. The pin and string method is very simple and if used the chart should be mounted on a stable and accurate surface with pins that do not bend and a line that is easily positioned between left and right pin position holes. Try to ensure as straight a line as possible occurs between the two points.

Using this method, plot on a control chart the EWMA forecasts simultaneously with the observed deviations e(t). Then as long as the EWMA values were within (± L) it would serve as a basis for the next forecast as shown in Figure 8. Then if the EWMA would go out of limits, corrective action would move the EWMA back to zero, or the target value.

COMPUTING STANDARD ERROR FOR EWMA

Using (σ) as the standard deviation of the observations the error equals

$$\sigma \text{ EWMA} = [(\lambda/2) - \lambda]^{0.5}\sigma$$

In our example, when the EWMA (λ) is set at ($\lambda = 0.2$) then the

(σ EWMA $= \sigma/3$).

This example sets the limits at ($L = \pm 0.10$) which was equal to (σ) and likewise equal to the three sigma limits for the EWMA. As with control chart plotting, whenever an observation fell outside of the three sigma limits, an assignable cause was sought.

Using the Box-Jenkins method, this example, whenever the EWMA forecast was outside of the three sigma limits the process was adjusted. The calculations for the data plot were developed in Table 3, Compulation of EWMA for Bounded Adjustment Chart.

Process control using the Box-Jenkins methodology, based on value selection can tightly control a process if the variables can be controlled, machine, raw materials, other inputs, etc. Analysis of the example yields, without any adjustments the sum of squares of deviation from the Target mean would have been,

$$\Sigma Z^2 t = 2.1472$$

Using the EWMA with $\lambda = G = 0.2$ with limits of ± 0.10 reduced the sum of squares to,

$$\Sigma e^2 t = 0.7738$$

with only three major adjustments as shown in Figure 8.

Setting tighter limits to reach Six Sigma such as ($L = \pm 0$) each observation would have required an adjustment.

The sum of squares of deviations from Target mean would then have resulted in

$$\Sigma\ e^2 t = 0.5920.$$

The process engineer and management must then decide how many adjustments to the process are required. With a reduction in the number of

Table 3. Computation of the EWMA for Bounded Adjustment Chart ($\lambda = 0.2$, $g = 0.8$, $L = \pm 0.1$).

t A		z(t) B	e(t) C	Forecast D	Error E	EWMA F	Need Now G	Total Need H	Control x(t) I	Total X(t) J
1	51	–0.02	–0.0200	–0.0000	–0.0200	–0.0040	0.0000	0.0000	0.0000	0.0000
2	52	0.16	0.1600	–0.0040	0.1640	–0.0288	0.0000	0.0000	0.0000	0.0000
3	53	0.24	0.2400	0.0288	0.2112	0.0710	0.0000	0.0000	0.0000	0.0000
4	54	0.29	0.2900	0.0710	0.2190	0.1148	–0.1148	0.1148	–0.1435	–0.1435
5	55	0.34	0.2252	0.0000	0.2252	0.0450	0.0000	–0.1148	0.0000	–0.1435
6	56	0.24	0.1252	0.0450	0.0801	0.0611	0.0000	–0.1148	0.0000	–0.1435
7	57	0.31	0.1952	0.0611	0.1341	0.0879	0.0000	–0.1148	0.0000	–0.1435
8	58	0.26	0.1452	0.0879	0.0573	0.0993	0.0000	–0.1148	0.0000	–0.1435
9	59	0.38	0.2552	0.0993	0.1658	0.1325	–0.1325	–0.2473	–0.1656	–0.3092
10	60	0.29	0.0427	0.0000	0.0427	0.0085	0.0000	–0.2473	0.0000	–0.3092
11	61	0.25	0.0027	0.0085	–0.0059	0.0074	0.0000	–0.2473	0.0000	–0.3092

Cell 1C = 1B	Cell 2C = 2B + 1H	Cell 2c: Copy 2C
Cell 1D = 0.0000	Cell 2D = 1F(ABS(1F > 0.1), 0, 1F)	Cell 3D: Copy 2D
Cell 1E = 1C – 1D	Cell 2E = 2C – 2 D	Cell 3E: Copy 2E
Cell 1F = 1D + (0.2)* 1E	Cell 2F = 2D + (0.2)*2E	Cell 3F: Copy 2F
Cell 1G = 1F(ABS(1F > 0.1), -1F,0)	Cell 2G = 1F(ABS(2F > 0.1), –2F,0)	Cell 3G: Copy 2G
Cell 1H = 0.0000	Cell 2H = 1H + 2G	Cell 3H: Copy 2H
Cell 1I = 1G/(0.8)	Cell 2I = 2G/(0.8)	Cell 3I: Copy 2I
Cell 1J = 0.0000	Cell 2J = 1J + 2I	Cell 3J: Copy 2J

(Adapted from reference [3]).

adjustments how much loss in control will result, a slight increase in the sum of squares of deviation about the target mean. Bounded adjustment will always increase the process variability around the target from the minimum possible, but only modestly.

This is another strong reason for ensuring all variables used in the process are held to their original specifications and not allowed to exceed their limits, as they will affect control of any process.

For other process control situations not adjusted within the next time/cycle interval (x(t)) refer to *Statistical Control by Monitoring and Feedback Adjustments* ((Box, G.E.P. & A. Luceno (NY, NY: John Wiley, 1997)).

When process dynamics are involved, process (variables) consequences will occur beyond the immediate time internal. The Box-Jenkins control charts can easily be modified to compensate for these situations. Under these process conditions both (e_t and e_{t+1}) are use to provide a forecast. The adjustments then become equivalent to proportional-integral control procedures.

There are now very accurate process control software programs with like controllers that can greatly assist in maintaining excellent control as long as the variables are within their specifications. The main issue is determining what processes lends itself to automatic control versus manual control as discussed.

The Box-Jenkins manual adjustment method and charts can prove of great value once a process is in control and also in developing a controlled process for manufacture.

The process engineer and management must then decide how many adjustments to the process are required. With a reduction in the number of adjustments how much loss in control will result, a slight increase in the sum of squares of deviation about the target mean. Bounded adjustment will always increase the process variability around the Target from the minimum possible, but only modestly.

WHEN SHOULD THE PROCESS BE ADJUSTED?

Adjustments will be necessary to bring a process into Six Sigma process control. Typically, after the first year of effort your control of the process is within five sigma control parameters By completing your pre-Six Sigma analysis correctly you now have all contributing variables as tight to their specifications as possible. This should minimize the degree of variable adjustments required to reach Six Sigma control.

The Six Sigma teams assignment now is determining where the process needs further adjustments to produce Six Sigma repeatable results and to further reduce supplier and customer risk to almost zero.

The process engineer must be able to recognize when a monitored process produces a time trace of deviations from the target that are statistically nearly identical. This is recognized as process stability, with mean zero random statistical independence and constant variance that is deemed normal. At this time the process should be left alone as it is in the best control possible based on all variable inputs over time.

Very often this is a difficult decision to make when trying to ascertain when the process is in its best process control settings. It is very difficult to know when an adjustment is unnecessary and can only cause the system to go out of control and to not improve. Deming's philosophy on a process in this state of operation is based on his famous funnel experiments was, "Leave the process alone".

Statistical process control requires both adjusting the process and monitoring the results. Both the Box-Jenkins manual adjustment chart and control charting the results are simple to use producing a proven dynamic method of adjusting a process into Six Sigma control.

Also, one needs to know if an unnecessary adjustment is made after every observation, how much is the variability of the process changed. This is best answered in G. Box and A. Luceno's book *Statistical Control by Monitoring and Feedback Adjustment* (1) (New York: John Wiley & Sons, 1997).

POTENTIAL WHEN AN ADJUSTMENT IS MADE

Variance Inflation

$$\sigma^2 e = \sigma^2[(1+\lambda)/(2-\lambda)]$$

λ is the EWMA parameter, $(0<\lambda<1)$ (Exponential weighted moving average)

If the process engineer makes a process adjustment when it should have been left alone of say,

$$\lambda = 0.20$$

the variance is inflated as follows:

$\sigma^2 e = \sigma^2[(1 + 0.2)/(2 - 0.2)]$ equals

σ^2 [1.11]

If a λ of 0.4 was chosen

$$\sigma^2 e = \sigma^2[(1 + 0.4)/(2 - 0.4)] = \sigma^2 \ [1.25]$$

Very minor variation results when; $\lambda = 0.2$ versus $\lambda = 0.4$, that results with: $\{[\sigma^2 = 1.25] - [\sigma^2 = 1.11]\} = 0.14$.

But, the real loss in performance occurs in not adjusting the process when ($\lambda > 0$).

Statistical process control requires both adjusting the process and monitoring the results. Both the Box-Jenkins manual adjustment chart and control charting the results are simple to use producing a proven dynamic method of adjusting a process into Six Sigma control.

IMPLEMENT CHANGE WITH SIX SIGMA METHODOLOGIES

Six Sigma quality operations require a new method for management and their employees to become empowered, to be creative, and implement changes to reduce risk in the company that will be productive for them, the company, and their suppliers and customers.

Creativity is more than thinking of new ways to improve your business. The thought process must be channeled, challenged, and encouraged to become creative.

Creativity is the focus for change, see Figure 10, funneling knowledge and experience to improve your business involves using what processes you have in place, analyzing them for improvement, and then when the ideas make good business sense for you, your suppliers and customers, to act on them.

Also, you must study or measure their effects to ensure the results are realized and do not cause problems. What improvement path is best for your company may not be the best for your customer should it require a modification in process or product. This is why QFD is such a valuable analytical tool for knowing what is required for you to improve based on feed back and input from the customer and your internal personnel, conducting their own QFD in your business and manufacturing departments in the plant.

A model for this method of the creativity improvement process is shown in Figure 11. It is based on obtaining new knowledge, TQM, ISO9000, QS-9000, and now Six Sigma methods to make improvements. The knowledge obtained is used to develop, test, and implement positive changes in the operations of the company.

Too often when new ideas are requested the thought process we have used is not connected to all members of the team. New members will rely on what they have accomplished before at other companies; and use these ideas with the group. Older members of the team, rely on what was tried and did or did not work. If it did not succeed they tend to be very negative and often shut down their thought process. The new black belt now has the task of getting their idea concept mechanism working to the benefit of the company and team.

The idea shown in Figure 12, needs to be discussed with the team members. The recognition of a concept under lying a specific idea and using it to follow a new direction are important parts of developing new and

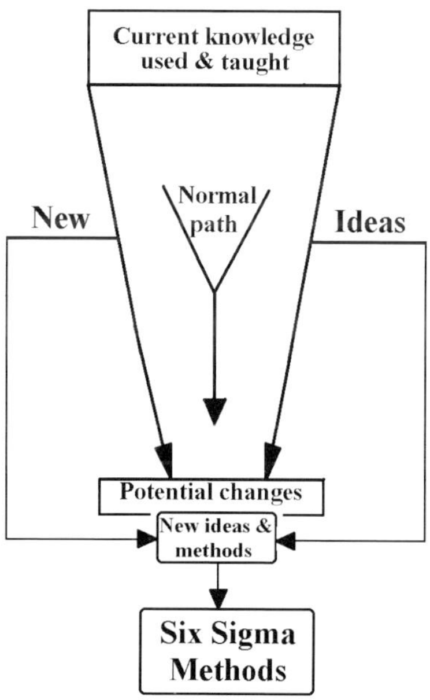

Figure 10. Provoking new thought patterns.
(Adapted from reference [4])

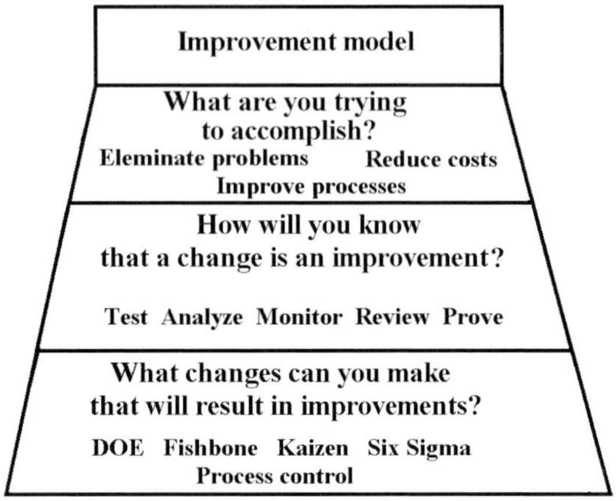

Figure 11. The model for improvement.
(Adapted from reference [4])

creative ideas for the Six Sigma team. Concepts are like forks in a road. They give the person a choice to travel a new route or stay in the same rut and not proceed successfully down the path of positive change. It is important to keep all ties to creativity strong and enforce these change concepts in the team members. Always ask the question "What if?" we tried this idea, what would we accomplish?

The "Concept Alternate", Figure 13, illustrates the creativity concepts of provocation (stimulating the thought process) and movement for developing new ideas and concepts for improvement for a specific purpose. It is used to separate our current ideas from the concepts they have in the past been

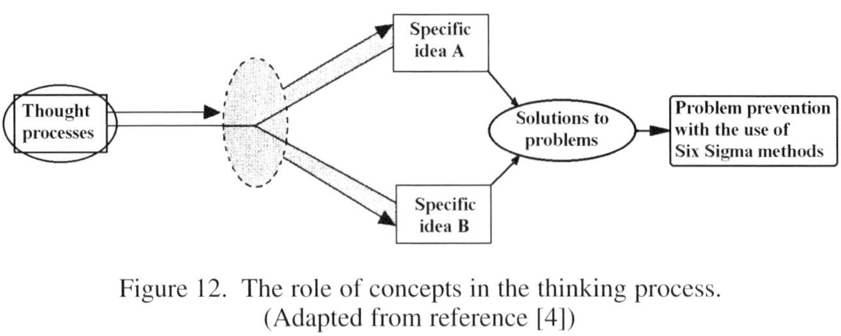

Figure 12. The role of concepts in the thinking process.
(Adapted from reference [4])

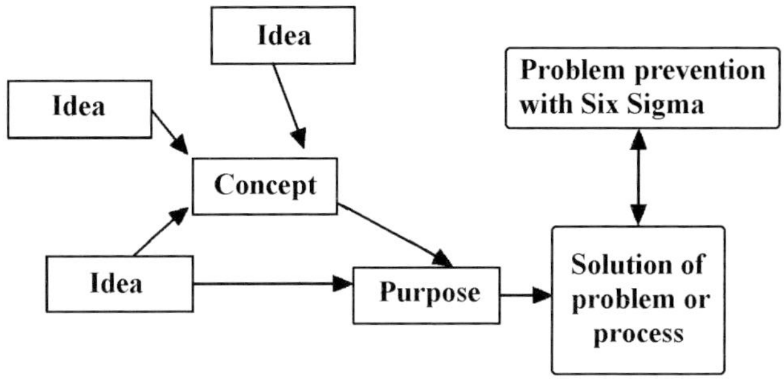

Figure 13. Alternate ideas for concept.
(Adapted from reference [4])

attached. This can produce new ideas for improving the old concept of doing business. This leads to more specific ideas that can carry out the same concept only better. Always ask yourself the question, "What if"? It opens up a whole new idea generating capability, as anything is now possible!

Customer service and technical support in a company can always be improved to give timely, correct, and good information feedback to their potential and current customers. This is the first contact a new customer has with your company. Impressions are formed early and if negative, very difficult to change later. An example of this was an improvement team developed to improve customer order and technical support services from the order entry department of a major plastic supplier company.

The team concept was simple, "Provide consistent and accurate responses to customers orders and technical questions and problems". Ideas generated from the team were:

1. Always express a helping attitude on the phone with an interested voice and "listen" to what is requested before replying.
2. Real Time order entry data tied to plant inventory and production schedules.
3. Hold training classes to educate CSR's (customer service representatives) in frequently asked technical product data questions. Provide answers for FAQ (frequently asked questions) in their computers database, easily accessible for on screen search for product data.
4. Provide each technician with a computerized database for frequently asked technical, design, process, and assembly type question and where

the customer can find it in the supplier's literature or other sources as the internet.

Questions like "How can this help improve customer loyalty?" and "What are we really trying to do to better assist our customers?" Can move the discussion into new areas (concepts) for improvement. The information in the data base led to improvements in the order entry data base for the CSR's to eliminate the chance of wrong data, product codes, package types, quality, being entered plus the true customer required delivery date. This led to discussions with plant production personnel on a daily basis to ensure all CSR's questions were being answered in a timely manner. The CSR's had a production, answers required screen, they could fill out with a few key strokes requesting production information, noting requests, order size, and package change questions. These questions were identified with a CSR's number and production responded directly back to their computer terminal with the answers to their individual questions.

This resulted in better and faster information flow, production schedule changes were reduced with "Real Time" input, and customers received the latest "Real Time" information in 24 hours or less that was the goal of the department.

This process is shown in Figure 14, with new ideas radiating out to improve the existing or creating a new conceptual way to improve business operations.

New concepts spread rapidly as do new ways and type of equipment to improve a process. In the early 1980's the concept of electrical drive, versus hydraulic, for injection molding machines was introduced. Many new ideas were introduced, lower operation (power costs), no noise (hydraulic pumps), no temperature problems with hydraulic fluids or leaks, very repeatable cycle movements, etc. The only initial negative was price.

Major companies bought these machines and used the older hydraulic machines as a benchmark to see if as advertised the electric drive machine costs and problems were reduced plus improvements in processing, capability and cycle reduction could be realized.

The results of the study were found to agree with supplier claims resulting in a gradual replacement of hydraulic machines to all electric drive. Other side effects were realized as the reduce cost of maintenance plus less shock, vibration, no oil fitting leaks, and excessive tool and machine movements during cycle operation that resulted in less wear of mold guides and tie bar bushings. Plus the elimination of the problem of hydraulic oil temperature control that resulted in the elimination of viscosity variation problems that changed injection speed, created a near perfect repeatable molding cycle. These electric machines were accepted by the technical staff as having fewer problems and as a result they proved very reliable in service.

A BETTER WAY TO GENERATE IDEAS FOR CHANGE

Ideas for change require an interest and willingness to find a better way of doing existing operations. A change concept is usually too general to use directly. Concepts such as "reduce or change the order of process steps" and "minimize handoff" must be applied to specific situations and then turned

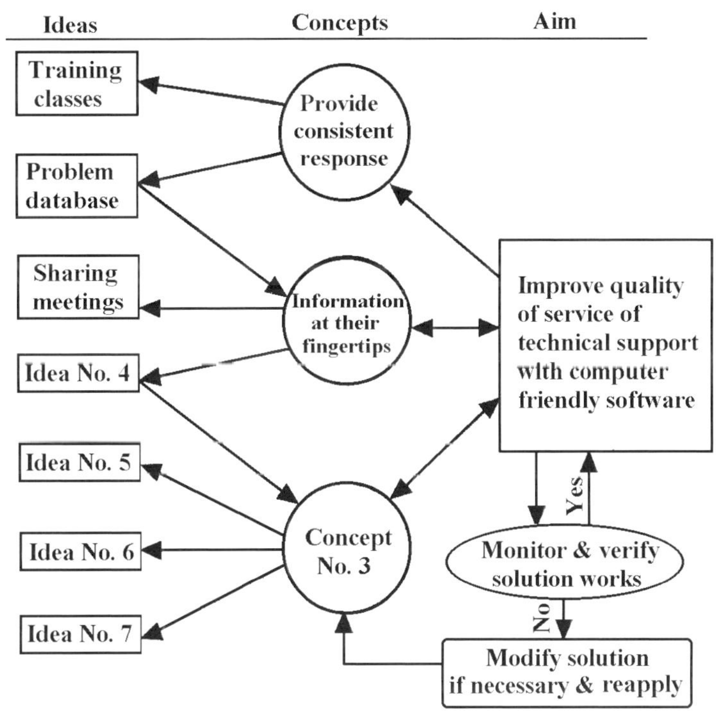

Figure 14. Concept fan structure.
(Adapted from reference [4])

into "How to ideas". The way ideas get tied to a concept are based on the skill of going back and forth between the general change concept and generated specific ideas as illustrated by the fan concept.

LEAN MANUFACTURING PRINCIPLES

Lean manufacturing is nothing new except the concept of pulling the work through the manufacturing process instead of pushing it to the next work station. Lean manufacturing requires product department rework stations that immediately evaluate, repair, and put the product back into the workflow. If a product is not repairable, the problem is identified as to where it originated and the department or work station where the problem originated is notified and the problem fixed.

Also, the reason it was allowed to continue on in the manufacturing flow is analyzed and corrected. The lot size is also designed for a smooth travel of product through each work station without backup resulting from stations that require more time to complete their operations. In this manner the work flow is controlled base on the management of time for completing each operation. When a problem results, it is immediately looked at, or an MRB called, to disposition the product, rework or be scrapped. Also, the operator immediately receives feedback on what the problem was and how to eliminate it in the future. Preventative actions are immediately implemented to permanently solve the problem. Sufficient information and training at each work station is supplied to ensure operators know what they are doing with parts available eliminating down time. The list can be expanded based on product requirements.

An example of a good lean manufacturing production line also employing JIT (just-in-time) manufacture is a Ford supplier for custom built truck seating. The company receives the seat orders from the Ford assembly plant and is required to supply the custom seating in the exact order the vehicle is being built on the assembly line in three hours. Three assembly lines are setup and the supplier, with work order routing is required to build each type of special seating on a designated assembly line. Each line builds a special type of seat.

The seating varies by model style, seat foam density, number and type of seats, front and rear cab, position control, fabric, and color. When the order is received each component is pulled from stock, staged and marked for a specific assembly line, work order requirements printed on the order, and delivered to the assembly floor in the correct flow for completion to then be put on the delivery truck to the Ford plant and their assembly line. All materials were bar coded, special parts staged at each work station, and personnel trained to build each type of seat to the customers specifications. The time allotted for receipt of order to loading on the delivery truck was one hour.

A list of change concepts is shown in Table 4, that can be used to promote ideas. The list is far from complete and a team can easily develop more concepts to aid them in their company and department improvements. Be sure to always document all new concepts.

How these change concepts work are described in the following four steps for team reference.

1. Select a change concept at random and solicit ideas. Since this is not a concept to be considered, random selection, this can lead to very innovative ideas. Remember, no idea is considered trivial as it may spark other more specific ideas in other team members. Criticism is not permitted! This is also a good mind warm up exercise before considering the main area to improve.
2. Begin by randomly selecting a change concept in the list that is applicable to your improvement program, to stimulate ideas from the team.
3. Study each change concept in the group and document the ideas generated. Use this information to develop changes or file it for further discussions when a new situation requires a concept change.
4. Document specific ideas based on a change concept if the ideas apply directly to your operation for improvement. Expand on these 71 concepts applicable to your company's method of operation. Keep them handy for "idea generation."

A company's rate of improvement is based on employee ideas related to a change concept. Change concepts are developed by relating to past experience and change concepts that worked at other companies where your employees worked.

Just when we believe there are no new ideas coming forth, there is a breakthrough. One method to develop new change concepts is to study the improvements that have been made in your company and expand on these and ask the following questions.

Table 4. Six Sigma Change Concepts.

A. Eliminate waste	
1. Eliminate things that are not used	7. Reduce classifications
2. Eliminate multiple entry	8. Remove intermediaries
3. Reduce or eliminate overkill	9. Match the amount to the need
4. Reduce controls on the system	10. Use sampling
5. Recycle or reuse	11. Use sampling
6. Use substitution	
B. Improve work flow	
12. Synchronize	18. Smooth work flow
13. Schedule into multiple processes	19. Do tasks in parallel
14. Minimize handoffs	20. Consider people as in the same system
15. Move steps in the process close together	21. Use multiple processing units
16. Find and remove	22. Adjust to peak demand bottlenecks
17. Use automation	23. Change the order of process steps
C. Optimize inventory	
24. Match inventory to predicted demand	26. Reduce choices of features
25. Use pull systems	27. Reduce multiple brands of same item
D. Change the work environment	
28. Give people access	34. Invest more resources in improvement to information
29. Use proper measurements	35. Focus on core processes and purpose
30. Take care of basics	36. Share risks
31. Reduce demotivating aspects	37. Emphasize natural and logical of pay system consequences
32. Conduct training	38. Develop alliance/cooperative relationships
33. Implement cross-training	

Table 4. Continued.

E. Producer/customer interface	
39. Listen to customers	43. Reach agreement on expectations
40. Coach customers to use product/service	44. Outsource for "free"
41. Focus on the outcome to a customer	45. Optimize level of inspection
42. Use a coordinator	46. Work with suppliers
F. Focus on time	
47. Reduce setup or start-up time	50. Extend specialist's time
48. Set up timing to use discounts	51. Reduce wait time
G. Focus on variation	
52. Standardize (create a formal process)	56. Develop contingency plans
53. Stop tampering	57. Sort product into grades
54. Develop operational definitions	58. Desensitize
55. Improve predictions	59. Exploit variation
H. Mistake proof	
60. Use reminders	62. Use constraints
61. Use differentiation	63. Use affordances
I. Focus on the product or service	
64. Mass customize	68. Influence or take advantage of fashion trends
65. Offer product/service any time	69 Reduce the number of component parts
66. Offer product/service any place	70. Disguise defects or problems
67. Emphasize intangibles	71. Differentiate product using quality dimensions

(Adapted from reference [4]).

1. What was the specific change made to prompt improvement?
2. What was the idea that sparked the change, where did it originate?
3. Where, from whom, or from what area of your company did it originate?
4. What change concept was the spark for this idea?

5. Can the idea be more generalized for use within other departments of the company?
6. Would a new concept better describe this new idea generated for change?

In 1992, approximately, a new method was introduced for the solving of inventive problems. These solution development concepts and knowledge were developed by Genrikh Altshuller, in 1946, a Russian inventor and creative thinker. He developed a body of principles and knowledge that became part of a data base for the development of solutions to difficult problems. Based on his work in the Soviet Union patent office where he reviewed new patents for recording, he found that the same problem had been solved in different technical fields using a core set of fundamental inventive principles. From this he developed a Russian acronym "TRIZ" meaning "the theory of the solution of inventive problems."[6] His process was further enhanced by his associates and engineers in Russia and recently introduce in the United States and work shops being presented to develop these ideas and principles to problem solving here.

Altshuller observed with his associates there were only 27 inventive principles or concepts behind all existing patents and that these principles address standard technical conflicts in design or problem solving. A list of these methods is presented in Table 5. There have been different versions of these principles and each is adopted to suit the specific problem solving technique used.

TRIZ has three thought processes that develop the idea patterns of the individuals in the teams searching for solutions. These are evolution, contradiction and ideality. The concept behind this thinking is described when a review of a new patent is made.

1. The analysis of a patent described the typical evolution of a technical system.
2. Mutually exclusive demands is the contradiction portion where a material must be both light and strong at the same time as a plastic material, glass reinforced, for example.
3. The ideal technical system theoretically does not exist. But, all of its functions are fulfilled.

TRIZ is a well-developed approach to using concepts to obtain creative solutions to both technical and everyday business type of problems. The principles are able to be used by all levels of company personnel. The list presented are more technical and science based than the change concepts described earlier in Table 4.

For complicated problem areas the process is found to be very effective and even the simple examples as the one that follows. For a customer service area, as previously presented, the following Table 6, of information can be used for concepts of improvement

To begin, an improvement team of company experts on the area or topic are convened. Each team develops a set of change concepts specific to the topic being discussed for change. These topic-specific change concepts are then taught to collaborative members and become the focus for developing specific ideas for improvement in the collaborative members team. Members use the Model for Improvement to test and implement specific changes developed for; the change concept workshop.

Changes in your company and department can spark changes for improvements in other company work areas. Do not be stopped by

Table 5. TRIZ: 27 Inventive Principles, Methods, Effects, and Tricks (Scientific example).

1. Do it	15. Macrostructure to microstructure
2. Change the state of the physical property	16. Effect of the "Corona discharge
3. Do it in advance	17. Curie point of ferromagnetic materials
4. Do a little less	18. Combination of various effects
5. Matreshka (nested dolls of Russia)	19. Geometric effect of the Moebius Ribbon
6. Separate conflicts in time or space	20. Geometric effect of the Rotating Hyperboloid
7. Replace special terms with simple words	21. Ideal final result (IFR)
8. Incorporation into one system	22. Introduction of a second substance
9. Fragmentation, consolidation	23. Utilization of soap bubbles and foam
10. Dynamization	24. Operator STC (Size, Time, Cost)
11. Add magnetic powder; apply a magnetic field	25. Model with Miniature Dwarfs (MMD)
12. S-field modeling	26. Make a copy and work with it
13. Self-service	27. Build a model of the problem
14. Heat expansion inversely	

(Adapted from reference [6]).

Table 6. Customer Service Change Concepts for Breakthrough, Series on Delays and Waiting Times.

Change concepts for system design

1. Do tasks in parallel
2. Remove or rearrange a step
3. Use customer service check list, add to computer screen
4. Use multiple process computer screens
5. Give timely feedback to supervisors
6. Use pull systems for information
7. Synchronize to a common point in time (real time production reports)

Change concepts for system design

1. Triage system when problem between departments occur
2. Combine services on computer screens
3. Smooth the flow of information into common data base
4. Software compatibility within plant(s)
5. Collect requests for information centrally and obtain answers

Change concepts for matching capacity to demand

1. Improve predictions (sales forecasting vs. plant)
2. Identify and manage the constraints of computer and production systems
3. Work down the inventory backlog of old products
4. Balanced centralized and decentralized capacity for inventory draw down
5. Use contingency plans when required.

(Adapted from reference [3]).

negativity such as, “We tried it once and it didn’t work!” Ask, what they tried and why they believe it did not work! This was found true at a company that did not believe in department meetings. They were considered a waste of time and non-productive.

IMPLEMENT MEETINGS OF SUBSTANCE

It has been proven that many companies have no idea of what a good meetings structure is composed. Too many meetings are called in haste with little or no personnel preparation. Spur of the moment meetings will occur without warning. These must be held to handle business and manufacturing problems. But, try to minimize these types as they are more disruptive to the business and personnel who have to work on answers for the meeting plus perform their daily duties. What often is forgotten by upper management are the more meetings personnel are requested to attend, the less time they have for their own departments and time to prepare for their next scheduled meeting. Always be mindful of the time a meeting takes up and try to get the maximum benefit from them and the personnel who attend.

The problem with their meetings the majority of companies have are that their meetings are not structured and not always planned for maximum utilization of resources, personnel and time. Once the following organization changes were made in meeting time (one hour), place (CEO board room), when (each Thursday at 10:00 am), and an agenda of items prepared and distributed. Each member received the agenda at least two days prior to the meeting and had the option of adding new discussion items to the agenda, time permitting after all other items were discussed unless the CEO wanted to discuss this item. The meetings were very productive with new ideas and methods of doing business discussed and some adopted. Problems were identified and solved or a team selected to gather data for the next meeting. The meetings worked when organized as they had never been done before!

This can also be taken to the other extreme as we have all experienced. Too many meetings were called with little or no preplanning by upper management. These often spur of the moment meetings caused problems due to the wrong personnel being involved, lack of adequate preparation and department inter-communication, personnel not knowing who was in charge of specific programs, lack of information on the customers requirements and manufacturing capability of the plant, lack of process control procedures and no enforcement of manufacturing work instruction on the assembly lines.

This resulted in so many daily meetings, management and staff never had the time to plan and prevent problems. A true case of crises management costing the company over $1,500,000.00 in scrap and waste in one years time. As a result of this high loss, a "Cost of Quality Program" was implemented. Each operating department developed a SWAT team for solving the five major problems in their department. This was all fine except the teams were only applying a band-aid on a festering problem that required major analysis before a cure could be considered or even attempted to be implemented.

EIGHT (8)-D PROBLEM SOLVING METHODOLOGY

The solution to these workstation problems was performed at this company using the 8-D Problem Solving Procedure. This problem solving procedure was developed, I was told at Ford Motor Company, in Detroit, Michigan. The problem solving plan was developed by Ford engineering to assist in solving their process and production line problems at their and supplier plants.

The procedure for performing an 8-D Problem solving is straight forward. A copy of the 8-D or Step Problem summary template is shown in Table 7. The procedure steps relate directly to the numbered items on the "8-D Step Problem Solving Summary"

Coupled with Table 7, 8-D Problem Solving are Figures 15, 16, and 17, 8-D form, Identification Problem/Project sheet, and Plant Managers Problem Solving Log for control of these documents for reference and use of this technique.

Some major problems were solved if they were work station or "problem specific." This means a solution could be implemented without requiring a cure or fix of a prior product or process problem on the product. But, quality assurance knew that to really solve the high scrap and rework problems a full analysis of the manufacturing operation was necessary. As a result each quality engineer was assigned a specific department for analysis, preparing Ishikawa variable diagrams for each machine, material, and manufacturing operation performed at each work station in the entire plant.

From this information manufacturing process control plans were developed that listed all variables, their control parameters, testing requirements, tolerances and what to do if an out of control condition occurred. These specific and individual manufacturing process control plans were never developed before with manufacturing and quality relying on ISO9001 work instructions to control the manufacturing of their products on multiple assembly lines. During reexamination it was discovered that the current work instructions were not current, being followed, and supervisors aware they were not up to current revision. Engineering did not follow through with ECR to the assembly line, only issued verbal orders to the line supervisors who passed this information to the assemblers and operators.

What was found were the manufacturing operations were completely out of control with no equipment setting, variables, or control settings documented and verified as correct for each operation. This meant that for each operation the conditions may never be the same and as a result high scrap rates occurred and when problems occurred the department had no history documentation to fall back on to know what might have caused the

Table 7. "8D" Problem Solving Sheet Procedure.

(1) Team Members/Phones
This is the group of people working on the problem and their phone numbers for contact

(2) Describe Concern
In this section write a brief concise statement of the problem. This should be in object defect form. This is to be supported by supplying the information used to determine a problem existed. (How is it a problem, what indicators, what areas are effected, charts, graphs, reports, pictures, etc.)

(3) Containment/Short Term Corrective Actions
Indicate here the immediate actions taken to contain the problem. These may be in the form of added inspection, sorting, changing materials, Kepner-Tregoe plan, PDCA (Plan, Do, Check, Act) Cycle, etc. Also attach how you determined the action was effective. How was it measured, attach results, data, etc. Complete a start date for the containment and an end date for the containment actions. An action plan should be developed at this point

(4) Root Cause(s) Identification
Attach the analysis and test information to determine the root cause of the problem. This could include Fishbone Charts, Is – Is Not analysis, DOE's with Results, FMEA's

(5) Corrective Action(s) Verified (Permanent Corrective Actions)
Eliminate the cause of the problem
Criteria a. Fixes The Problem at the Root Cause Level
b. Generates no additional problems
c. Has been verified to work

(6) Permanent Corrective Action(s) Implemented
Plan and Implement the Selected Permanent Corrective Actions, and to remove the Containment/Short Term Actions, and develop monitors for long-term results. Tools used during this step include Kepner-Tregoe planning, PDCA (Plan, Do, Check, Act) methods, Control Plans, Charts, and Graphs

(7) Actions to Prevent Recurrence
Show what procedures, policies, practices are modified to prevent recurrence of the issue.
Useful tools include Questioning to the void, FMEA, Control Plans, Kepner-Tregoe planning, Subject matter experts, Routings, Work Instructions

(8) Actions taken to Recognize Team
The purpose is to convene the team and recognize individual and group contribution and to celebrate a job well done

Problem Number	Assigned		Date	
#	By	To	Opened	Closed

(1) Team Members / Phones	(2) Describe Concern (attach supporting data, graphs, charts, etc.):

(3) Containment / Short Term Corrective Actions Verified	Dates	
	Committed	Completed

(4) Root Causes Identified	(5) Corrective Actions Verified

(6) Permanent Corrective Actions Implemented (related to root causes)	Dates	
	Committed	Completed

(7) Actions To Prevent Recurrence (modifications to management systems, operating systems, practices, and procedures)	Dates	
	Committed	Completed

(8) Actions Taken To Recognize Team

Summary				Copy Final Summary To
Closed By				
Closed By				
	Name	Signature	Date	

Attach action plans, flow charts, cause/effect, is-is not, time lines, and other supporting data and documents.

Figure 15. 8-D Step problem solving summary.

problem. If the problem continued, a new one developed or they went away with the next lot of product and no one knew why.

Identify Problem/Project: Plan Revision Date:

ACTION	COMPLETION		RESPONSIBLE	STATUS	VERIFIED
	Planned	Actual	Name/Phone	CONCLUSIONS	Name/Date

Figure 16.

Problem	Date Opened	Code	Note	Date Closed

Problems must be closed by the plant and the quality managers.

Figure 17. Plant manager's problem solving log.

To solve the problems, as there were many, the quality department went back to the basics to develop the information lacking in all process, equipment, and material areas. Once this information was developed, selected in calibration testing machines were chosen to perform in process testing on the product to see what could be causing the problems associated in each department. When the analysis was completed, the machine variables were known, setting established to product good product each time and all other machines inspected and calibrated. It was then determined the state of manufacturing was capable and verified as capable for repeatable manufacture.

The assembly line personnel and their supervisor were also retrained and certified in the work operations performed at each station. The line supervisors were required to inspect each operators work on a regular basis. Each operator was also required to signoff on all work they performed, accepting responsibility for their work operations. Then if a problem was discovered, the line supervisor knew where the problem occurred and a solution could be developed in training, materials, or equipment and implemented immediately.

This program was not completed overnight. Using the quality metrics that have been discussed and with a lot of hard work and training the program was a success. It took over one and a half years to get the plant running in a manner that was able to bring the cost of quality down from 5% of sales to an acceptable level of management of less than 1%. Then the plant was operating within three sigma limits for the first time in ten years.

Managers must be willing to accept and implement new concepts and ideas for improvement. They need to reorganize their thought process and move into the new company culture of change with improvements being required for continual growth, program improvement, and profitability.

Providing management with an idea and the methods for achieving change requires backup and data that it (the idea) has or is capable of working to get them to consider the change. A way to convince some managers is by studying improvements in your organization and other companies that were both successful and also not successful and ask the questions:

Improvement Questions

1. What specific change were made or intended?
2. What was the idea that sparked the change?
3. What change concept generated the idea?
4. Can the idea be carried over to your area?
5. Would a new concept be better used to describe this idea?
6. Where or from whom was this idea generated?
7. Can the change be implemented?
8. Cost in assets and training of personnel?
9. Work or Business improvements?
10. How will they be accepted?
11. What caused a change to fail?
12. How could it have been better implemented?

Recognition of idea generators, the thinkers, can be helpful but could also become a deterrent based on personal bias of some personnel. Look out for this problem in your organization and always try to eliminate it. Rewarding new and creative ideas that work is one-way team recognition and idea generation is improved. Use words of recognition as, “this was a team idea”, as in a team environment it always is no matter who’s idea it was originally. Remember it is and was the team!

Use employee creativity to build your business since they do the work and often only a mention of a change concept can spark new ideas and rewards to the company in increased profits, more efficient use of assets, and benefit both employee and customer alike. Creativity leads to developing new ideas that move to meaningful improvements.

Use the creative thinking methods in training, process improvements, procedures and work instructions, work station layout and flow, and business discussions with employees and customers. The three thinking methods that focus on concepts are; concept triangle, concept fan, and change concepts.

Creative thinking leads a company to improvements and will always be the method used for continued quality improvements.

ORGANIZATIONAL CHANGE EXPERIENCE FOR EXCELLENCE

Management must early on make the business decision to lead the company into quality improvement and delegate responsibility to their chosen black belt team or teams for implementing the quality improvements in their company. They must champion the process in their department and in the company with assets and support to make the programs successful.

Six Sigma uses a trained black belt to lead the company to excellence and cost savings through project improvements and elimination of defects and waste. Why Six Sigma is now successful is asked with the answer simply stated as; “Dedicated management implementing a company culture for improvement.”

Company organizations must be focused on taking action and to be responsible for the action taken. Management must ensure quality and process improvements remain the focal point for continued improvement. This was the goal with Jack Welch at General Electric to have “all” of their employees trained in Six Sigma methods.

All management is required to buy-in and have a common work attitude in their daily operations. They continue to improve and validate the systems and processes to remain in Six Sigma control, once deemed in control by the black belt team. Quality process control must not just be accepted, but proven with monitoring and documentation that the system is truly in Six Sigma control and capability.

The black belt teams goal is to have all of the companies business and manufacturing operations Six Sigma capable. The major operations first and then to the other important programs, will be considered as free time for all programs becomes available for investigation of improving all business and manufacturing department and systems. Remember, one of the requirements of ISO9000-2000 is for continual improvement of the business and manufacturing systems of an ISO certified company quality program.

Using Six Sigma methodology the daily correcting of reoccurring problems will be replaced with better efficiency of operations to eliminate the risk and prevent the possibility of any reoccurring daily problems.

New personnel must be trained in Six Sigma methods to carry on the program and current employees frequently retrained to reinforce the knowledge and methods learned to maintain control of their operations. This is done by monitoring and tracking the variables in the process on a continuous basis of operations.

TRACKING SIX SIGMA CONTINUOUS IMPROVEMENT

Once a business or manufacturing operation is in control, never relax your monitoring of the process. Rescheduling the frequency of inspections, tests, or other reviews of the process are acceptable as long as the operation is proven reliable or stable enough to remain in control. Train your operators using procedures and work instructions to recognize when change occurs and what to do if the quality or specifications are affected by any change in material, equipment, or process. This includes both business and manufacturing operations. Look for new and better ways to perform the tasks as suggested in creative thinking and idea generation for preventative action programs. Often it takes additional time to both look and listen to the operator performing the operation. This involves seeing how it is performed and to be told some of the problems that have occurred so new ideas can be generated to eliminate a potential, and yet unsolved problem.

The last method described is used in the Kaizen quality improvement program discussed earlier. New eyes looking at existing areas with the authority to reorganize the work area, reassign personnel, combine operations to improve work flow, efficiency, and productivity. An offshoot from this should also be improved quality, but this is not always the end product of this technique.

IMPROVED WORK FLOW, PULLED VERSUS PUSHED

It has been documented that pull versus push is the correct method to move product through your plant. It is the responsibility of management to always be aware of new and better methods for improving the work flow and resulting quality within their organizations. The difficulty often observed by outside consultants is their reluctance to make change and recognize better methods of performing their operations. There was also apathy discovered within their work force for embracing change even when change was necessary for keeping their business alive and competitive within their business sector. This does not always mean they do not want to change only that they must be led with the person in charge a proven leader. The leader will assume the responsibility to make the change while inspiring and delegating the authority to his subordinates who must implement the change. It has been proven both in big and little companies that the right leader can overcome all forms of disagreement by developing a workable plan and explaining it in detail to their employees that change is necessary for the health and well being of the employees.

Personnel who are charged with the ongoing responsibility of continual quality improvement will continue to implement new processes and products. This is the best sign for continued growth in customer base and market share. Therefore, frequently track customer satisfaction (QFD), competitive reactions, and what your departments are doing to continually meet their customer's requirements. Each department has a high stake in how successful the Six Sigma program matures within the company. Continue to monitor operations and processes for improvements and increased efficiency. Keep your personnel focused on doing the job as directed in their procedures and work instructions. Develop, if not already in place, job descriptions and responsibilities and publish this information using your storyboard information display. You will find many employees, even long time ones, did not know who does what in their organization. If a title can be used to assist in identifying the department personal specific responsibilities, assign them this title.

Also, be sure this person is capable of doing the task and accepts the responsibility. Be sure each person knows who to go to for information and the decision points in their department or work area within the company. Also, who is their alternate should a decision need to be made when their supervisor is absent.

COMMUNICATION, BEFORE AND AFTER SIX SIGMA IMPLEMENTATION

A major, under estimated, benefit of Six Sigma is an improvement of communications within the company. Discussed earlier was how communications, at the start of a new program should be conducted. This is accomplished by getting all interested parties together at the start of any current or new program. All involved departments and personnel making decisions must rely on every ones input to ensure the program is as problem free as possible from the start to finish. This is communications at it's very best and at its earliest, when important decisions must be made that affect all departments and the customer. It is then the most productive and cost saving to the company. This is the major concept of program and quality problem team preventative action for bringing a new product or process in on time and within projected cost.

DEPLOYING SIX SIGMA TO CUSTOMERS AND SUPPLIERS

Companies with Six Sigma success programs want to share their stories and success with their customers and suppliers. The benefits of having your suppliers Six Sigma are significant as your product risk is decreased for your cost of incoming quality inspection.

The elimination of having to perform incoming quality inspection is typically a price of quality you pay for in the products purchase price from the supplier. The savings on your ledger sheet may be four or more times greater if their part is always good and in specification so it can be used in your product without worry or cost of inspections on your time. These products are often know as "dock to stock" purchased items. They are inspected by incoming inspection on a set basis agreed to by supplier and your company. This is typically yearly when re-approval of the supplier usually occurs. This type of business relationship builds up trust, loyalty, and long-term agreements, even when price must be increased, for supplier and customer to remain in business.

There can also be beneficial cost savings for both supplier and customer. General Electric has achieved these monetary and product quality rewards by training their supplier base quality personnel in Six Sigma at their training facilities. The rewards have been cost reductions, reliability improvements and on-time delivery. The use of the Six Sigma training has achieved rewards to General Electric and also their other business base customers. When taken across the company organization, Six Sigma can have a significant influence on business, output, and growth for a company.

Marketing and sales can use these success stories and their positive results to keep and obtain new customers. This can be done through case studies, company visits, audit results, and awards of high quality and low, on-time, product risk for their customers. Companies frequently audit their suppliers to access how their services measure up to their competition. Continue this and be sure your evaluation system is a real measure of your suppliers service and quality. It should not be a numbers game. The number is the rating and how you audit and analyze their experience, delivery, and quality of product and service is the real importance for an operation.

A major resin supplier conducted a customer survey for their customer service department to determine how their customer service representatives (CSR's) were performing for their customers. The survey determined they selected the most responsive, service oriented, and accurate in their order processing and supplying "Real Time" product information versus their other resin suppliers. This information was obtained by an independent customer survey of molders, purchasing agents, engineers, technicians, and management response with specific supplier related for service, delivery accuracy and quality of purchasing questions they answered as to which resin suppliers CSR's were, in their opinion, "the best they worked with on a continuing basis".

The CSR's won this award because they had the Real Time product information tools, motivation, and personalities plus the training, organizational support (production, sales, and technical) to provide accurate and timely answers and information to specific questions asked by their customers. Each CSR had a specific industry segment she was responsible for and dealt with the same customer personnel, often on a daily basis. A business relationship developed between each person of trust, responsibility, and friendship that proved to be so successful they were often invited to visit the customer's plant to help them do a better job in planning and other order operations. This is true Six Sigma in operation.

Company management should also encourage and even assist, when possible, their suppliers to initiate Six Sigma programs. The better control your suppliers have on their incoming products can only make your program easier to remain in Six Sigma process control. This includes inter-company departments that pass on product after they have performed their respective department operations.

Even before a problem develops, offer the services of your black belt to assist them in solving problems and implementing correct preventative action plans to eliminate future problems. This can result in a savings to your company to eliminate a major supplier problem before it becomes your incoming inspection problem at your plant.

It is less expensive and more productive to work with your current supplier to improve their quality than to qualify an unknown new supplier. Management decisions at these value points are difficult to decide what is best for the company and your customer.

In conjunction with assisting your suppliers, you often act as their consultant in how they can improve their business and quality of operations. Many major corporations do this with their suppliers and it has turned into a win-win arrangement for each company.

The emphasis, rewards, and improvements now being obtained by major suppliers is fast moving down to medium size and small companies. When used with ISO/QS-9000 it will only assist these companies to further improve their performance and quality. The awakening of Six Sigma quality methodology, even considering it's initial and ongoing costs will stimulate more companies to be the best they can be as long as their management remains on this course of proven quality improvement.

Six Sigma is truly the first 20th century quality program that will continue into the 21st century that will succeeded early in its implementation. Management now has full approval of the Six Sigma program values with an understanding of the results that can be attained for long-term success. The main reason is the program is designed and requires programs show a positive reward in monetary savings that management can see and verify the Six Sigma system is succeeding.

They can also see the positive result obtained when the Cost of Quality falls below 1% of their Cost of Sales that is the true cost and improvement result of a quality program and a true measure of the success of the Six Sigma program.

REFERENCES

1. Harrold, D., "Optimize Existing Processes to Achieve Six Sigma Capability." Control Engineering March 1999: 87–90.
2. Hunkar, D. B., "An Engineering Approach to Process Development and the Determination of Process Capability." Hunkar Laboratories, Cincinnati, Ohio, 1991, Document No. 228.
3. Hunter J. S., "The Box-Jenkins Bounded Manual Adjustment Chart." Quality Progress August 1998: 129–137.
4. Provost, L. P. et. al., The Importance of Concepts in Creativity and Improvement." Quality Progress March 1998: 32–38.
5. Langley, G. et al, "*The Improvement Guide: A Practical Approach to Enhancing Organizational Performance*, (San Francisco, CA: Jossey-Bass, 1996).
6. H. Altov, *The Art of Inventing: And Suddenly the Inventor Appeared,* translated and adapted by Lee Shulyak (Worcester, MA: Technical Innovation Center, 1994).

Glossary

Abrasion The wearing away of some surface area by its contact with another material.

Absorption Moisture.

Accelerated aging Aging by artificial means to obtain an indication on how a material will behave under normal conditions over a prolonged period.

Accelerated Weathering Duplicating or reproducing weather conditions by machine-made means.

Acceptable quality level The minimum quality level at which a product will be accepted or rejected

Acetal resin A crystalline thermoplastic material made from formaldehyde. Trade names: Delrin and Celcon.

Acrylics The name given to plastics produced by the polymerization of acrylic acid derivatives, usually including methyl methacrylate. An amorphous thermoplastic material that is clear.

Acrylonitrile, Butadiene, Styrene (ABS) A thermoplastic classified as an elastomers-modified styrene.

Additive A material added to resin prior to molding or forming to add a desired property or characteristic to the finished product or to assist in the processing of the material.

Adsorption *See* Moisture.

Aesthetics Referring to the external surface appearance of a plastic product.

Aging The change of a material over time under defined natural or synthetic environmental conditions, leading to improvement or deterioration of properties. *See also* Accelerated aging: Artificial aging.

Air vent A small outlet or area around the periphery of a mold cavity used to prevent entrapment of gases within the mold cavity.

Algorithms A mathematical procedure and series of equations used in a computer with support software to produce a result based on input data.

Ambient temperature Temperature of the medium surrounding an object. Used to denote prevailing room temperature.

Amorphous Plastic materials that have no definite order of Crystallinity.

Amortized The cost of an item spread out equally over time or a specified number of parts. Frequently used when estimating finished product costs if the cost of a mold or capital equipment is spread out and added into the piece part cost.

Analog Refers to a needle on a scale readout that is almost instantaneous from the input signal to the output readout. Determined by the design of the circuitry. Gives an approximate reading based on the detail of the readout scale.

Angular welding See ultrasonic sealing.

Anneal (1) To head a molded plastic article to a predetermined temperature and slowly cool it to relieve stresses. (Annealing of molded or machined parts may be done dry, as in an oven, or wet, as in a heated tank of mineral oil). Often done with the part in a holding fixture. (2) To heat steel to a predetermined temperature above the critical range and slowly cool it to relieve stresses and reduce hardness.

Antioxidant A substance added to a material to inhibit oxidation.

Antistatic agents (antistats) Agents when added to the molding material or applied on the surface of the molded part, make it less able to conduct electricity (thus hindering the fixation of dust).

Approved supplier A product supplier who has been rated satisfactorily on previous jobs. May involve a detailed analysis of manufacturing and quality capability to be sure it meets customer requirements.

Artificial aging The accelerated testing of plastic specimens to determine their changes in properties. Carried out over a short period of time, such tests are indicative of what may be expected of a material under service conditions over extended periods. Typical investigations include those for dimensional stability; the effect of immersion in water, chemicals, and solvents; light stability; and resistance to fatigue among others.

ASTM Abbreviation for the American Society for Testing and Materials.

Attribute Unlike a property it is a quality that is less precisely known and is only ascribed to someone or something.

Automatic mold A mold or die in injection or compression molding that repeatedly goes through the entire cycle without human assistance.

Auxiliary equipment Refers to equipment, other than the injection molding machine and mold, required to ensure the manufactured product would be made correctly, including for example, dryers, chillers, material and part conveyors, and robots.

Balanced mold A mold laid out with runner and mold cavities spaced and sized for uniform flow, fill, and packing pressure throughout the system.

Bar coding The electronic/optical bar recognition system for identification, storage, printout, and retrieval of specified data and information.

Barrel (extruder) In injection molding, extrusion, or bottle-blowing equipment. It is the hollow tube in which the plastic material is gradually heated and melted and from which it is extruded into a die or rammed into the mold cavity under pressure.

Bench marking The base line or starting point for a program that all future measurements or contributing comments are referenced from or to.

Blanket Purchase Order (BPO) A purchase order placed with a supplier for materials over a set time period. Customer then releases material as required or as specified.

Blend Any combination of mixtures of a base resin with additives or modifiers. The base resin has been modified.

Blow molding (1) A molding process primarily used to produce hollow objects. (2) A molding process in which a hollow tube (a parison) is forced into a shape of the mold cavity using internal air pressure. The two primary types are injection blow molding and extrusion blow molding.

Blush The tendency of a plastic to turn white or chalky in areas that are highly stressed, such as gate blush.

Boss Projection on a plastic product designed to add strength, facilitate alignment during assembly, and provide for a point to fasten or screw the product together.

Branched chains In polymer chemistry, side chains attached to the main original polymer chain.

Bubble A spherical, internal void; globule or air or other gas trapped within a plastic. *See* Void.

Burned A carbonized condition showing evidence of thermal decomposition through some discoloration, distortion, or localized destruction of the surface of the plastic. Usually caused by poor venting of the mold cavity.

Burning (1) Overheating the resin in the barrel causing discoloration and, if long enough, charring the material. (2) Caused by trapped gasses in a poor or non-vented area of the mold. The gasses may ignite, due to pressure and temperature, as in a diesel engine, and discolor or char the product.

C of C, Certificate of Compliance A letter or form furnished by a supplier who states the material meets company or predetermined customer requirements.

Capable process A process able to make a high percentage of product within specification (ie, 99.5%).

Caprolactam A cyclic amide compound containing six carbon atoms. When the ring is opened Caprolactam is polymerizable into a nylon resin known as type-six (6) nylon or polycaprolactam.

Carbon black A black pigment produced by the incomplete burning of natural gas or oil. It is widely used as filler, particularly in the rubber industry and wire/cable applications. Because it possesses useful ultraviolet protective properties, it is also use in molding compounds intended for outside weathering applications.

Cavity Depression in the mold that usually forms the outer surface of the product. Depending on number of such depressions, molds are designated as a single cavity mold, a multi-cavity mold, or a family cavity mold.

Cavity number A sequential number engraved in a mold cavity and reproduced on the molded part for later reference in case a problem ever occurs with the part. Used in multi-cavity molds of similar parts.

Cellular plastics Foamed plastic products.

Cementing A process of joining two similar plastic materials to themselves or to dissimilar materials by means of solvents.

Chalking Dry, chalk-like appearance or deposit on the surface of a plastic.

Change request A written request, often called an (Engineering Change Request [ECR]) to modify or alter the dimensions, material, tolerances, or manufacture or a part now in or soon to be in production. Use to ensure all interested and involved department personnel are informed and can comment and approve or disapprove of the pending change.

Check list A list of written guides or instructions that must be completed before the next step of an operation is begun with the items completed in a set order so the operation or act is completed as specified to obtain the desired results.

Chemical resistance Ability of a material to retain utility and appearance following contact with chemical agents.

Chiller A refrigeration unit used to supply cooling water in a closed loop system for a mold or equipment used to regulate temperature.

CIM Computer integrated manufacture, the use of computer technology to manufacture and control a product.

Clamping area The largest rate molding area an injection or transfer press can hold closed under full molding pressure.

Clamping force (Clamping pressure) In injection molding, the pressure is applied to the mold to keep it closed despite the fluid pressure of the compressed molding material within the cavity and runner system.

Clarity Material clearness or lack of haze.

Closed loop System used with microprocessor for control of a machine's cycle. *See* Feedback.

Closed loop continuous feed back process control A system collecting operation variable manufacturing data, analyzing the data in "Real Time" and using software to adjust the machine variables for the next cycle when required. Used to adjust and perform in a continuous operation, cycle to cycle.

Coefficient of expansion The fractional change in a specified dimension (sometimes volume) of a material for a unit change in temperature. Values for plastics range from 0.01 to 0.2mils/in.°C (ASTM D 696).

Color concentrate A mixture of a measured amount of dye or pigment and a specific plastic material base. A more precise color can be obtained using concentrates than using raw colors. *Note:* Care should be taken to verify that the color concentrate base is compatible with the plastic it is to color. Color concentrate is normally used at 1 to 4% of the plastic material to be colored.

Colorfast The ability to resist change in color.

Color standard The exact color a plastic resin or product must match to be acceptable. Resin suppliers often submit color chip samples of the matched resin color to be compared to the molded part. The color chip, or standard, is usually 2×3 inch. With one polished surface and various textured surfaces on the opposite side. Suppliers use similar standards to verify the color of each lot of resin shipped to their customers.

Combination mold *See* Family mold.

Commodity resin Usually associated with the higher-volume lower-priced plastics, with low-to-medium physical properties. Examples are PE, PP, PS, and acrylic, PVC, EVA, and ABS. Used for less critical applications.

Compound A mixture of polymer(s) with all materials necessary for the finished product.

Compression ratio In the extruder of an injection/blow molder screw, the ratio of volume available in the first flight at the hopper to the last flight at the end of the screw.

Compressive strength Crushing load at the failure of a specimen divided by the original sectional area of the specimen (ASTM D 695).

Concentricity (1) The relationship of all circular surfaces with the same center. (2) Relationship Of all inside dimensions to all outside dimensions. Usually, as with diameter, expressed in thousandths of an inch (F.I.M. = FULL INDICATOR MOVEMENT). Deviation from concentricity is often referred to as run out.

Conditioning The subjection of a material to a stipulated treatment so that it will respond in a uniform way to subsequent testing or processing. The term is frequently used to refer to the treatment given before testing. ASTM standard conditions for a plastic testing laboratory are 23°C (73.4°F + 3.6°F) and 50 + 5% relative humidity.

Conditioning chamber An enclosure used to prepare parts for their next step in the assembly or decoration process. Parts can be stress relieved, humidity or moisture conditioned. or impregnated with another element.

Consigned material Material given over to another supplier for care and use in manufacturing a customer's product.

Contamination Any foreign body in a material that affects or detracts from the parts quality.

Control plan A written plan that lists step-by-step procedures describing how a specific operation will be conducted and followed.

Controllers The instruments, timers, and pressure controls used to control and regulate the molding cycle or any manufacturing cycle.

Cooling channels Channels or passageways within the body of a mold through which a cooling or heating medium can be circulated to control temperature on the mold surface. May also be used for heating a mold by circulating hot water, steam, hot oil, or other heated fluid through channels as in molding thermoplastic materials.

Cooling time The time period required after the gate freezes for the part to solidify and become rigid enough for ejection from the mold cavity.

Copolymer A polymer produced by polymerization of two or more monomers. Can also be done as a secondary compounding operation on an extruder.

Core (1) Male element in die that produces a hole or recess in a part. (2) Part of a complex mold that molds undercut parts. Cores are usually withdrawn to one side before the main sections of the mold open. (3) A channel in a mold for circulation of a heat-transfer medium. (4) The central member of a laminate.

Corona treatment Exposing a plastic part to a corona discharge increasing receptivity to inks, lacquers, paints, and adhesives. *See also* Surface treatment.

Creep The dimensional change with time of a material under load, following the initial instantaneous elastic deformation. *See also* Cold flow (ASTM D 674).

Critical path A method of charting/scheduling for identifying key elements in the path to completion of a program.

Cross-linking The chemical combination of molecules to form thermally stable bonds within a polymer, not broken by heating.

Crystallinity A state of molecular structure in some resins that denotes uniformity and compactness of the molecular chains forming the polymer. Normally attributed to the formation of solid crystals with a definite geometric form. High Crystallinity causes a polymer to be less transparent or opaque.

Cure That portion of the molding cycle during which the plastic material in the mold becomes sufficiently rigid or hard to permit ejection.

Curing time The time between the end of injection pressure and the opening of the mold.

Cycle The complete, repetitive sequence of operations in a process or part of a process. In molding, the cycle time is the period, or elapsed time, between a certain point in one cycle and the same point in the next.

Degradation A deleterious change in the chemical structure, physical properties, and/or appearance of a plastic, usually caused by exposure to heat.

Density Weight per unit volume of a substance, expressed in grams per cubic centimeter or pounds per cubic foot.

Design of Experiments (DOE) A problem-solving technique developed by Taguchi using a testing process with an orthogonal array to analyze data and determine the main contributing factors in the solution to the problem.

Design stress A long-term stress, including creep factors and safety factors that is used in designing structural fabrication.

Destaticization Treating plastic materials to minimize their accumulation of static electricity and subsequently the amount of dust picked up by the plastics because of such charges. *See* Antistatic.

Destructive test Any test performed on a part in an attempt to destroy it, often performed to see how much abuse the part can tolerate without failing.

Deterioration A permanent change in the physical properties of a plastic as evident by impairment of these properties.

Die A metal form in making or punch in plastic products. It is used interchangeably with mold. In extrusion it refers to the tooling forming the plastic shape the molten plastic is forced through.

Drop test *See* Impact test.

Dry as molded (DAM) Term used to describe a part immediately after it is removed from a mold and allowed to cool down. All physical, chemical, and electrical property tests are performed on non-conditioned test bars and results recorded on the data sheets. Parts and test bars in this state (DAM) are felt to be their weakest in some properties, as they have not had time to condition or relieve any molded-in-stresses.

Dry coloring Method commonly used to color plastic by tumble blending uncolored particles of the plastic materials with selected dyes and pigments.

Dryers Auxiliary equipment used to dry resins before processing to ensure that surface properties are within manufactured specifications. There are several styles of dryers, including ovens, microwave, and hot-air desiccant bed and refrigeration types.

Ductility The extent to which a solid material can be drawn into a thinner cross-section without breaking.

Durometer hardness The hardness of a material as measured by the Shore Durometer.

Dyes Intensely colored synthetics or natural organic chemicals that are soluble in most common solvents and dissolve in the plastic substrate while imparting color. Characterized by good transparency, high-tincturial strength, and low specific gravity.

Economic order quantity Ordering a product in a quantity for cost savings and for projected use in a predetermined time period.

Ejection The removal of the finished part from the mold cavity by mechanical means.

Ejection time The time in the cycle when the mold opens, the part is ejected, the mold closes and clamping pressure is applied.

Elastic deformation A deformation in which a substance returns to its original dimensions on release or the deforming stress.

Elasticity That property of a material by virtue of which it tends to recover to its original size and shape after deformation. If the strain is proportional to the applied stress, the material is said to exhibit Hookean or ideal elasticity.

Elastomer A material that at room temperature can be stretched repeatedly under low stress to at least twice its length and snaps back to the original length upon release of the stress.

Elongation The increase in length of a material under test, expressed as a percentage difference between the original length and the length at the moment of the break.

EMI/EMP Electromotive Interference and electromotive protection, terminology to describe electronics vulnerable to electrical radiation interference, which can affect and damage solid-state devices.

Endothermic An action or reaction that absorbs heat.

End use The function the part or assembly was originally designed and manufactured to perform.

Engineering resin Associated with plastics having medium to high physical properties used for structural and demanding applications. Examples are nylon, acetal, PBT, PET, PC, PPS, AND LCP.

Environmental stress cracking (Esc) The susceptibility of a thermoplastic article to crack or craze under the influence of certain chemicals or aging, weather, and stress. Standard ASTM test methods that include requirements for environmental stress cracking are indexed to ASTM standards.

Ethylene-vinyl acetate A plastic copolymer made from the two monomers, ethylene and vinyl acetate. This copolymer is similar to polyethylene, but has considerable increased flexibility.

Exothermic Pertaining to an action or reaction that gives off heat.

Extrusion The plasticizing of a material in an extruder (barrel-and-screw or plunger assembly) and forcing of the molten material or extrudate through a die or into a mold. The initial part of the molding process.

Fabricate To work a material into a finished form by machining, forming, or other operations. In the broadest sense, it means to manufacture.

Factor An independent variable that will be controlled in the designed experiment. (ie. mold temperature).

Fractional Factorial A type of designed experiment that investigates all primary effects of factors as well as some of the interactions between factors.

Failure mode and effects analysis (FMEA) A quality assurance tool analyzing all manufacturing operations in a continuous step-by-step manner to determine any variables in an operation that can affect the operation. Once these are determined develop ways to control the variability and selection of control methods for the control of these variables to produce a repeatable good product, cycle-to-cycle.

Family mold (1) A multi-cavity mold in which each cavity forms a part that often has a direct relationship in usage to the other parts in the mold. Family molds can have more; than one cavity making the same part, but they will still always have that same direct relationship; to usage. (2) A multi-cavity mold in which each cavity forms one of the component parts of the assembled object. The term often applied to molds in which parts from different customers are grouped together in one mold for economy of production. Sometimes called a combination mold.

Feedback Information returned to a system or process to maintain the output within specified limits.

Fiber Thin strands of glass used to reinforce both thermoplastic and thermosetting materials. One-inch-long fibers are occasionally used. but the more common lengths are 0.25 to 0.50 inches long and often less than 0.100 inches long.

Fill rate The pressure-tie relationship used to described the filling of the mold cavity.

Filler An inert substance added to plastics for the purpose of improving physical properties or processability or reducing the cost of the material.

Finish (1) To complete the secondary work on a molded part so that it is ready for use. Operations such as filling, deflashing, buffing, drilling, tapping, and degating are commonly called finishing operations. (2) The plastic forming the opening of a bottle, shaped to accommodate a specific closure. The ultimate surface structure of a part. (3) *See* Surface finish.

Finite element analysis (FEA) A stress analysis technique of a part using a computer-generated model that can take finite sections of the part for analysis of the forces and loads the part will experience in service. It generates a part-section analysis that shows the force concentrations in the section and determines if the material selected will be suitable for the part by calculating the stresses in the material.

First surface The front surface of a plastic part, nearest the eye.

Fishbone diagram (Ishakawa Diagram) A problem analysis technique used to list all the variables and steps in the solution to a problem. All contributing elements are associated with each factor and taken back to their starting point to ensure that all variable elements are considered.

Fixture Means of holding a part during a machine or other operation.

Flame retarded A resin modified by flame-inhibiting additives so that exposure to a flame will not burn or will self-extinguish. Some resins will not burn as thermosets; others can be modified to meet agency flame/burning specifications; and others, depending on their base materials, may not be able to be modified.

Flammability Measure of the extent to which a material will support combustion.

Flex bar An ASTM specified test bar used to develop physical property data for plastic materials. Usually sized at $4 \times 1/2 \times 1/8$ inches or thicker, depending on the ASTM specification.

Flexural strength Ability of a material to flex without permanent distortion or breaking (ASTM D 790).

Flow (1) A qualitative description of the fluidity of a plastic material during the process of molding. (2) A quantitative value of fluidity when expressing a melt flow index. *See* Melt index.

Flow-chart A line chart that traces a process from start to finish.

Flow length The actual distance a material will flow under a set of molding machine conditions. Influenced by the processing and mold design variables, the composition of the polymer, and any additives in the polymer.

Flow line A mark on a molded piece made by the meeting of two flow fronts during molding. Also called weld line or weld mark. *See* Weld line.

Flow marks Wavy surface appearance on a molded object caused by improper flow of the material into the mold. *See* Splay marks.

Fluoropolymer A generic name given to fluorine based plastic, trade named Teflon®, plastic material.

Foamed plastics Resins in sponge form. The sponge may be flexible or rigid, the cells closed or interconnected. The density anything from that of the solid parent resin down to, in some cases, 2 pound per cubic foot.

Force (1) (Physics) that which changes the state of rest or motion in matter, measured by the rate of change of momentum. (2) That portion of the mold that forms the inside of the molded part. *See* also Core; Plunger.

Freeze off Refers to the gate area when it solidifies as well as any area in the resin flow system when the melt becomes too cool to flow and solidifies.

Friction welding A means of assembling thermoplastic parts by melting them along their line of contact through friction. *See* also Spin welding.

Full Factorial A designed experiment that determines all the primary effects of the factors as well as all the interactions between factors.

Full indicator movement (F.I.M.) A term in current use to identify tolerance with respect to concentricity. "Former practices" terms are Full Indicator Reading (F.I.R.) and Total Indicator Reading (T.I.R.) runout.

Fusion bond (1) The joining of two melt fronts that meet and solidify in a mold cavity. (2) The bond formed during the assembly operation where the joint line is melted prior to assembly. *See* Hot plate welding; Induction welding; Ultrasonic welding.

Gardner A type of drop-weight impact test. *See* Impact test.

Gardner test *See* Impact test.

Gas assisted "injection" molding (GAM) An injection molding process that introduces a gas (usually nitrogen) into the plasticized material, to form voids in strategic locations.

Gate In injection and transfer molding the orifice through which the melt enters the mold cavity.

Gauges The measuring instruments used to determine if the part meets customer specification, including go/no go plugs, micrometers, and vernier calipers.

Generic Descriptive of an entire type or class of plastic resins. The base resin is one of a family of polymers, but there may be hundreds of product combinations.

Heading The mechanical, thermal, or ultrasonic deformation of a pin to form a locking attachment to retain whatever is under the deformed material.

Heat-distortion point An arbitrary value of deformation under a given set of test conditions. In ASTM Test D 648, it is defined as a total deflection of 0.010″ in a rectangular bar supported at both ends under a load of 66 or 264 psi. while submersed in oil. The temperature is raised 2 °C per minute until this deflection is reached.

Heat sealing A process of joining two or more thermoplastic films or sheets by heat and pressure.

Heat stability The resistance of a plastic material to chemical deterioration during processing.

Heat stabilizer An ingredient added to a polymer to improve its processing or end-use resistance to elevated temperatures. The term is used in different contexts depending on the benefit to be derived from the additive. For processing k it retards changes in resin color. For end-use, it protects the surface of the part exposed to elevated temperature from oxidation. It does not imply that a resin can be use beyond its recommended end-use temperature rating if it is heat stabilized.

Heater bands The only heat source for the barrel and nozzle temperature control divided usually, into read, middle, and front, and nozzle temperature control sections. They are very accurate resistance heaters with high heat output.

Heating chamber Injection molding, that part of the machine in which the cold feed is reduced to a hot melt. Also called heating cylinder or barrel.

Hermetic As in seal, to form a bond that is pressure tight, so that air or gasses cannot enter or escape.

Histogram A bar chart used to determine the pattern of variation of a single variable.

Holding pressure The pressure maintained on the melt after the cavity is filled and until the gate freezes off. *See* Packing pressure.

Holdup time *See* Residence time.

Homopolymer The product of the polymerization of a single monomer.

Hot/heated manifold mold A thermoplastic injection mold in which the portion of the mold that contains the runner system has its own heating elements to keep the molding material in a plastic state ready for injection into the cavities, from which the manifold is insulated.

Hot plate welding The use of a heated tool to cause surface melting of a plastic part at the joint are. It is then removed prior to the joint surfaces being pressed together to form a fusion bond.

Hot-runner mold A thermoplastic injection mold in which the runners are insulated from the chilled cavities and remain hot so that the center of the runner never cools in a normal cycle operation. Runners are not usually ejected with the molded pieces. Called insulated runner molds when heating elements are not used in the mold. Note: A heated manifold mold is a hot-runner mold that is both heated and insulated; and insulated mold is a hot-runner mold that does not contain heaters.

Hydrolysis Chemical decomposition of a material involving the addition of water.

Impact strength (1) The ability of a material to withstand shock loading. (2) The work done in fracturing, under shock loading, a specified test specimen in a specified manner. (3) The relative susceptibility of plastic articles to fracture under stress applied at high speeds.

Impact test Often associated with the Gardner (ball or falling dart) test, with a known weight falling at a known distance and hitting a part, thereby subjecting it to an instantaneous high load. Could also be a pendulum-type of impact test. ASTM impact tests for material properties are the Izod, Charpy, and Tensile Impact tests.

Induction welding The use of radio, magnetic, or electrical energy to form a melt through the application of a foreign medium at the joint line to form a fusion bond.

Inert pigment A pigment that does not react with any components of a paint.

Initiator Any foreign additive mixed in a material to cause a chemical or physical reaction in the melt or liquid stage.

Injection molding A molding procedure whereby a heat-softened plastic material is fed into a cavity (mold), which gives the article the desired shape using a screw and ram. Used with both thermoplastic and thermosetting materials.

Injection pressure The pressure in the mold during the injection of plasticized material into the mold cavity. Expressed in psi, with the hydraulic system pressure being used to indicate changes, when there are no sensors in the mold.

Injection time The time it takes for the screw's forward motion to fill the mold cavity with melt.

Inorganic A mineral compound not composed of carbon atoms.

Insert An integral part of plastics molding. It consists of metal or other material, which may be molded into position or may be pressed into the molding after the molding is completed. Also, a removable or interchangeable component of the mold.

Ishakawa Developed the "fishbone diagram" method of analysis.

ISO9000 International Organization of Standardization, the current world class recognized quality standard for all businesses in the world.

Izod A type of pendulum impact. *See* Impact test.

Izod impact test An impact test in which a notched sample bar is held at one end and broken by a blow. This is a test for shock loading. *See* Impact test.

Just-in-time (JIT) A practice developed to minimize customer inventory by the Japanese. The supplier provides the product, at predetermined intervals, so that it can proceed directly to the customer's assembly line. This practice demands excellent quality control and production schedules. Customers who use JIT must demand the same care and treatment from their own suppliers. Suppliers and customers are usually located within a few hours shipp9ng time of each other to make it work effectively.

Kaizen A Japanese developed quality assurance method where in a group of consultants are brought in to a plant with complete control to change, modify, and replace the current manufacturing line. The new methods, procedures, and manufacturing positions are developed for more efficient, economical, and quality proficient operation without management's permission while working with the employees of the company.

Laminar flow Laminar flow of thermoplastic resins in a mold is accompanied by solidification of the layer in contact with the mold surface that acts as an insulating tube through which material follows to fill the remainder of the cavity. This type of flow is essential to duplication of the mold surface.

Land (1) The horizontal bearing surface of a semi-positive or flash mold by which excess material escapes. (2) The bearing surface along the top of the flights of a screw in an extruder. (3) The surface of an extrusion die parallel to the direction of the melt flow. (4) The bearing surfaces of any mold. (5) The gate, when entering a part, has either one or two dimensions. There is always one more dimension involved, which is the length of the gate itself. This would be called the land. On a round gate, it is the second dimension. On a rectangular or square gate, it is the third dimension.

Level The value of a factor (ie, mold temperature - 230° F).

Locked-in-stress *See* Residual stress.

Lot number A number assigned to a specific lot of material or parts. Used for traceability and accountability by the supplier and customers on all paperwork for the product.

Lubricants (1) A processing aid to assist material flow in the barrel of an injection molding machine or extruder. Can be a solid, such as sodium or zinc styrate, or a liquid usually compounded into the base material. (2)

Internally lubricated resins that use oils, Teflon, molybdenum disulfide, or other materials to give the molded part a lower coefficient of friction.

Macbeth A lighting system used for checking color.

Manifold A pipe channel, or mold, with several inlets or outlets.

Master curve The acceptable or required curve that all subsequent test curves must match.

Material review board (MRB) A panel of representatives from departments of the company who are involved with a product. It decides if the material and or product meet customer requirements if a question or problem about quality arises.

Matrix Refers to the base resin or material used for a molded product.

Melt front The exposed surface of molten resin as it flows into a mold. The melt front advances as the molten resin is continuously pushed through its center section.

Melt generation capacity Ability of the injection molding machines barrel and screw to produce the required melt quantity for the size of the barrel and screw combination required for the molding cycle. Used to size the molding machine based on polystyrene melt generation capacity and listed as ounces of melt generation capacity.

Melt index (MI) or melt flow index (MFI) The amount, in grams, of a thermoplastic resin that can be forced through a 0.0825″ orifice when subjected to the prescribed force (grams) in 10 minutes at the prescribed temperature (°C) using an extrusion plastomer (ASTM D 1238).

Melt strength The strength of the plastic while in the molten state.

Melt temperature (1) The temperature at which a resin melts or softens and begins to have flow tendencies. (2) The recommended processing temperature of resin melt for correct processing. (3) The temperature of the melt when taken with a pyrometer melt probe.

Meter SI length unit equal to 100 centimeters or 39.37 inches.

Metering equipment A machine or system to accurately meter additives or regrind to the machine's hopper or feed throat. Comes in many sizes and types to suit each particular application, including augers, shuttle plates, photoelectric eyes, and positive or negative weight loss belt feeders.

Metering screw An extrusion or injection molding screw that has constant shallow depth and pitch section, usually over the last three to four flights.

Methyl methacrylate An amorphous thermoplastic resin. A common name is acrylic resin.

Microprocessor Computer system that stores, analyzes, and adjusts the controls of a machine based on the parameters established during the operation of the machine it is controlling. Only operates within preset limits. Continuously analyzes output data to adjust and maintain the machine's cycle within programmed limits. Can also store data and output it as directed by programming.

Migration of plasticizer Loss of plasticizer from an elastomers plastic compound with subsequent absorption by an adjacent medium or lower plasticizer concentration.

Mil English unit of length equal to 0.001 inch or 0.00254 centimeters.

Milestone chart Usually identified as a go/no go decision point in a critical path flow chart or schedule.

Modifiers Any additive to a resin that improves the processing or end-use properties of the polymer. An example would be plasticizers added to PVC resin to make it soft and pliable and improve its impact strength. All PVC resins use different modifiers to meet desired product requirements. This is true of almost all plastic resins currently manufactured.

Modulus of elasticity The ratio of stress to strain in a material that is elastically deformed (ASTM D 790).

Moisture (1) ABSORPTION, The pickup of moisture from the atmosphere by a material that penetrates the interior. (2) ADSORPTION, Surface retention of moisture from the atmosphere.

Moisture conditioning A method to ensure a product has a predetermined amount of moisture absorbed by the product. Typically done by placing parts in a moisture conditioned and maintained enclosure at ambient or elevated temperature to drive the action of absorption. The moisture level is maintained at the required moisture level for absorption. The operation is timed and parts are weighed to determine the correct amount of moisture absorption, a percentage of part weight.

Moisture vapor transmission rate (MTR) The rate at which water vapor permeates through a plastic film or wall at a specified temperature and relative humidity (ASTM E 96).

Mold (1) (noun) A medium or tool designed to hold a cavity form to make a desired shape and/or size of product. (2) (verb) To process a plastic material using an injection molding process.

Mold deposits Material build up on a cavity's surface due to plate out of resin, usually in a gaseous state. Can also be attributed to additives in a resin adhering to the mold's surface.

Mold open time *See* Ejection time.

Mold release (1) A lubricant used to coat a mold cavity to prevent the molded piece from sticking thereby facilitating its removal from the mold. (2) Additives put into a material to serve as a mold release. Also called a release agent.

Molding A group of plastics processes using molds.

Molding cycle The period of time required to complete a cycle and produce a product.

Molding material Plastic material in varying stages of granulation often comprising plastic or resin filler, pigments, plasticizers, and other ingredients, ready for use in the molding operation. Also called, molding compound or powder.

Molding pressure (1) The pressure applied directly or indirectly on the compound to allow the complete transformation to a solid dense part. (2) The pressure developed by a ram or screw to push molten plastic into a mold cavity. *See* Injection pressure.

Molding shrinkage The difference in dimensions, expressed in inches per inch, between a part and the mold cavity in which it was molded. Both the part and the mold cavity are at normal room temperature when measured. Also called mold shrinkage and contraction.

Molecular weight (MW) (Average Molecular Weight) The sum of the atomic masses of the elements forming the molecule, indicating the relative size typical chain length of the polymer molecule.

Monomer A low-molecular-weight-reactive chemical that polymerizes to form a polymer.

Morphology The study of the physical form and structure of a material.

Mottle A mixture of colors or shades giving a complicated pattern of specks, spots, or streaks.

Multi-cavity mold A mold having more than one cavity or impression for forming finished items during one machine cycle.

Node A single point on a FEA model. A node is the starting and connection points of a mesh. All nodes connect to each other in a 2-D or 3-D geometric analysis.

Nonrigid plastic A plastic that has a modulus of elasticity (either in flexure or in tension) of not over 10,000 psi at 25 °C and over 505 relative humidity (ASTM D747).

Normal Distribution A pattern of variation that looks like a bell shaped curve.

Notch sensitive A plastic material is said to be notch sensitive if it will break when it has been scratched, notched, or cracked. Glass is considered to be highly notch sensitive.

Nucleation (nucleator) With crystalline polymer, any foreign additive that assists or acts as a starting site for Crystallinity within the resin. These initiators can reduce cycle time by speeding up the crystalline formations, there by causing the part to solidify faster so its ejection from the mold can occur sooner.

Nylon A generic term for polyamides. A crystalline thermoplastic.

Olefin plastics Plastics produced from olefins (polyolefins). Examples are polyethylene and polypropylene.

Opaque A material that will not transmit light and is not transparent.

Optical comparator An inspection machine using optics to compare the outline of a part to its required dimensions on a graphic screen.

Organic Refers to the chemistry of carbon compounds.

Orientation The alignment of the crystalline structure in polymeric materials so as to produce a highly uniform structure. Can be accomplished by cold drawing or stretching during fabrication.

Pack time The amount of time that packing pressure is kept on the screw until the gate freezes off. Occurs immediately after the initial injection stroke ends.

Packing pressure The pressure applied just before the part cavity fills, and is maintained to keep melt flowing into the mold cavity to compensate for in-mold material shrinkage and until the gate freezes off. Packing pressure must be the same as the injection pressure so that the mold cavity is not depressurized during the final packing period.

Pareto analysis An analytical and statistical technique used to determine part defect type and quantity. Ranks each type of defect as a percentage of the total number of defects found, based on the quantity of each type of defect.

Parison Term given the extruded molten material, usually hollow, to form a product in an extrusion or blow molded operation.

Piece part price The calculated finished part cost based on material, processing, assembly, decoration, and packaging, including productivity and overhead costs.

Pigment Imparts color to plastic while remaining a dispersion of undissolved particles.

Pigmented Color pigments are added to a resin to produce a desired color in the plastic resin after molding. Pigments can be either organic or inorganic based material. The inorganic pigments are usually heavy metals that are carcinogenic and no longer used.

Plastic (1) (noun) One of the high polymeric materials, either natural or synthetic, exclusive of rubbers, which either melt and flow with heat and pressure, as with a thermoplastic, or chemically "set", as with a thermoset material. (2) (verb) Capable of flow under pressure or tensile stress.

Plastic deformation The deformation of a material under load that is not recoverable after the load is removed. Opposite of elastic deformation.

Plastic memory A phenomenon of a plastic to return, in some degree, to its original form upon heating.

Plasticate To soften by heating or kneading.

Plasticity A property of plastics that allows the material to be deformed continuously and permanently without rupture upon the application of a force that exceeds the yield value of the material.

Plasticize To make a material soft and moldable with the addition of heat and/or pressure or a plasticizer.

Polyallomers Crystalline thermoplastic polymers made from two or more differed monomers, usually ethylene and propylene.

Polyamides A group of crystalline thermoplastics, of which nylon is typical.

Polycarbonate resin An amorphous thermoplastic material. It is transparent and can be injection molded, extruded, thermoformed, and blow molded. It is known for its high impact force retention capabilities but is solvent sensitive. The material is amorphous.

Polyethylene A crystalline type thermoplastic material made by polymerizing ethylene gas.

Polyimide Classified as a thermoplastic, it cannot be processed by conventional molding methods. The polymer has rings of four carbon atoms tightly bound together. It has excellent resistance to heat.

Polyliner (1) A perforated, longitudinally ribbed sleeve that fits inside the cylinder of an injection-molding machine. Used as a replacement for conventional injection cylinder torpedoes (older machines). (2) A plastic bag placed inside a carton or box to prevent moisture and foreign material contamination during shipment of resin to a customer.

Polymer A high molecular weight organic compound – natural or synthetic – whose structure a repeated small unit, the MER, can represent. Examples are polyethylene, rubber, and cellulose. Synthetic polymers are formed by addition of condensation polymerization of monomers. Some polymers are elastomers and some are plastics.

Polymerization A chemical reaction in which the molecules or a monomer are linked together to form large molecules whose molecular weight is a multiple of the original substance. When two or more monomers are involved, the process is called copolymerization. Addition and condensation are the two major types of reactions.

Polyphenylene oxide (PPO) An amorphous thermoplastic. This material is noted for its useful temperature range from –275 to 375 °F.

Polystyrene An amorphous thermoplastic made by polymerizing styrene.

Polysulfone An amorphous thermoplastic noted for its high strength.

Polyvinyl chloride (PVC) A thermoplastic material made by the polymerization of vinyl chloride with peroxide catalysts. The pure polymer is brittle and difficult to process. It yields a flexible material when compounded with plasticizers.

Post annealing Stress relieving of molded parts by external means, hot air, or oil, humidity chambers, or submersion in a fluid.

Post mold shrinkage The shrinkage occurring after a part has been removed from the mold. Influenced by the material and chemical properties of the resin and its molding conditions. Also influenced by end-use conditions and environmental conditions.

Postforming A process used to impart a shape to a previously molded article.

Potentiometer An electrical control device that senses changes in voltage or a potential difference by comparison to a standard voltage and can transmit a signal to a control switch.

Preplastication Technique of premelting injection molding powders in a separate chamber, then transferring the melt to the injection cylinder. Device used for preplastication is commonly known as a preplasticizer.

Pressure drop The decrease in pressure on a fluid attributed to the number of turns it has to make and the distance it must flow to fill a cavity.

Pressure gradient lines A hypothetical set of pressure lines in a part created by the material's pressure drop as the part is filled. The further the material flows from the gate, the lower the packout pressure.

Procedure A set of established steps or methods for conducting the affairs of a business or for manufacture of a product or providing a service. A way of performing an operation.

Process A series of actions, functions, or operations that result in the completion of an act or process such as an end or result.

Process control procedures A separate document, often included as an attachment to the quality control manual, which is a detailed description of the methods to be followed in the manufacture of a product. A copy may be attached to the work order for reference and revised as required should changes in the product occur.

Product certification The certificate or letter stating that the material or product meets or exceeds customer requirements. Values are often listed for the tested or measured results. Signed by a key representative of the company to verify accuracy.

Projected surface area The exposed resin area of a mold on the parting line that transmits the injection pressure on the closed mold halves. Includes part, runner, and sprue surfaces expressed in inches squared of the surface area.

Prototype mold A simplified mold construction often made from a light metal casting alloy or from an epoxy resin in order to obtain information for the final mold an/or part design.

Property Designates a specific quality that is basic to a thing and often makes it act in a certain or specific way.

Pyrometer An electrical thermometer for measuring high temperatures. Unit comes with two probes to measure melt and surface temperatures.

QS-9000 Automotive harmonization of Dammler/Chrysler, Ford, and General Motors with input from the trucking manufacturers. An add on to ISO9000 requiring documentation and verification in greater depth and detail to the automotive suppliers specifications and requirements. Soon to be combined with a world wide automotive specification in or about 2002.

Quality assurance A separate department established to direct the quality function of the business and systems responsibility areas. Major concentration is direct to assisting and auditing the activities of the quality control department in their efforts to ensure that quality products are produced.

Quality circles A quality analysis group consisting of employees with specific departmental knowledge used to provide suggestions and ways to solve a procedural or manufacturing quality problem. If found acceptable, the groups findings and solutions are ten passed on to upper management for implementation.

Quality control A department set up to be technically involved in the control of product quality. Involved in the principal inspection and testing of a product, with limited systems responsibility.

Quality control manual A document that states the company's quality objectives and how they will be implemented, documented, and followed in the manufacture and conducting of business with their customers.

Quality Function Deployment Method of obtaining the required information from a customer, supplier, or your own company personnel for solving a problem, improving a product, or providing a required or necessary service to a customer.

Quality rated *See* Approved supplier.

Quench A method of rapidly cooling thermoplastic molded parts as soon as they are removed from the mold. Submerging the parts in water generally does this.

Quick mold change An efficient method of quickly changing over to a new molding program, often staging mold, equipment and tools at the machine to reduce setup charges and program cost.

RFQ, Request for quote A request for a supplier to furnish a price and delivery quote to a customer with in a specified time period as defined by the instructions in the quote.

Real Time An action or operation occurring in present time.

Reciprocating screw A combination injection and plasticizing unit in which an extrusion device with a reciprocation screw is used to plasticizer the material. Injection of material into a mold can take place by direct extrusion into the mod, by reciprocation the screw as an injection plunger, or by a combination of the two. When the screw serves as an injection plunger this unit, the screw and barrel, acts as a holding measuring and injection changer.

Recycled plastics A plastic material prepared from previously used or processed plastic materials that have been cleaned and reground.

Regrind (1) Waste plastics that are recovered and processed for reuse. (2) Plastics that have been ground or palletized at least twice, non-virgin resin pellets.

Reinforced molding compound A material reinforced with special fillers to meet specific requirements, such as glass fibers, mineral, or other reinforcing modified medium.

Release agent *See* Mold release.

Residence time The amount of time a resin is subjected to heat in the barrel of an injection-molding machine.

Residual stress The stresses remaining in a plastic part as a result of thermal or mechanical treatment.

Resin (1) Any of a class of solid or semisolid organic products of natural or synthetic origin, generally of high molecular weight with no definite melting point. (2) In a broad sense any polymer that is a basic material for plastics. *See* Polymer.

Rib An object designed into a plastic part to provide lateral, longitudinal. or horizontal support and additional strength to the section it is added.

Rockwell hardness A common method of testing materials for resistance to indentation in which a diamond or steel ball, under pressure, is used to pierce the test specimen (ASTN D 785).

Runner In an injection or transfer mold, the channel that connects the sprue with the gate to the cavity. The channel through which the molten plastic flows into the mold cavity.

Runner system With plastics, the sprue, runners, and gates that lead the material from the nozzle of an injection-molding machine to the mold cavity.

Salt and pepper blends Resin blends of different concentrate additives, in pellet form, mixed with virgin resin to make a different product. Usually associated with color concentrate blends, that, when melted and mixed by the injection molding machine's screw, yield a uniform colored melt for a product.

SAN An abbreviation for styrene-acrylonitrile copolymers.

Scrap A product or material that is out of specification to the point of being unusable.

Screening Experiment A designed experiment that evaluates only the primary effects of factors. Used in the initial stages of process experimentation.

Screw The main component of the "reciprocation screw" injection-molding machine. Has various sizes, lengths, and compression ratios to feed, compress, melt, and meter for injection into the mold cavity. Basically divided into three major sections but there can be more. *Feed Section* – deep screw depths to convey the resin into the next screw's section. *Transition Section* – Gradually decreasing screw depths when resin is compressed, forced against the barrel's surface, and melts. *Metering Section* – The molten melt is further compressed in a shallow, uniform screw depth conveying forward as the screw turns past the check ring at the front of the screw.

Screw flights The circular groves cut into the screw whose size, depth, and shape convey the pellets down the barrel compressing and melting them while preparing the melt for the next molding cycle.

Screw plasticating injection molding *See* Injection molding.

Secondary finishing operations Operations performed on a product after it has achieved its primary form required to complete the manufacturing of the product, i.e.: decorating assembly, packaging, etc.

Semi-automatic molding machine A molding machine in which only part of the operation is controlled by direct human action. The machine according to a predetermined program controls the automatic part of the operation.

Setup charge A monetary amount calculated to cover the expense of preparing a machine for the next molding operation, time, material, equipment, and labor. A set fee, actual cost or percentage of overhead.

Shear Stress developed because of the action of the layers in the material attempting to slide against or separate in a parallel direction.

Shear heat The rise in temperature created by the compression and longitudinal pressure on the resin in the barrel by the screw's pumping and turning action.

Shelf life The time a material, such as an additive for a molding compound, can be stored without losing any of its original physical or functional properties.

Shore hardness A method of determining the hardness of a plastic material. This device consists of a small conical hammer fitted with a diamond point and acting in a glass tube. The hammer is made to strike the material under test and the degree of rebound is noted on a graduated scale. Generally, the harder the material, the greater the rebound (ASTM D 2240).

Shot The yield from one complete molding cycle, including sprue, runner and flash.

Shot capacity The maximum volume of material that a machine can produce from one forward motion of the plunger or screw. All machines are rated using polystyrene as the melt standard and presented in ounces or pounds of melt per cycle.

Shot volume *See* Shot capacity

Shot weight The amount of molten resin generated in the barrel and injected into the mold cavity on a typical molding cycle to fill and packout the mold cavity to the correct part weight.

Shrink fixture *See* Cooling fixture.

Shrinkage In a plastic, the reduction in dimensions after cooling.

Shrinkage allowance The additional dimensions that must be added to a mold to compensate for shrinkage of a plastic material on cooling.

SI units Systems International Units.

Silicone (1) Chemical derived from silica used in molding as a release agent and general lubricant. (2) A silicon-based thermoset plastic material.

Six Sigma The new quality term and methodology for identifying a process control technique to control a process within Six sigma limits that reduces defects to 3.4 defects per million, a reduction of 20,000 times.

Snap fit An assembly of two mating parts, with one or both parts deflecting until the mating parts are together. They then return to their as-molded condition or nearly so, depending on the design of the attachment. Parts can be under high to low stress after assembly.

Solvent Any substance, but usually a liquid, that dissolves other substances.

Solvent welding (solvent cementing, solvent bonding) A method of bonding thermoplastic articles of like materials to each other by using a solvent capable of softening the surfaces to be bonded. Thermoplastic materials that can be bonded by this method are ABS, acrylics, cellulosics, nylons, polycarbonate, polystyrene, and vinyls.

Sonic bonded High frequency vibrations generated by a transducer and transmitted in a tuned horn that contacts a part. The vibration energy generates heat and while pressure is applied to form a seal/connection or shape.

Specific gravity The density (mass per unit volume) of a liquid or solid material divided by that of water (ASTN D 792).

Specification A written statement that dictates the material, dimensions, and workmanship of a manufactured product.

Spin welding The process of fusing two objects by forcing them together while one of the pair is spinning, until frictional heat melts the interface. Spinning is then stopped and pressure held until they are fused together.

Spiral flow test A method of determining the flow properties of a thermoplastic or thermoset material, in which the resin flows along the path of the spiral cavity that is circular in design from the sprue. The length of the material that flows into the cavity and its weight gives a relative indication of the flow properties of the resin.

Splay marks or splay Marks or lines found on the surface of the part after molding that may be caused by overheating the material, moisture in the material, or flow paths in the part. Usually white, silver, or gold in color. Also called silver streaking.

Spot welding The localized fusion bonding of two adjacent plastic parts. Does not require a molded protrusion or hole in the parts. To be effective, used where two parallel and flat surfaces meet.

Statistical process control The gathering of variable data using quality control methodology and charting the results to monitor and control a process.

Stereolithography A three-dimensional modeling process that produces copies of solid or surface models in special plastic resins. This process uses a moving laser beam, directed by computer; to copy or draw sections of the computer generated drawing or model onto the surface of photo-curable liquid plastic. After each pass the model indexes down into the resin for the next layer to be developed.

Storage life *See* Shelf life.

Strain The dimensionless numbers (or units of length/length, i.e. inch per inch) that characterize the change of dimensions of a test specimen during controlled deformation. In tensile testing, the elongation divided by the original gage length of the test specimen.

Strength of material Refers to the structural engineering analysis of a product to determine its strength properties.

Stress The force applied to produce a deformation in the material. The ratio of applied load to the original cross-sectional area of a test specimen (psi).

Stress concentration Sections or areas in a part where the molded-in or physical forces are very high or magnified by a force or action.

Stress crack External or internal cracks in a plastic caused by tensile stresses less than its short-term mechanical strength can withstand.

Styrenic Indicates a group of plastics materials that are polymers, either whole or partially polymerized from styrene monomer.

Surface finish Finish of a molded product. Refer to the SPI-SPE Mold Finishes Comparison Kit, available from DME Corporation, Detroit, Michigan.

Surface treatment Any method of treating a material so as to alter the surface and render it receptive to inks, paints, lacquers, and adhesives such as chemical, flame, and electronic treatments.

Taguchi *See* Design of experiments.

Temperature gradient The slope of a graphed temperature curve. An increasing or decreasing temperature profile on the barrel of he molding machine is an example.

Tensile impact test A test whereby the sample is clamped in a fixture attached to a swinging pendulum. The swinging pendulum strikes a stationary anvil causing the test sample to rupture. This is similar to the Izod test. *See* Impact test.

Thermal expansion The linear rate at which a material expands or contracts due to a rise or fall in temperature. Each material is unique and has its own rate of expansion and contraction. Expressed in (in/in °F, mm/mm °C).

Thermal stress cracking (TSC) Crazing and cracking of some thermoplastic resins that results from overexposure to elevated temperatures.

Thermocouple A thermoelectric heat-sensing element mounted in or on machinery and the mold to transmit accurate temperature signals to a control and readout unit.

Thermoelastomers *See* Elastomers.

Thermoplastic (TP) (1) (adjective) Capable of being repeatedly softened by heat and hardened by cooling. (2) (noun) A material that will repeatedly soften which heated and harden when cooled. Typical of the thermoplastic family are the styrenic polymers and copolymers, acrylics, cellulosics, polyethylene, polypropylene, vinyls, nylons, and the various fluorocarbon materials.

Thermosets (TS) A material that undergoes or has undergone a chemical reaction by the action of heat and pressure, catalysts, ultraviolet light, etc., leading to a relatively infusible state. Typical of the plastics in the thermosetting family are the aminos (melamine and urea), unsaturated polyesters, alkyds, epoxies, and phenolics. A common thermoset goes through three stages. *A-Stage* – an early stage when the material is soluble in certain liquids, fusible, and will flow. *B-Stage* – an intermediate stage at which the material softens when heated and swells in contact with certain liquids, but does not dissolve or fuse. Molding compounds resins are in this state. *C-Stage* – the final stage is the TS reaction when the material is insoluble, infusible, and cured.

Three Standard Deviation A band of variation that encloses three standard deviations should account for 99% of the variation in future products.

Timers Analog or digital timers used to accurately control the molding cycle operations of occurrences.

T.I.R. (Total Indicator Reading) An abbreviation used to identify tolerances with respect to concentricity. *Note:* The term T.I.R. is a "former practices" term; the more acceptable current term is F.I.M. (Full Indicator Movement).

Tolerance A specified allowance for deviation in weighing and measuring or for deviations from the standard dimensions of weight (SPI Guidelines of Plastic Custom Molders).

Tool *See* Mold.

Translucent The quality of transmitting light without being transparent.

Transparent A material with a high degree of light transmission that can be easily seen through.

Treatment Combination A trial with a set of factors run at high and low levels. The eight run screening experiment uses eight treatment combinations.

Ultimate strength Strength (measured in stress as psi (pounds per square inch)) at the break point in tensile test.

Ultrasonic sealing or bonding A method in which sealing is accomplished through the application of vibratory mechanical pressure at ultrasonic frequencies (20 to 40 kc.). Electrical energy is converted to ultrasonic vibrations through the use of either a magnetostrictive or piezoelectric transducer. The vibratory pressures at the interface in the sealing area develop localized heat losses that melt the plastic surfaces effecting the seal.

Unbalanced mold A nonuniform layout of mold cavities and runner system, fill rate, packing pressure, and part quality will vary from cavity to cavity. Used only for noncritical, stand-alone parts.

Universal testing machine A machine used to determine tensile, flexural, or compressive properties of a material in test bar form.

UV (ultraviolet) stabilizer Any chemical compound that when added to thermoplastic material, selectively absorbs UV rays. Carbon black is a natural UV absorber and used extensively in plastic materials.

Variation The differences between items that are supposed to be exactly equal.

Vendor A company or person who sells or supplies a part or service to another for a price.

Vibration welding *See* Ultrasonic sealing.

Vinyl Usually polyvinyl chloride, but may be used to identify other polyvinyl plastics.

Virgin plastics or virgin material Material not previously used or processed and meeting manufacturer's specifications.

Viscosisty A measurement of resistance of a material to flow.

Volume Synonym for capacity or displacement.

Weigh packing A method, often automated, which packs product in a container, based on individual part weight of combination of it. Often weight is compared to part count for small parts.

Weld line *See* Flow line.

Welding Joining thermoplastic parts by one of several heat-softening processes. Butt fusion; spin welding, ultrasonic, and hot gas or plate welding. Each is different and unique but accomplish the same end result.

Yield value (1) (yield strength) In tensile testing, the stress, usually in psi at which there is no increase in stress with a corresponding increase in strain: usually the first peak on the curve. (2) (yield point) The specific limiting deviation from the proportional stress-strain curve.

Young's modulus *See* Modulus of elasticity.

Zero defects A quality control method where anyone in the production cycle who discovers a quality problem can stop the assembly line or manufacturing process until it is corrected. The problem associated with this method is that upper management is often never made aware that a problem occurred. This lack of knowledge may prevent a complete repair from being initiated and the problem continues to occur.

Appendix A

Check Lists for Business and Manufacture

GORDON & ASSOCIATES
www.qualityplasticconsult.com

No. 1 PRODUCT DEVELOPMENT CHECK LIST

DATE:

CUSTOMER:
ADDRESS:
CONTACT: PHONE: FAX: E-MAIL:
ALTERNATE:

CUSTOMER-/-IN-HOUSE PART DEVELOPMENT NUMBER:
MARKET ESTABLISHED: BENEFITS TO MARKET:
MARKET SIZE: ESTIMATED VOLUME/YEAR:
USERS OF PRODUCT:

ANTICIPATED SALES PRICE: HOW CALCULATED: HOW SOLD:
MANUFACTURE IN-HOUSE: OUTSIDE SUPPLIER: JOINT:
OUTSIDE SUPPLIER(s):

PURCHASED PARTS REQUIRED:
PARTS:
SUPPLIERS:
COST:

NEW PART: EXISTING: REDESIGN REQ'D.: METAL REPLACEMENT:
COMPETITION: WHO: SALE PRICE:
MARKET SIZE: SHARE DESIRED: ESTIMATED SELL PRICE:
PATENTABLE: APPLIED FOR: DATE:
PATENT NO.:

PROGRAM ASSETS AVAILABLE: REQUIRED:
MARKET INTRODUCTION DATE ANTICIPATED:
ASSISTANCE REQ'D.: TYPE:
WHOM: WHAT AREAS:
PROBABILITY OF PROGRAM SUCCESS:
ESTIMATED COMPLETION DATE:

PROJECT TEAM LEADER: ALTERNATE:
PROJECT START DATE:
DECISION DATES: ____________ ____________ ____________
ALL REQ'D. INFO. AVAILABLE ASSETS AVAILABLE START DATE

DEVELOPMENT TEAM MEMBERS:
CUSTOMER IF DESIGNATED:
SALES:
ENGINEERING:
DESIGN:
PRODUCTION:
TOOLING:
QUALITY:
PURCHASING
FINANCE:
MANAGEMENT:
SUPPLIERS:

PART DEVELOPMENT (TEAM ANALYSIS OF PRODUCT)

Continuation of Part Development Check List:

PART REQUIREMENTS (GENERAL, SPECIFIC, LIABILITY ITEMS), BE SPECIFIC:
1.
2.
3.
BENEFITS TO USER:
LIMITATIONS OF CURRENT PRODUCT:
COMPETITIONS PRODUCT EVALUATION:
QUALITY REQUIREMENTS:
IMPROVEMENTS POSSIBLE:
POSSIBLE TO COMBINE FUNCTIONS:
POSSIBLE TO CHANGE MATERIAL:

CHECK LIST ANALYSIS:

CUSTOMER REQUIREMENTS:	DATE:
SALES/CONTRACT:	DATE:
ENGINEERING:	DATE:
PROBLEM:	DATE:
DESIGN:	DATE:
MATERIAL:	DATE:
PROGRAM SCHEDULING	DATE:
MANUFACTURING:	DATE:
TOOLING:	DATE:
PURCHASING:	DATE:
SUPPLIERS:	DATE:
PRICE ESTIMATION:	DATE:
DEVELOPMENT:	DATE:
ASSEMBLY:	DATE:
DECORATION:	DATE:
PACKAGING:	DATE:
SHIPPING:	DATE:

AGENCY AND CODE REQUIREMENTS:
WHO: WHAT:
CUSTOMER REQUIREMENTS:
WHAT:
PROGRAM STATUS, CONTINUE: NEED MORE INFORMATION:
WHAT:
TERMINATE: REASONS:

CONTINUE: APPROVED BY: DATE:

PART DESIGN & MATERIAL SELECTION:
DESIGN CHECK LIST COMPLETED TO SUIT REQUIREMENTS:
ADDITIONAL INFORMATION REQUIRED:
WHAT:
FROM WHOM: REQUIRED BY:
AVAILABLE: NEEDS TO BE DEVELOPED: BY WHEN:
HOW:
BY WHOM:

PART/DESIGN ANALYSIS:
DESIGNER:
TYPE-CALCULATIONS: FEA: SLA/SLS: MODEL: PROTOTYPE:
TIME ESTIMATED TO COMPLETE: COST:
COMPLETION DATE:

MATERIAL CANDIDATES:

Continuation of Part Development Check List:

A: WHY:
B: WHY:
C: WHY:

SUPPLIER A:
SUPPLIER B:
SUPPLIER C:

SUPPLIER PROPERTY DATA AVAILABLE: REQUESTED: DATE:
REQUIRED DATA:

IF NOT AVALIABLE, CAN IT BE DEVELOPED: BY WHOM: BY WHEN:
SUPPLIER CONTACT: PHONE: FAX; E-MAIL:

PROTOTYPE: TESTABLE: TYPE:
WHO PROVIDES: WHEN:
SPECIFIED IN CONTRACT:
COST ESTIMATE:
FULL SIZE: MATERIAL:

PROTOTYPE TESTABLE: TYPE OF TESTS REQUIRED:
ACTUAL END USE CONDITIONS: SIMULATED:
CONDITIONS:
REQUIREMENTS TO PASS:
WHAT IDENTIFIES FAILURE/PASS:
WHO DETERMINES:
AGENCY/CODE REQUIREMENTS TESTABLE:
WHAT:
TESTING TIME:
TEST COST:
NUMBER OF TESTS: SAMPLES REQUIRED:
SUPPLIER TEST DATA REQUIRED:
PROCEDURE DEFINED: PROCEDURE NUMBER:
WHO DOES TESTING:
CONTACT: PHONE: FAX: E-MAIL:
WHO EVALUATES DATA:]
DOCUMENTATION REQ'D.:
CERTIFICATION REQ'D.:
TEST RESULTS IN WHAT FORM:

PASS/FAIL:
COMMENTS:

PROJECT STATUS CHECK POINT:
CONTINUE: TERMINATE: NEED MORE DATA:
WHAT:
BY WHOM:

DESIGN FINALIZED:
CUSTOMER APPROVED: BY WHOM: DATE:
TITLE:

MATERIAL SELECTED:
SUPPLIER: PRODUCT CODE:
ALT. SUPPLIER: PRODUCT CODE:
CAN EITHER BE SUBSITUTED AT WILL:
DECISION AUTHORIZED BY ONLY:

Continuation of Part Development Check List:

EACH MATERIAL MUST BE END USE TESTED BEFORE FINAL APPROVAL:
SUPPLIER ON CERTIFIED SUPPLIER LIST:
IF NOT, WHEN:
WHO APPROVED:

QA APPROVAL STATUS OF SUPPLIERS:
SUPPLIER CERTIFICATION TYPE REQUIRED:
SPECIFIC LOT DATA:
TYPICAL LOT DATA:
SPECIAL REQUIREMENTS:
APPROVAL STATUS:

PURCHASED PARTS REQUIRED: WHAT:
SUPPLIERS:
CERTIFICATION REQ'D.:
VENDOR AUDITED FOR QUALITY: WHEN: STATUS OF AUDIT:

CRITICAL DIMENSIONS:
DRAWING AVAILABLE FOR DISCUSSION: DRAWING NO.:
DIMENSIONS ATTAINABLE: PLASTIC TOLERANCES: NUMBER OF CRITICAL TOLERANCES:
WHERE 1: 2. 3.

INSERTS USED: TYPE:
IN MOLD: AT ASSEMBLY:
SCREWS USED: TYPE:

OTHER ASSEMBLY METHODS:
SNAP/PRESS FIT:
SONIC:
THERMAL:
SOLVENT/ADHESIVES:

QUALITY REQUIREMENTS: (SEE QUALITY CHECK LIST):
WHAT MAJOR REQUIREMENTS:
WHO DETERMINES:
WHEN VERIFIED:
BY WHOM:
TEST EQUIPMENT REQUIRED:
WHAT:
AVAILABLE:
SUPPLIED BY WHOM:
COST OF TESTING:
CUSTOMER TO VERIFY TESTS: ONLY DATA: BOTH:
PROCEDURE NUMBER:
DOCUMENTATION REQUIRED:
WHAT:
HOW TO REPORT:

MANUFACTURING METHOD (SEE CHECK LIST)
METHOD:
TOOLING:
SPECIAL REQM'TS.:
WHERE: BY WHOM:
CONTACT: PHONE: FAX: E-MAIL:
CAPABILITY OF EQUIPMENT EVALUATED: CP: CpK: CR:
PERSONNEL TRAINING REQ'D.:

Continuation of Part Development Check List:

PROCESS CONTROL USED: CLOSED LOOP FEEDBACK: OTHER:
REAL TIME PROCESS CONTROL USED:
MANUFACTURING PROCEDURE DOCUMENTED:
PROCEDURE NUMBER:
WORK INSTRUCTIONS:

MOLD DESIGN (SEE CHECK LIST)
COMPLETED:
ALL TEAM MEMBERS APPROVED DESIGN:
IF NOT, WHO DISAGREES:
WHY:
HOW RESOLVED: BY WHOM: DATE:

SPECIAL REQUIREMENTS: WHAT:
MOLD TYPE: NUMBER OF CAVITYS:
BALANCED RUNNER SYSTEM: REPLACEABLE GATE BLOCK:

SUPPLIER:
CONTACT: ALTERNATE:
PHONE: FAX: E-MAIL:
ESTIMATED PRICE: DELIVERED WHEN:
MOLD SPECIAL FEATURES: WHAT:
CORE PULLS/UNSCREWING:

MOLD FLOW ANALYSIS: BY WHOM: RESULTS:
MOLD COOL ANALYSIS: BY WHOM: RESULTS:

MOLD TRYOUT: WHERE: BY WHOM:
PRESS SIZE: OUNCES: TONS OF CLAMP:
PLATTEN SIZE: MOLD FIT WITHIN:
PROCESS CONTROL USED FOR TRYOUT:

MATERIAL: GRADE: LOT NUMBER:
REGRIND ALLOWED: USED: PERCENTAGE: HOW BLENDED:
TRIAL DATE: LENGTH OF TRIAL:
GOOD PARTS PRODUCED:
IF NOT, WHAT WAS PROBLEM:
HOW WILL IT BE CORRECTED:
BY WHOM:
WHEN:
RETRIAL OF MOLD SCHEDULED: WHEN: WHERE:
MOLD TRIAL RESULTS:

FINAL MOLD TRIAL DATE:
LENGTH OF TRIAL: TIME: CYCLES: GOOD PARTS PRODUCED:
PARTS MEET CUSTOMER REQUIREMENTS:
IF NOT, WHAT WAS LACKING:
FIXABLE: PROCESSING: MOLD: MATERIAL:
TOOL APPROVAL: DATE: BY WHOM:

EXISTING TOOLING: LAST MOLDED AT:
MAINTENANCE PERFORMED TO MEET PRODUCT REQUIREMENTS:
QUALITY ASSURANCE APPROVED ALL DIMENSIONS:
DATE:
CONTACT:
REASON FOR TRANSFER:
TOOL DRAWINGS AVAILABLE:

Continuation of Part Development Check List:

PARTS LIST AVAILABLE:
KNOWN PROBLEMS WITH TOOL, DOCUMENTED:
CORRECTED:
BY WHOM:
VERIFIED: BY WHOM: TITLE: DATE:
MOLD APPROVED FOR PRODUCTION:
BY WHOM: TITLE: DATE:

ASSEMBLY METHOD (SEE CHECK LIST)
COMPLETED:
ASSEMBLY REQUIRED: FIXTURES REQ'D.: AVAILABLE:
MUST BE DEVELOPED: BY WHEN: BY WHOM: WHO PAYS:
ASSEMBLY DRAWING AVAILABLE: DRAWING NUMBER:
TYPE OF ASSEMBLY:
PRESS FIT: SNAP FIT: SONIC: THERMAL: SOLVENTS: ADHESIVES:
SCREWS: OTHER OR COMBINATION OF METHODS:
REPAIRABLE: TYPE ALLOWED:
SEALED UNIT: TYPE:
HAND/MACHINE ASSEMBLY: REQUIRED ASSEMBLY RATE:
PROCESS/SPECIFICATIONS DEFINED: DOCUMENT NUMBER:
PART CLEANING REQUIRED: HOW: WITH WHAT:
MUST KEEP PART DRY AS MOLDED: HOW: WITH WHAT:
STORED WHERE:

ASSEMBLY TESTING REQUIRED: WHAT:
PROCEDURE NUMBER: TESTING SPECIFICATIONS:
SPECIFICATION NUMBER:

ASSEMBLY EQUIPMENT REQUIRED: IS IT CAPABLE:
MUST IT BE CALIBRATED: TO WHAT:
SPECIFICATIONS:

ADHESIVE/SOLVENTS USED: SYSTEM:
SUPPLIER: CONTACT: PHONE:
MSDA SHEETS REQUIRED WITH ORDER:
OSHA REQUIREMENTS:

PURCHASED PARTS IN ASSEMBLY:
WHAT:
SUPPLIER:
CONTACT: PHONE: FAX: E-MAIL:
QUALITY RATING: APPROVED SUPPLIER:
INSPECT BEFORE ASSEMBLY: SPECIFICATION:
BY WHOM: WHEN: WHERE:

DECORATION METHOD (SEE CHECK LIST)
COMPLETED:

PACKAGING METHOD (SEE CHECK LIST)
COMPLETED:

PIECE PART COST ANALYSIS (SEE PRICE CHECK SHEET)
COMPLETED: BY WHOM: DATE:
APPROVED BY MANAGEMENT: DATE:
APPROVED BY PRODUCTION: DATE:

MATERIAL COST:

Continuation of Part Development Check List:

MANUFACTURING COST:
MOLD COST:
HOW PAID FOR:
BY WHOM:
AMORTIZED OVER PRODUCTION RUN:

ASSEMBLY COST:
PURCHASED PART COSTS:
DECORATION COST:
PACKAGING COST:

TOTAL COST OF PROGRAM:

FINAL PROGRAM ANALYSIS:
CONTINUE:
TERMINATED:
COMMENTS:

APPROVED BY:
DATE:
CUSTOMER REPRESENTATIVE:
DATE:

PROGRAM START DATE ANTICIPATED:

GORDON & ASSOCIATES
www.qualityplasticconsult.com

No. 2 SALES & CONTRACTS CHECK LIST

DATE:

CUSTOMER:
ADDRESS:
CONTACT: PHONE: FAX: E-MAIL:

APPLICATION:
VOLUME/YEAR: ANTICIPATED PART PRICE:
RELEASE QUANTITY: FREQUENCY:
PART SIZE (SQ. IN.)
DRAWING/SKETCH/PROTOTYPE AVAILABLE: TYPE: DWG. NO.
WHO DESIGNS PRODUCT:

REQUIREMENTS:
PART DEVELOPMENT CHECK LIST USED:
PART IS NEW/EXISTING/REDESIGN:
USERS:
AGENCY/CODE APPROVAL REQ'D.:
SPECIAL SITUATION:
SUPPLIER CERTIFICATION REQ'D.: WHAT:

TYPE OF MANUFACTURE:
ANTICIPATED MATERIAL: TYPE:
SUPPLIER:
IS COMPANY CAPABLE OF SUPPLYING:

PURCHASED PARTS USED: WHAT:
WHO FURNISHES: INVENTORY REQM'TS.:
ASSEMBLY REQ'D.: TYPE: CHECK LIST USED:
DECORATION REQ'D.: TYPE: CHECK LIST USED:
PACKAGING & SHIPPING REQM'TS.:

MOLD/TOOLING:
WHO SUPPLIES:
WHO DESIGNS:
TYPE: NUMBER OF CAVITIES: BALANCED:
SPECIAL IN MOLD REQM'TS.: WHAT:

EXISTING MOLD:
CONDITION:
REASON FOR TRANSFER:
LAST MOLDER:
CONTACT: PHONE: FAX: E MAIL:
IN-HOUSE TRIAL TO ACCESS CONDITION:
WHEN: WHERE:
MOLD DRAWINGS AVAILABLE: BILL OF MATERIALS:
WHO BUILT CURRENT MOLD:
CONTACT: PHONE: FAX: E-MAIL:
SPECIAL EQUIPMENT REQ'D. TO RUN MOLD: WHAT:
WHO FURNISHES:
PROCESS CONDITION RECORDS AVAILABLE: WHERE;
PARTS AVAILABLE:
MATERIAL USED:
SUPPLIER:

Continuation of Sales & Contract Check List:

GRADE: AMOUNT ON HAND:

NEW TOOL:
WHO DESIGNS:
WHO OWNS:
ANTICIPATED COST:
CUSTOMER PAYMENT METHOD, DIRECT ON APPROVAL, WITH PARTIAL PAYMENTS:
AMORTIZED OVER PRODUCTION RUN AS PARTS ARE DELIVERED:
WHO APPROVES TOOLING:
WHO APPROVES FIRST PARTS OFF TOOLING:
CONTACT: PHONE: FAX: E-MAIL:

MOLD CHECK LIST USED:
MAINTENANCE REQUIREMENTS:
WHO APPROVES REPAIRS:
WHO PAYS: APPROVAL REQ'D. BEFORE REPAIR:
CONTACT: PHONE: FAX: E-MAIL:

QUALITY REQUIREMENTS:
QUALITY CHECK LIST USED:
INCOMING MATERIAL TESTS REQ'D.:
PURCHASED PARTS TESTS REQ'D.:
REQUIREMENTS SPECIFIED:

PROTOTYPE TESTING REQUIRED:
TYPE, MODEL/MOLDED PART/SLA/SLS MODEL, OTHER:
WHO FURNISHES:
TIME REQUIREMENTS TO PROVIDE:
COST ANTICIPATED:
IN-PROCESS TESTING REQ'D.: WHAT:
REQUIREMENTS:
EQUIPMENT REQUIRED:
WHO FURNISHES: WHAT:

END USE TESTING REQUIRED:
WHAT:
WHO PERFORMS:
REQUIREMENTS:
TEST EQUIPMENT REQUIRED: WHO FURNISHES:

TESTING DOCUMENTATION (SPC) REQUIRED:
INCOMING:
PRODUCTION:
ASSEMBLY:
DECORATION:
SPECIAL DOCUMENTATION REQ;D.:
WHAT:
CUSTOMER REQUIRED DOCUMENTATION PRIOR OR AT TIME OF SHIPMENT:
WHAT:
PROBLEM RESOLUTION:
CONTACT: PHONE: FAX: E-MAIL:

CUSTOMER TESTS AT INCOMING: WHAT TESTING:
PROCEDURES DOCUMENTED: WHAT:
WHO FURNISHES:
CONTACT: PHONE: FAX: E-MAIL:

Continuation of Sales & Contract Check List:

PRODUCTION:
FIRST ARTICLE WHO APPROVES:
REQUIREMENTS (QUALITY CHECK LIST OR OTHER USED):
FORM/FIT/FUNCTION:
AESTHETICS:
DIMENSIONAL:
COLOR APPROVAL:
SPECIAL SPECIFICATIONS:

ANTICIPATED RELEASE QUANTITIES PER ORDER:
JUST-IN-TIME PRODUCTION REQ'D.: FREQUENCY:
SHIPMENT DISTANCE:
INVENTORY REQUIREMENTS: WHO PAYS:
PAYMENT TERMS/METHOD:
DIRECT: AMORTIZED: OTHER:

CONTRACT TERMS SPECIFIED:
RELEASE ON PURCHASE ORDERS: TIMED RELEASES: WHO SPECIFIES:
CUSTOMER APPROVAL TO SHIP: WHO:
SHIPPER SPECIFIED:
SHIPPING PAID BY: TERMS:
ON RELEASE:
CONTRACT TERMS USED: WHAT TERMS:

CUSTOMER QUOTE TO:
ADDRESS:
QUOTE DUE DATE: TIME:
QUOTE DELIVERED BY: HOW:

QUOTE ITEMS REQUIRED PIECE PART PRICE:
SCHEDULING:
MATERIAL:
TOOLING:
ASSEMBLY:
DECORATION:
PACKING:

SPECIAL TESTING:
SPECIAL REQM'TS.:
WHAT:

QUOTE NUMBER:
QUOTED BY:
APPROVED BY:
SALES CONTACT: PHONE: FAX: E-MAIL:
SUBMITTED DATE"
CUSTOMER RESPONSE DATE:

ACCEPTED/REJECTED BY: WHY:
ANY SPECIAL TERMS REQUIRED:
WHAT:
REQUOTE ALLOWED IF REJECTED:
TIME LIMITS:
DUE BY DATE:

CONTRACT SUBMITTED DATE:
CUSTOMER APPROVED DATE:
CUSTOMER OFFICER APPROVAL:

Continuation of Sales & Contract Check List:

CONTRACT SUPPLIER APPROVED BY: DATE:

GORDON & ASSOCIATES
www.qualityplasticconsult.com

No. 3. PRODUCT DESIGN CHECK LIST

DATE:

CUSTOMER:
ADDRESS:

CUSTOMER CONTACT: ALTERNATE:
PHONE: FAX: E-MAIL:
PROGRAM START DATE: EST. COMPLETION DATE:
JUST-IN-TIME PROGRAM: ANTICIPATED QUANTITY/SHIP FREQUENCY: ______/______

PART NAME: JOB NO.:
DRAWING AVAILABLE: DRAWING NO.:
PART AVAILABLE: MODEL/PROTOTYPE:

PART DESCRIPTION NEW: EXISTING: COMPETITORS:
FUNCTION OF PART:

EXISTING MATERIAL: PROPOSED MATERIAL:
PART WEIGHT & SPECIFIC GRAVITY OF MATERIAL:________/________

NUMBER PARTS IN ASSEMBLY, EXISTING: PROPOSED:
FUNCTION OF ADJACENT PARTS:
ABLE TO COMBINE FUNCTIONS:
WHAT:

PURCHASED PARTS USED: TYPE:
TYPE:
TYPE:

CUSTOMER INCENTIVE FOR PROJECT: COMPENSATION:
PERFORMANCE IMPROVEMENT: REDESIGN:
WEIGHT SAVINGS: COST REDUCTION:
MEET NEW REQUIREMENTS: ALTERNATE SUPPLIER NEEDED:
OTHER CONSIDERATIONS:

PRODUCTION INFORMATION:
MANUFACTURING METHOD:
VOLUME (PARTS/YEAR):
SUPPLIER, IN-HOUSE: OUTSIDE: WHO:

DESIGN CONSIDERATIONS (OBTAIN SKETCH OF FORCES ACTING ON PART)
PART FUNCTION:
CUSTOMER LIABILITY IF FAILURE OCCURS:

OPERATING CONDITIONS:	NORMAL	MAXIMUM	MINIMUM
TEMPERATURE:	:	:	:
SERVICE LIVE (HOURS)	:	:	:
FORCES (LBS/TORQUE/ETC.)	:	:	:
DURATAION OF FORCES TIME ON:	:	:	:
TIME OFF:	:	:	:
MAXIMUM DEFLECTION ALLOWED:	:	:	:

LOAD BEARING APPLICATION:
BUCKLING CONSIDERED:

IMPACT FORCES: WHERE ON PART (SKETCH): TYPE:

Continuation of Design Check List:

REPEATED: ONE TIME: FREQUENCY:
DROP HEIGHT: LOAD: IMPACT ENERGY:

VIBRATION EFFECTS CONSIDERED:
VIBRATION INPUT:
WEIGHT OF ASSEMBLY: WEIGHT OF INTERNAL COMPONENTS:
MOUNTING OF COMPONENTS (METHOD):
ATTACHES TO ANOTHER ASSEMBLY OR PART: PART:
FUNCTION OF THIS PART:
OPERATING SPEED (RPM):
EXCITING FREQUENCY (CPS):
DISPLACEMENT (INCHES/MM):
ACCELERATION (G FORCES):

ENVIRONMENTAL CONDITIONS:
CHEMICAL EXPOSURE: TYPE: CONCENTRATION:
CHEMICAL MAKEUP:
MOISTURE (HUMIDITY): PERCENT: TEMPERATURE:
WATER EXPOSURE TYPE (FRESH/SALT/BOILING/STEAM):
TEMPERATURE:
RADIATION: TYPE: EXPOSURE LEVEL: TIME:
SUNLIGHT: EXPOSURE TYPE (DIRECT/INDIRECT): TIME:
UV PROTECTION REQUIRED: COLOR FADING A FACTOR:
AMBIENT TEMPERATURE NOT OPERATING: OPERATING:

ELECTRICAL REQUIREMENTS:
MAXIMUM CURRENT SUPPORTED:
MAXIMUM VOLTAGE:
INSULATION PROPERTIES REQUIRED: TYPE:
EMI/EMP PROTECTION REQUIRED: VALUES REQUIRED:

FLAMMABILITY REQUIREMENTS REQUIRED: TYPE:
SMOKE GENERATION LIMITS REQ'D.: OXYGEN LEVEL LIMITS:
MUST MEET UL FLAME REQUIREMENTS: VO/V1/V2/HB/V5

AGENCY REQUIREMENTS (UL/CSA/NSF/FDA/ETC): AGENCY:
REQUIREMENTS:
CUSTOMER REQUIREMENTS:
OTHER REQUIREMENTS:

ABRASION & WEAR REQUIRED:
TYPE:
MATING PART MATERIAL:
COEFF. FRICTION REQ'D.: STATIC: DYNAMIC:
LUBRICATION REQ'D.: TYPE: CHEMICAL COMPOSITION:
SELF LUBRICATING: INTERNAL: EXTERNAL:
TYPE LUBRICATION ALLOWED:

SAFETY FACTOR REQUIREMENTS:
PART LIABILITY: SEVERITY IF FAILURE OCCURS:
DEGREE OF LIABILITY TO SUPPLIER:
PART FAILURE IMPACT ON PRODUCT/APPLICATION:
CONSUMER/INDUSTRY APPLICATION:
INSTRUCTIONS ON PRODUCT TO OPERATE:
WARNING LABELS REQUIRED:
WHAT INFORMATION ON LABEL:
WHO SUPPLIES:
AGENCY REQUIREMENTS/TESTING REQUIRED: WHAT: WHO:

Continuation of Design Check List:

QUALITY REQUIREMENTS:
CRITICAL TOLERANCES:
HOW MANY: WHERE: TOLERANCE:
IMPACT ON PART FUNCTION IF NOT MET:
PART REQUIREMENTS (FLASH/WARPAGE/SINK/POROSITY):
SPECIFICATIONS ESTABLISHED: WITHIN SUPPLIER CAPABILITY:
MATERIAL/PARTS INCOMING TESTING REQ'D.:
TESTING SPECIFIED:
TESTING REQUIREMENTS:
CUSTOMER REQUIREMENTS ESTABLISHED:
REQUIREMENTS:

WILL CUSTOMER TEST INCOMING PRODUCT: HOW: WHERE:
TESTS REQUIRED TO MEET:
TEST EQUIPMENT:
PROCEDURE ESTABLISHED:
PERSONNEL TRAINED:
SPECIAL EQUIPMENT REQ'D.:
CUSTOMER CONTACT: PHONE:
SUPPLIER WITNESS TESTS: MUST SCHEDULE:
IF FAILURE OCCURS, HOW ARE DISPUTES SETTLED:
BY WHOM:

INTEGRATION OF COMBINING PART FUNCTIONS CONSIDERED:
WHAT OPERATIONS CAN BE COMBINED:
MATERIAL CAPABLE:
ASSEMBLY METHODS CONSIDERED:

PART REQUIRES ASSEMBLY:
SNAP/PRESS FIT:
MECHANICAL FASTENERS:
WELDING (THERMAL/SONIC):
ADHESIVES/SOLVENTS:
OTHER:
PLANT HAS EQUIPMENT TO DO ASSEMBLY: WHAT IS REQUIRED:
PERSONNEL TRAINING REQUIRED: WHAT:
IF NO, USE OUTSIDE COMPANY:
EQUIPMENT REQUIRED:
COST FACTOR ON PRODUCT:

DECORATION REQUIREMENTS:
COLORED: COMPOUNDED: S & P: CONCENTRATES:
COLOR: REQUIREMENTS: COLOR SAMPLE:
MUST MATCH MATING PART: COLOR/MATERIAL/PAINT:
PIGMENT TYPE ALLOWED: WILL IT AFFECT MATERIALS PROPERTIES:
TESTING REQUIREMENTS: WHAT:
SPECIAL CLEANING REQUIRED: WHAT: AFFECTS ON MATERIAL:
PAINTED: PAINT TYPE: PRIMER: NUMBER COATS:
OSHA REQUIREMENTS INVOLVED: WHAT:

METALIZED: TYPE OF METAL: THICKNESS:
PLATED: TYPE OF METAL: THICKNESS:
ETCHING OF MATERIAL REQUIRED: WHAT: CONDITIONS:
IN-HOUSE: OUTSIDE: WHO:

PRINTING: TYPE: MOLDED IN: ON SURFACE:
SURFACE PREPARATION REQUIRED: WHAT:
FOILS/DECALS: TYPE: SUPPLIER:

Continuation of Design Check List:

PRINTING:
INFORMATION REQUIRED:
WHO FURNISHES:
FINISH ON PART: CLASS A: TEXTURED: DEPTH:
FINISH TYPE: HIGH POLISH: SAMPLE AVAILABLE:
MUST MATCH MATING PART: SAMPLE AVAILABLE:

TESTING REQUIREMENTS:
TESTS TO MEET:
PROCEDURE:
PROTOTYPE TESTED:
PRODUCTION TESTS:
END USE REQUIREMENTS:

PROTOTYPE:
MOLDED:
MODELED:
SLA/SLS/OTHER:

CUSTOMER INFORMATION:
DESIGN DEADLINE: EXTENSION TIME AVAILABLE:
ARE ALL DESIGN REQUIREMENTS/END USE INFORMATION AVAILABLE:
IF NOT, WHAT IS MISSING: WHEN AVAILABLE:
END USE TEST AVAILABLE: WHO TESTS:
WHERE: HOW MANY CYCLES:
CONDITION OF PART: CONDITIONING REQUIRED: WHAT:
CONTACT: PHONE:
WHO GIVES FINAL APPROVAL OF DESIGN:

PRELIMINARY COST FIGURES COMPLETED:
ANTICIPATED PIECE PART COST:
QUOTE NUMBER:

HAS ALL DEPARTMENTS BEEN CONTACTED FOR THEIR DESIGN INPUT:
SALES:
ENGINEERING:
PURCHASING:
MANUFACTURING:
QUALITY:
TOOLING:
ASSEMBLY:
DECORATION:
PACKAGING:
SHIPPING
MATERIAL SUPPLIERS:
OUTSIDE SOURCES REQUIRED:
UPPER MANAGEMENT APPROVAL:

ADDITIONAL INFORMATION REQUIRED, NOT LISTED TO ASSIST IN UNDERSTANDING COMPLETELY THE FUNCTION, MANUFACTURING, QUALITY, ASSEMBLY, AND ANY OTHER ABUSE OR REQUIREMENTS THE PART MUST WITHSTAND OR ENVIRONMENTAL STRESSES NOT LISTED:

DESIGNER:
DESIGN TEAM SIGNOFF:
COMPANY REPRESENTATIVE: APPROVAL DATE:

GORDON & ASSOCIATES
www.qualityplasticconsult.com

No. 4 MATERIAL CHECK LIST

DATE:

CUSTOMER:
ADDRESS:

PART NAME: JOB NUMBER:
PRODUCTION START DATE: PRODUCTION SUPERVISOR:
MATERIAL: PRODUCT CODE: VOLUME:
ALTERNATE SUPPLIER: PRODUCT CODE:
CRITICAL PARTS REQUIRING USE OF SAME LOT NUMBER OF MATERIAL; DUE TO COLOR, DIMENSIONS:
PART NUMBERS:
ALTERNATE PARTS PRODUCTION START DATE:

PRODUCT VOLUME: PRODUCT WEIGHT: MATERIAL REQ'D. LBS.:
ORDER SIZE, LBS.: ALL ONE LOT NUMBER OR MIXED: YES/NO CONFIRMED:

MATERIAL CERTIFICATION REQUIRED*: YES/NO CERTIFICATION TO:
SPECIAL REQUIREMENTS, MATERIAL VALUES, COLOR, PROPERTIES, SPECIFICATION:
VALUES REQUIRED:
CERTIFICATION REQUIRED WITH EACH SHIPMENT: YES/NO PRIOR TO RECEIPT OF MATERIAL: YES/NO
TEST VALUES ON MATERIAL REQUIRED:
VALUES REQUIRED:

PACKAGE TYPE: BAGS-DRUMS-GAYLORDS-BULK:
PRICE PER POUND/KILO: VOLUME DISCOUNT:
COLORED MATERIAL: YES/NO METHOD: COMPOUNDED-S/P-CONCENTRATE (TYPE)
CONCENTRATE SOURCE:
COLOR SAMPLE REQUIRED**: YES/NO TYPE: COLOR CHIP-SURFACE TYPE-RESIN
PIGMENT CHANGES PERMITTED: YES/NO MUST NOTIFY IF REQUIRED: YES/NO
NOTIFY WHO:

SPECIAL INCOMING MATERIAL TESTING REQUIRED: YES/NO CONTACT:
TESTS:

QC CONTACT AT RECEIVING:
PRODUCTION CONTACT:

Continuation of Material Check List:

MATERIAL ROUTING ON RECEIPT: WAREHOUSE-SILO-PRODUCTION-OUTSIDE MOLDER
HOLD TILL TESTING COMPLETED: YES/NO CONTACT:
DISPOSITION IF MATERIAL FAILS INCOMING TESTS:
NOTIFY CONTACTS IN: QC:
PRODUCTION:
SALES:
PURCHASING:

OTHER PARTS REQUIRED FOR PRODUCT SALE:

PRODUCT NAME :	PART NUMBER:
SUPPLIER:	CONTACT:
DATE REQUIRED:	

* SEE INSPECTION & MATERIAL FLOW SHEET NO.:
** SEE COLOR MATCH REQUEST FOR VERIFICATION:

PURCHASING REPRESENTATIVE:

GORDON & ASSOCIATES
www.qualityplasticconsult.com

No. 5 PURCHASING CHECK LIST

DATE:

CUSTOMER:
ADDRESS:
CONTACT: PHONE: FAX: E-MAIL:

CONTRACT NUMBER:
JOB NUMBER:
PRODUCTION START DATE:
JOB SCHEDULE COMPLETED:
APPROVED BY: DATE:

PURCHASING REQUIREMENTS:
BUYER: PHONE:
BUYER: PHONE:

PROTOTYPES: TYPE:
SUPPLIER:
CONTACT: PHONE: FAX: E-MAIL:
PURCHASE ORDER REQ'D.: P.O. NO.: DATE:
DUE BY: RECEIVING CONTACT:

NOTIFY DEPARTMENT MANAGER(s) WHEN SPECIFIC MATERIALS ARE RECEIVED:
PURCHASING: PHONE:
PRODUCTION: PHONE:
ENGINEERING: PHONE:
ASSEMBLY: PHONE:
DECORATION: PHONE:
QUALITY: PHONE:
PACKAGING: PHONE:

MATERIAL & FINISHED GOODS AND PARTS:
PRIME MATERIAL SUPPLIER:
GRADE:
CERTIFICATION REQUIRED: WHAT: SPECIFIC LOT DATA:
REQUIRED PRIOR TO OR WITH RECEIVING DOCUMENTS:
TEST RESULTS REQUIRED: WHAT: SEND TO:
MSDS REQUESTED WITH ORDER:
QUANTITY ORDERED: POUNDS: PACKAGE TYPE:
PURCHASE ORDER NUMBER: ORDERED DATE:
CONTACT: PHONE:
DUE TO ARRIVE ON OR BEFORE:
SHIPPER PRO NUMBER REQUIRED: WHAT IS NUMBER:
TRUCKING COMPANY:
RECEIVING NOTIFY ON RECEIPT:
PURCHASING: PHONE:
PRODUCTION: PHONE:
QUALITY: PHONE:

INCOMING TESTING REQUIRED:
QUALITY CONTACT: PHONE:
PRODUCTION: PHONE:

TESTING RESULTS:
QUALITY APPROVED BY: DATE:

Continuation of Purchasing Check List:

INVENTORY PLACEMENT:
WHERE: BAR CODED TO INVENTORY: HEATED AREA REQ'D.: TEMP. REQ'D.:

IF REJECTED, REASON: BY WHOM: DATE:
SUPPLIER QUALITY CONTACT: PHONE: E-MAIL:
DISPOSITION:
SEGREGATED FROM CURRENT INVENTORY: WHERE: BY WHOM:
SPECIAL REJECTION LABEL ON PACKAGING:

REORDER REQUIRED:
PURCHASING NOTIFIED: WHEN: BY WHOM:
NEW P.O. NUMBER:
ABLE TO MEET PRODUCTION START DATE:
PRODUCTION NOTIFIED: WHO: WHEN:
SCHEDULING REQUIRED TO BE ADJUSTED:
SALES NOTIFIED: WHO: WHEN:
CUSTOMER CONTACTED BY SALES: WHEN:
COMMENTS:

TOOLING:
SUPPLIER:
CONTACT: PHONE: FAX: E-MAIL:
CONTRACT/P.O. NO.: DATE ENTERED:
DUE BY DATE: RECEIVED ON DATE:
PRODUCTION NOTIFIED: WHOM: DATE:
TOOLING NOTIFIED: WHOM: DATE:
QUALITY: WHOM: DATE:
ENGINEERING: WHOM: DATE:

PURCHASED PARTS FOR MOLDING PRODUCT:
PARTS:
PURCHASE ORDER NUMBER: ENTERED:
QUANTITY ORDERED:
DUE IN BY:
CERTIFICATION REQUIRED:
WHAT:
SUPPLIER:
ON APPROVAL LIST: NEEDS APPROVAL: WHO/WHEN APPROVED:
CONTACT: PHONE: FAX: E-MAIL:

INCOMING TESTING REQUIRED:
WHAT:
QUALITY CONTACT: ALTERNATE:
ACCEPTED/REJECTED: BY WHOM: DATE:
REASON:
REORDER: P.O. NUMBER: REQUIRED BY:
DUE IN:

PURCHASED PARTS FOR ASSEMBLY:
PARTS:
QUANTITY ORDERED:
PURCHASE ORDER NUMBER: DATE:
CERTIFICATION REQUIRED:
WHAT:
SUPPLIER:
ON APPROVAL LIST: NEEDS APPROVAL: WHO/WHEN APPROVED:
CONTACT: PHONE: FAX: E-MAIL:

Continuation of Purchasing Check List:

INCOMING TESTING REQUIRED:
WHAT:
QUALITY CONTACT: ALTERNATE:
ACCEPTED/REJECTED? DATE:
REASON:
REORDER: P.O. NUMBER: REQUIRED BY:
DUE BY:

PARTS:
QUANTITY ORDERED: DATE:
PURCHASE ORDER NUMBER:
CERTIFICATION REQUIRED:
WHAT:
SUPPLIER:
ON APPROVAL LIST: NEEDS APPROVAL: WHO/WHEN APPROVED:
CONTACT: PHONE: FAX: E-MAIL:

INCOMING TESTING REQUIRED:
WHAT:
QUALITY CONTACT:
ACCEPTED/REJECTED: DATE:
REASON:
REORDER: P.O. NUMBER REQUIRED BY:
DUE BY:

PURCHASED PARTS FOR DECORATION:
PARTS:
QUANTITY ORDERED:
PURCHASE ORDER: DATE:

PACKAGING:
PURCHASE ORDER:
QUANTITY:
SPECIAL REQUIREMENTS:
REQUIRED BY:

RECEIVING TO NOTIFY:
PURCHASING: PHONE:
PRODUCTION: PHONE:

SPECIAL MANUFACTURING EQUIPMENT REQUIRED:
WHAT:
SUPPLIER:
CONTACT: PHONE: FAX: E-MAIL:
P.O. NUMBER: ORDER DATE: DUE DATE:
NOTIFY:
PURCHASING: PHONE:
PRODUCTION: PHONE:

GORDON & ASSOCIATES
www.qualityplasticconsult.com

No. 6 QUALITY CHECK LIST

DATE:

CUSTOMER:
ADDRESS:
CONTACT: PHONE: FAX: E-MAIL:

PART NAME: JOB NUMBER:
MANUFACTURING START DATE:
PRODUCTION SUPERVISOR:
MATERIAL: PRODUCT CODE:
SUPPLIER:

QUALITY PROCEDURES PER ISO9000/QS9000/OTHER
QUALITY INSPECTOR:

CUSTOMER QUALITY REQUIREMENTS KNOWN:
DOCUMENT: REVISION:
ENGINEERING CHANGE ORDERS RECEIVED: WHAT:
INCORPORATED INTO PRODUCTION: WHEN: BY WHOM:
ANY DEVIATIONS ALLOWED: WHAT:
WHO APPROVED AT CUSTOMER: WHEN: TITLE:

PART REQUIREMENTS:
PHYSICAL: DOCUMENT: DRAWING:
CHEMICAL: DOCUMENT:
ELECTRICAL: DOCUMENT:
AGENCY REQUIREMENTS: WHAT:
CODE REQUIREMENTS: WHAT:

PART DESIGN DOCUMENTED: DRAWING NO.:
MATERIAL DOCUMENTED: CERTIFICATION RECEIVED: RESULTS:
INCOMING INSP/TEST RESULTS:
CONFIRMED BY: DEPT: TITLE:

REVIEW OF PROCEDURES BY: ALL CURRENT:
MANUFACTURING: REVIEWED BY: DATE:
DECORATING: REVIEWED BY: DATE:
ASSEMBLY: REVIEWED BY: DATE:
FINAL TESTING: REVIEWED BY: DATE:
PACKAGING: REVIEWED BY: DATE:
SHIPPING: REVIEWED BY: DATE:

MATERIAL SAFETY DATA SHEETS AVAILABLE & CURRENT:
MANUFACTURING EQUIPMENT MAINTENANCE CURRENT:
TOOLING MAINTENANCE CURRENT:
AUXILIARY EQUIPMENT MAINTENANCE CURRENT:

PROCESS CONTROL LIMITS ESTABLISHED: BY WHOM:
PART QUALITY LIMITS DOCUMENTED: BY WHOM: DOCUMENT:
TEST & INSPECTION EQUIPMENT DOCUMENTED: BY WHOM:
PROCEDURE:
STATISTICAL PROCESS CONTROL DATA REVIEWED:
PROCESS CONTROL:
DOCUMENTED FOR RECORDS:

Continuation of Quality Check List:

QFD, ANALYSIS COMPLETED WITH CUSTOMER: DATE:
FMEA, ANALYSIS COMPLETED: DATE:
FISH BONE ANALYSIS COMPLETED: DATE:
MEASUREMENT TOOL ANALYSIS: DATE:
METRIC REQUIREMENTS COMPLETED: DATE:
SPC, REQUIREMENTS ESTABLISHED: DATE:
SIX SIGMA ANALYSIS COMPLETED: DATE:
ALL MEASUREMENT ITEMS IN CERTIFICATION: DATE:

TEST EQUIPMENT AVALIABLE:
WHAT IS REQUIRED:
INSPECTOR: ALTERNATE:

CUSTOMER ON SITE INSPECTION REQUIRED: DURING MANUFACTURE: FINAL:
CUSTOMER INSPECTOR: ALTERNATE:
PHONE: FAX: E-MAIL:
SAME INSPECTION EQUIPMENT USED BY CUSTOMER:
WHAT IF NOT:
WHO SUPPLIES:
AGENCY TESTING REQUIRED: CONTACT:
ADDRESS:
NUMBER OF PARTS TO SEND: SENT:
INFORMATION DUE BACK WHEN:
INFORMATION/FORMS REQUIRED:

WAS "REAL TIME" PROCESS CONTROL USED DURING MANUFACTURE:
COMPUTER OUTPUT SAVED AND FILED:
FILE NAME:

QUALITY RECORDS REVIEWED AND SIGNED OFF FOR SHIPMENT:
BY WHOM: DATE:

GORDON & ASSOCIATES
www.qualityplasticconsult.com

No. 7 Design and Development Schedule Checklist

Outline program: Start Date:

A. Design Team:

1. Primary Members: Alternate: Responsibility:

2. Secondary Members: Alternate: Responsibility:

B. Check Lists: Start: Finish:

1. Sales & Contract

2. Part Development

3. Part Design

4. Material

5. Purchasing

6. Mold, Customer Requirements

Mold, Design & Materials

7. Pricing

8. Vendor Survey

9. Scheduling for Manufacture

10. Manufacturing

11. Quality

12. Assembly

13. Decorating

14. Packaging & Shipping

15. Problem solving (if required)

C. Design Reviews

1. Preliminary-Program review with primary and secondary team input.

2. Design analysis and Reviews from check lists.

a. Part consolidation/function incorporation/value added extras.

b. Material possibilities/selection.

c. Material supplier inputs/data availability.

3. Design layout of part/parts, system.

4. Review of initial design/cost, projections/assembly/decorating, and design team input.

5. Review of mold requirements, type, functions, tolerances, and cost.

6. Manufacturing capability study.

a. Injection molding machine

b. Mold

c. Auxiliary equipment

d. Shipping

e. Packaging

f. Material/certify/test/verify

g. Plant support facilities

h. Personnel training required

I. Purchased parts/suppliers

J. Assembly/decorating

7. Secondary Operations.

. Assembly/type/source/equipment/training

b. Decoration

c. Packaging and shipping

8. Quality Requirements.

a. In-house – material, mold, process, product, and tests

b. Supplier requirements/certifications

9. Finalize Preliminary Design.

10. Prototyping, Part/System.

a. Method/type/source/schedule

b. Product requirements

c. Testing required, code, customer, and agency

d. Part function/aesthetics review

11. Customer Feedback/Analysis/Conclusions.

12. Finalize Design.

13. Tooling/Mold Design Review with CheckList.

a. Schedule, in-house/outside builds

14. Outside Support Equipment/Services Required.

a. Define, cost/schedule

15. Evaluate Production Tooling on Manufacturing Machines.

a. Process capability requirements

b. Quality system to be capable and in control

c. Finalize total quality process control procedure to produce zero defects and monitor in "Real Time".

16. Final Customer Approval/Sign off/Begin Program

GORDON & ASSOCIATES
www.qualityplasticconsult.com

No. 8 PRICE ESTIMATING CHECK LIST

CUSTOMER: DATE:
ADDRESS:
CONTACT: PHONE: FAX: E-MAIL:

PART NAME: JOB NUMBER:
DRAWING NO.:

PIECE PART COST ESTIMATING PER 1000 PARTS

A. MATERIAL	:	:	:	:
B. RESIN COST ($/LB)	:	:	:	:
C. SPECIFIC GRAVITY (Sg)	:	:	:	:
D. PART WEIGHT (lbs)	:	:	:	:
E. PART WEIGHT (D x 1000)	:	:	:	:
F. MATERIAL COST (B x E)/0.95	:	:	:	:
G. CYCLE TIME (CT)	:	:	:	:
H. NUMBER OF CAVITIES (NC) *a	:	:	:	:
I. PARTS/HOUR (H/G x 3600)	:	:	:	:
J. CAVITY AREA (PROJECTED) *b	:	:	:	:
K.CLAMP FORCE *c (CF) TONS x (J x MATERIAL FACTOR)	:	:	:	:
L. SHOT WEIGHT (oz) (D x H x W *d x 16 oz/lb)	:	:	:	:
M. MACHINE HOUR COST (RATE x (MC *e)	:	:	:	:
N. PROCESSING COST ($/1000 PARTS) M/I x 1000	:	:	:	:
O. ADJUSTED PROCESSING COSTS *f [N/(0.95)(0.80)]	:	:	:	:
TOTAL COST (PROCESSING PER 1000 PARTS	:	:	:	:

*a Assumed three shifts/day, 6 days/week (*f), one years production produced
*b Projected cavity area & runner/sprue, mold cavity in square inches x number of cavities, plus runner and sprue area of mold surface in square inches
*c 80 % to 20% maximum shot weight of resin, use material clamp factor to estimate tons of clamp required
*d Use reference chart for shot weight Figure A.
*e Use machine hour rate chart Figure B, adjust for current machine rates
*f Assumes 95% yield and 80% utility of molding process

GORDON & ASSOCIATES
www.qualityplasticconsult.com

No. 9 PROGRAM SCHEDULING FOR MANUFACTURE CHECK LIST

CUSTOMER: DATE:
ADDRESS:
CONTACT: PHONE: FAX: E-MAIL:

QUOTE NUMBER: COMPLETION DATE: REVIEW DATE:
QUOTED BY:
SUBMITTED TO CUSTOMER:
ACCEPT/REJECT: REASON:
CONTRACT SIGNED: BY WHOM: TITLE: DATE:

PROGRAM CHECK LISTS COMPLETED, DATE:
SALES & CONTRACT: MANUFACTURING:
PART DEVELOPMENT: QUALITY:
PART DESIGN: ASSEMBLY:
MATERIAL: DECORATION:
PURCHASING: PACKING & SHIPPING:
MOLD: WARRANTY PROBLEM SOLVING:
PRICING:

MANUFACTURING DOCUMENTS COMPLETED:
JOB TRAVELER:
JOB NUMBER FOR TRACKING:

BAR CODING USED: LABELS SPECIFIED: TYPE:
INFORMATION REQUIRED:

MOLD EXISTING: STATUS:
REPAIR REQ'D.: WHAT;
BY WHOM: WHO PAYS:
MOLD TRIAL DATE: WHERE:
BY WHOM: MATERIAL:
ACCEPT/REJECT MOLD: REASON:
CAN MOLD BE MODIFIED TO MAKE PARTS: BY WHOM:
WHEN:
MODIFICATIONS REQUIRED:
MOLD ACCEPTED BY PRODUCTION: BY WHOM: DATE:

NEW MOLD: WHO DESIGNS:
WHO PAYS: HOW:
START: FINISH:
SCHEDULE DETERMINED FOR MANUFACTURE: WHEN:
MOLD TRIAL DATE: WHERE: BY WHOM:
MATERIAL:
RESULTS:
MODIFICATIONS REQUIRED:
WHEN COMPLETED:
MOLD ACCEPTED BY PRODUCTION: BY WHOM: DATE:

PURCHASING:
RESIN:
PO/DATE: DUE IN: PACKAGE: LBS:
STORED WHERE:
BAR CODED INTERNALLY:

Continuation of Program Development Check List:

FINISHED PARTS: WHAT:
PO/DATE: DUE IN: INSPECTION REQ'D.:
WHAT:
PROCEDURE NO.:
STORED WHERE:
SPECIAL STORAGE REQUIRED:
WHAT:

SPECIAL EQUIPMENT: WHAT:
PO/DATE: DUE IN: INSPECTION REQ'D.:
WHAT:
PROCEDURE NO.:

MANUFACTURING:
START DATE: FINISH DATE:
QUANTITY TO PRODUCE: TO ORDER: MONTHLY: QUARTERLY:
MOLD NUMBER:
MACHINE NUMBER:
PROCEDURE DOCUMENTED: DOCUMENT NUMBER:

NEW MOLD SETUP: DOCUMENT DATA REQUIRED:
BY WHOM:
COMPLETED: DOCUMENT NUMBER:

AUXILIARIES:
DRYER: CHILLER: FEEDER:
CONVEYOR: PART SEPARATOR: ROBOT:
GRINDER: WEIGH SCALE: PACKER:
BLENDER: OTHER:

ASSEMBLY:
START DATE: FINISH DATE:
SPECIAL EQM'T.: WHAT:
PO/DATE: DUE IN:
SPECIAL PARTS:
WHAT:
PO/DATE: DUE IN: QUANTITY: INSPECTION REQ'D.:

DECORATION:
START DATE: FINISH DATE:
SPECIAL EQM'T.: WHAT:
PO/DATE: DUE IN:
SPECIAL PARTS:
WHAT:
PO/DATE: DUE IN: QUANTITY: INSPECTION REQ'D.:

PACKING:
START DATE: FINISH DATE:
SPECIAL PACKAGING REQ'D.:
WHAT:
PO/DATE: DUE IN: QUANTITY:

PART/MATERIAL TESTING REQUIREMENTS:
WHAT:
WHERE:
WHEN:
BY WHOM:
TEST PROCEDUE:

Continuation of Program Development Check List:

CUSTOMER TO VERIFY:
CERTIFICATIONS REQUIRED:
WHAT:

INCOMING:
MATERIAL: TEST:
MATERIAL: TEST:

IN PROCESS:
PART: TEST:
PART: TEST:

ASSEMBLY:
PART: TEST:
PART: TEST:

FINAL:
TEST: TYPE: INSPECTION:

SPECIAL EQM'T. REQUIRED:
WHAT:
CUSTOMER SUPPLIED/PURCHASED:
PO/DATE: DUE IN:

PRODUCT CERTIFICATION REQ'D.:
WHAT:
DUE TO CUSTOMER: HOW: WHEN:
TO WHOM:

INVOICING: DATE:
INVOICE NUMBER:
AMOUNT INVOICED:
QUANTITY SHIPPED:
QUANTITY ORDERED:
OVER/UNDER % ALLOWED: PERCENT:
DISCOUNTS:
TERMS:

REORDER ANTICIPATED:
WHEN:
QUANTITY:
PRICE:

GORDON & ASSOCIATES
www.qualityplasticconsult.com

No. 10 MANUFACTURING CHECK LIST

DATE:

CUSTOMER:
ADDRESS:

PART NAME: JOB NUMBER:
MANUFACTURING START DATE:
PRODUCTION SUPERVISOR:
SET UP TIME & DATE: SET UP TECH.:

MACHINE NUMBER/SIZE SCREW TYPE: CHECK RING:
NOZZLE TYPE: MOLD INSULATED FROM PLATTENS:
PROCESS PROCEDURE: PROCESS CONTROL CHART NO.:
PRODUCT SPECIFICATION SET: DOCUMENT NO.:
SPECIFICATION LIMITS ESTABLISHED: MEAN: UCL: LCL:

MOLD NUMBER/SIZE:
OWNERSHIP:
SPRUE BUSHING FITS NOZZLE: SPECIAL BUSHING REQUIRED: PART NO.:
MAINTENANCE COMPLETED: SIGNED OFF BY: DATE:
SPECIAL REQUIREMENTS:
INSTALLATION PROCEDURE: NUMBER:
QUICK CHANGE: MOLD PREHEAT REQ'D.: TEMPERATURE: TIME:
MOLD RELEASE ALLOWED: TYPE:

PRODUCTION EQUIPMENT:
DRYER: DRY TO % MOISTURE: TEMPERATURE: DRY TIME:
HOPPER/CENTRAL/SIDE DRYER TYPE: FILTERS CLEAN: DESICANT GOOD: CLEAN:
DESICANT BED DRIED: START BEFORE MATERIAL ADDED: TEMP.: TIME:

MOLD CHILLER NO: TEMPERATURE SETTING: FLOW GP: PRESSURE:
COOLING MEDIUM: SPECIAL HOSES REQ'D.: TYPE: FITTINGS:
SETUP PROCEDURE AVAILABLE: DOCUMENT NO.:

GRINDER NO.: LAST INSPECTED: FILTER & UNIT VACUUMED CLEAN:
BLADE SHARPENED: LAST MATERIAL GROUND:
SCREEN SIZE IN HOPPER:

ROBOT NO.: SET UP PROCEDURE: DOCUMENT NO.:
SET UP BY:
SPECIAL INSTRUCTIONS REQUIRED: DOCUMENT NO.:

PART HANDLING:
OPERATOR: GLOVES REQUIRED: TYPE:
PROTECT PART SURFACE: HOW: ELECTRO-STATIC:
CONVEYOR: MACHINE NO.:
SPRUE PICKER: MACHINE NO.:
MOLD SWEEP: MACHINE NO.:

SPECIAL OPERATIONS AT PRESS: WHAT: EQUIPMENT REQ'D.:
SPECIAL OPERATOR TRAINING: WHAT:

PACKAGE PRODUCT AT PRESS: HOW:
SPECIAL REQUIREMENTS: PACKAGING SUPERVISOR:

Continuation of Manufacture Check List:

MATERIAL:
SUPPLIER:
PACKAGE TYPE: HOPPER LOADING METHOD: HOPPER CAPACITY:
DRYING TIME REQUIRED: SAMPLE TEST PRIOR TO START: % MOISTURE ALLOWED:
REGRIND ALLOWED: PERCENT: HOW BLENDED AT HOPPER:
PROCEDURE NO.: WHO ADDS TO HOPPER: FREQUENCY:

PROCESS CONTROL LIMITS ESTABLISHED: DOCUMENT NO.:
QUALITY CHECKS AT PRESS: WHO APPROVES:
PROCEDURE NO.:
TEST EQUIPMENT:
VERIFY WHAT VARIABLES AT START UP:
PROCEDURE NO.:
WHO APPROVES SAVING FIRST PRODUCTION PARTS:
SAMPLES SAVED: HOW MANY: QUALITY CHECKS:

ANY SECONDARY OPERATIONS AT PRESS:
WHAT:
SPECIAL PART HANDLING REQUIRED: WHAT:

PRODUCTION PROBLEMS CONTACT:
QUALITY ASSURANCE CONTACT:
MAINTENANCE CONTACT:
MATERIAL HANDLING CONTACT:

PARTS BOXED/PALLETIZED/COUNTED/WEIGH COUNTED/WHAT

PARTS TO STORAGE/STATION/HOLDING POINT:
PARTS PROTECTED: HOW & WITH WHAT:

ANY SPECIAL INSTRUCTIONS:

GORDON & ASSOCIATES
www.qualityplasticconsult.com

No. 11 ASSEMBLY CHECK LIST

DATE:

CUSTOMER:
ADDRESS:

PART NAME: JOB NUMBER:
MANUFACTURING START DATE:
PRODUCTION SUPERVISOR:
ASSEMBLY START DATE: ASSEMBLY SUPERVISOR:

ASSEMBLY DRAWING NO.:
PART NUMBERS:
TYPE OF ASSEMBLY REQUIRED:
PROCEDURE AVAILABLE: PROCEDURE NUMBER:
SPECIAL INSTRUCTIONS:
SPECIAL EQUIPMENT REQUIRED:
EQUIPMENT:
SPECIAL MATERIALS REQUIRED:
MATERIALS:
OSHA REQUIREMENTS:
OPERATOR TRAINING REQUIRED: WHAT: PROVIDED BY WHOM:
COLOR OR TEXTURE MATCH REQUIRED: TYPE:
WHO APPROVES:
PROCEDURE: PROCEDURE NUMBER:

PURCHASED PARTS REQUIRED: PURCHASE ORDER NO.:
RECEIVED IN-HOUSE: WHERE IN INVENTORY:
PART NUMBERS:
QUANTITY REQUIRED:
QUALITY APPROVED FOR ASSEMBLY:

TOTAL ASSEMBLY IN HOUSE: OUTSIDE SUPPLIER:
SUPPLIER:
CONTACT: PHONE: FAX: E-MAIL:

FINISHED TESTING OF ASSEMBLY REQUIRED:
REQUIREMENTS:
WHO APPROVES:
TEST PROCEDURE NO.:
WHO DOES TESTING:
REJECTS SALVAGEABLE. HOW: PROCEDURE NO.:
WHO APPROVES:
DISPOSITION OF REJECTS:

PACKAGING REQUIREMENTS:
PROCEDURE: PROCEDURE NO.:

JUST-IN-TIME PRODUCT:
SPECIAL INSTRUCTIONS:
DOCUMENTATION REQUIRED: WHAT:
DOCUMENTATION TO WHOM:
SHIPPING CONTACT:

GORDON & ASSOCIATES
www.qualityplasticconsult.com

No. 12 DECORATING CHECK LIST

DATE:

CUSTOMER:
ADDRESS:

PART NAME: JOB NUMBER:
MANUFACTURING START DATE:
PRODUCTION SUPERVISOR:
DECORATING START DATE: DECORATION SUPERVISOR:

DECORATION REQUIRED: TYPE:
DRAWING NUMBER:
BEFORE OR AFTER ASSEMBLY:
PROCEDURE NO.:
COLOR MATCH/TEXTURE REQUIRED:
APPROVAL BY WHOM:
PROCEDURE NO..:
REJECT HANDLING PROCEDURE:
SALVAGEABLE:
PROCEDURE NO.:

PART SURFACE PREPARATION REQUIRED: TYPE:
PART SURFACE TESTING REQUIRED:
TEST REQUIREMENTS:
EQUIPMENT REQUIRED:
EQUIPMENT
PROCEDURE NO.:
IN-HOUSE DECORATION: OUTSIDE: WHO:
TRAINING REQUIRED: WHAT:
FIXTURES REQUIRED: WHAT:
SPECIAL: ORDERED: PURACHASE ORDER NO.: RECEIVED:

DECORATING MATERIALS ORDERED:
SUPPLIER:
MATERIALS:
PURCHASE ORDER NO.:
SPECIAL REQUIREMENTS:
CERTIFICATION REQUIRED: WHAT: BY WHOM:

OSHA REQUIREMENTS TO BE MET:
REQUIREMENTS:
SPECIAL EQUIPMENT REQUIRED: WHAT: ORDERED:
PURCHASE ORDER NO.: RECEIVED:

DECORATED PARTS TO:
SPECIAL HANDLING REQUIRED:
PARTS TO STORAGE/STATION:

JUST-IN-TIME PRODUCT:
SPECIAL INSTRUCTIONS:
DOCUMENT:
PACKAGING CONTACT:

GORDON & ASSOCIATES
www.qualityplasticconsult.com

No. 13 PACKAGING AND SHIPPING CHECK LIST

DATE:

CUSTOMER:
CONTACT: PHONE: FAX: E-MAIL:
ADDRESS:

PART NAME: JOB NO.:
PRODUCTION SUPERVISOR:
MANUFACTURING START DATE: QUANTITY:
PACKAGING REQUIRED: TYPE: SPECIAL ORDER:
PURCHASE ORDER ISSUED: P.O. NO.: WHEN:
LEAD TIME TO ORDER PACKAGING:
PACKING DUE IN: NOTIFY WHOM:

PART PROTECTION REQ'D. BEFORE PACKING: WHAT:

JUST-IN-TIME MANUFACTURE USED:
SPECIAL PACKAGING REQUIRED FOR SHIPMENT: IF NOT, DUNNAGE AVAILABLE:
WHAT TYPE: WHO FURNISHES: SPECIAL REQUIREMENT:
SUPPLIER:
REQUIREMENTS:
REUSABLE: HOW RETURNED TO SUPPLIER: BY WHOM:

PART PACKAGING PERFORMED WHERE: BY WHOM: SPECIAL TRAINING REQ'D.:
WHAT:

NUMBER OF PARTS PER PACKAGE:
NUMBER OF PARTS PER CARTON:
NUMBER OF CARTONS PER PALLET:
ARE PALLETS STACKABLE: HOW MANY PALLETS HIGH ALLOWED:

STORAGE REQUIRED BEFORE SHIPPING:
SECURED AREA:
QS9000 INSPECTION REQ'D.: WHAT:
PROCEDURE NUMBER:
WHO PAYS: HOW LONG:

BAR CODING REQUIRED: WHO SUPPLIES:
BAR CODE SPECIFIED:
SPECIAL INSTRUCTIONS REQUIRED: WHAT:
LOT NO.:
DATE OF MANUFACTURE:
PRODUCT NAME:
PRODUCT CODE:
OTHER INFORMATION:
SPECIAL PACKAGING REQUIRED: WHAT:
WHO SUPPLIES:

PACKAGING PRODEDURE DOCUMENTED:
DOCUMENT NUMBER:

SPECIAL TRUCKING REQUIRED FOR SHIPMENT: WHAT:
SHIPPER: CONTACT: PHONE:

GORDON & ASSOCIATES
www.qualityplasticconsult.com

No. 14 WARRANTY PROBLEM CHECK LIST

DATE:

CUSTOMER:
ADDRESS:
CONTACT: PHONE: FAX: E-MAIL:
INDUSTRY/MARKET OF USE:

PROBLEM:

PROBLEM REPORTED BY:
REPORT OF PROBLEM SUBMITTED: PROBLEM.:

OCCURRED AT DEVELOPMENT: PROTOTYPE: FINAL DESIGN: PRODUCTION:
ASSEMBLY: DECORATION: END USE: OTHER:

FAILURE DEFINED AS:

DESIGN-MATERIAL-PURCHASED PARTS-ASSEMBLY-DECORATION-PACKAGING:
SHIPPING-OTHER AREA, DESCRIBE IN DETAIL:

FAILURE OCCURRENCE – ONCE: SEVERAL TIMES: REPEATABLE: NO. TIMES:
SAME POINT OR AREA ON PART: VARIABLE: WHERE:
SKETCH SHOWING LOCATIONS:

SAMPLE OF FAILED PARTS AVAILABLE:
SENT TO: WHEN: BY WHOM:

FAILURE OCCURRED AT:
MANUFACTURE: ASSEMBLY: WAREHOUSE: SHIPPING: END USE:

OCCURRED IN WINTER: SUMMER: SPRING: FALL:
TROPICAL: DRY AREA: OTHER CONDITIONS, DESCRIBE:
SECTION OF COUNTRY:

SERIOUSNESS:
LIABILITY INVOLVED: WHAT EXTENT:
STATUS OF FAILURE: KNOWN: MUST BE INVESTIGATED:

MOLDED PART FAILURE ANALYSIS:
MATERIAL: SUPPLIER: PRODUCT NO.: LOT NO.:
CERTIFIED BY SUPPLIER: WHAT CERTIFICATIONS:
INCOMING TEST RECORD: DATE TESTED: TEST RESULTS:
REGRIND USED: PERCENTAGE: NUMBER OF PASSES ALLOWED:

CHEMICAL DATA AVAILABLE:
PHYSICAL DATA AVAILABLE:
SAMPLE OF PART RETAINED:

Continuation of Warranty Problem Check List:

ENGINEERING CHANGE ORDER: NUMBER: DATE:
APPROVED BY:
CUSTOMER APPROVAL REQUIRED: WHOM:
GRANTED BY: ON DATE:
ALL COMPANY DEPARTMENTS NOTIFIED OF CHANGE ORDER AND THEIR SIGNATURE ON APPROVAL SIGN-OFF SHEET:
CONFIRMED BY: TITLE: DATE:

INCORPORATED INTO PART: WHEN:

AGENCY/CODE APPROVAL GRANTED: BY WHOM: DATE:

PART REQUIRED AGENCY/CODE CERTIFICATION:
WHAT:
CONTACT: PHONE: FAX: E-MAIL:

COLORED MATERIAL: BLENDED WHERE: CONCENTRATE:
SUPPLIER:
LOT/P.O. NUMBER:
CONTACT: PHONE: FAX: E-MAIL:
ANY CHANGES IN PIGMENT SYSTEM INGREDIENTS DURING MANUFACTURE:
WHAT:
LOT SAMPLES AVAILABLE: TEST RESULTS FROM SUPPLIER:

MOLDED PART: PART NUMBER: DATE MFG'D.:
PURCHASED PART: PART NUMBER: DATE MFG'D.:
SUPPLIER:
MOLD NUMBER:
CAVITY NUMBER: CONSISTANT WITH FAILED PARTS:
MOLD NUMBER: DRAWING OF MOLD AVAILABLE:
SUPPLIER INCOMING INSPECTION RECORD: DATE: TEST NO.:

PART FAILURE ANALYSIS:

MECHANICAL FAILURE:
FAILURE TYPE:
DESCRIBE TYPE OF FAILURE AND IF DURING USE:
CUSTOMER:
SEVERITY:
REPAIRABLE:
HOW:

ELECTRICAL FAILURE:
USED AS:
FAILURE TYPE:
CUSTOMER:
FREQUENCY OF OCCURANCE:
SEVERITY:
REPAIRABLE:
HOW:

QUALITY ASSURANCE:
ANY TESTING SHOWED PROBLEMS:
METHOD OF TESTING BASED ON FAILURE:
ANY REPORTS OF PRIOR FAILURES OF THIS TYPE:
CUSTOMER REACTION:
SEVERITY:

Continuation of Warranty Problem Check List:

MATERIAL PROBLEM:
MOLDING PROBLEM:
END USE APPLICATION TO SEVER FOR PART/MATERIAL:
ANALYSIS OF FAILURE:

MANUFACTURED IN HOUSE: OUTSIDE MOLDER: WHO:
CONTACT: PHONE: FAX: E-MAIL:
PRODUCTION DATE: LOT NO.: SHIFT:
PROCEDURE FOR MANUFACTURE FOLLOWED:
PROBLEMS NOTED DURING PRODUCTION: WHAT:
PROCESS CONTROL RECORDS REVIEWED: IN REAL TIME: BY WHOM:
MOLDING PRESS NO.:
MOLD NUMBER:
MAINTENANCE PERFORMED LAST:
MAINTENANCE RECORDS AVAILABLE:
SPC PROCESS DATA AVAILABLE:

PART ANALYSIS/TEST RESULTS:
VISUAL INSPECTION:
ANALYTICAL RESULTS:
PHYSICAL:
CHEMICAL:
DSC:
TGA:
IR:
OTHER:
MATERIAL SUPPLIER ANALYSIS/INPUT:

SOLUTION TO PROBLEM:

CORRECTIVE ACTION RESPONSE ASSIGNED TO:
DATE:
TIME ESTIMATED TO RESOLVE:
ESTIMATED COST TO COMPANY:

Appendix B

DOE (Design of Experiments) Statistical Troubleshooting Process Screening for Reducing the Number of Variables

Once a process is believed to be in control and the optimum manufacturing parameters controlled, there often occurs the additional problem that products may still not meet customer specifications. Something has changed in the process or environment. The material and/or machine parameters though not noticeable may have changed or are different. The plant manufacturing environment, plus plant provided services are varying or the tooling has worn sufficiently to now cause a problem. As a result any number of multiple manufacturing variables are varying but not enough to be easily detected by the operator or system control. This is often known as the process was in "control" but no totally "capable" of producing the desired product.

What do you do now? The old method was to guess and begin by holding all variables constant but one. Then through trial and error and typically a lot of time, find the elusive controlling variable by running tests on the manufacturing machine and process. Then once the controlling variable(s) are determined they can be adjusted and brought into control to make good parts to the required specification. This technique works but often the troublesome variable(s) are not easily or quickly detected and the operator has to constantly adjust the system to make acceptable products. This is wasteful, expensive, and often causes a delay in shipping acceptable product to the customer.

There has to be an alternate method available! There is and this is a rapid screening technique using a statistically designed series of trials where in many variables are changed at the same time in the manufacturing process. This rapid screening technique developed by Dr. Genichi Taguchi with the name for his method known as "Design of Experiments" or DOE. The intent of the DOE is to rule out non-significant variables by modifying an array of variables and combining them in specific pattern. In this manner valid statistical information on the impact value of each variable in the process, on the product, can be obtained and evaluated with other variables.

The Taguchi experiments are used once a process is in control, the process parameters optimized, but the parts still do not meet or drift in and out of specifications. The process can be affected by the material, machine settings, tooling changes, plant environment, or any number of items n the manufacturing system. The manufacturing process is in control but not yet capable.

The Taguchi method utilizes orthogonal arrays – rows of experiments (factors or variables) versus the trial (runs) to be performed. The variables for each factor and run are established using quality control problem analysis techniques. These factors are then tested with each run using a different combination of these variables.

Taguchi's orthogonal arrays capture the most significant variables combination levels for testing. For example, to test independently seven variables at two setting levels, high and low, the complete orthogonal array would be large: 2^7, or 128 separate testing sites. This means to test each variable, at its high and low level, in combination with all the other variables, at each of their levels, would take 128 separate trial runs. There are many combinations of arrays and levels of testing. The tester determines which levels to test and the amount of time the plant can spend to find the elusive variable.

To solve the problem, the Taguchi variable screening analysis is applied. This is a mass-screening technique that uses a statistically designed series of trial runs. Many process variables are changed at the same time in a controlled test environment during the manufacturing process. The goal is to rule out insignificant variables by modifying an array of variables and combining them in specific patterns. In this manner, valid statistical information on the impact of each variable on the process and the product can be obtained. These variables are rapidly evaluated and concentration placed on the highly probable causes of the problem. Determining when to use this technique is not always clear. It is based on the customer requirements and the capability of the manufacturing process, including machinery, tooling, material, and personnel.

A process is in control when the variation in the product, plotted on a bar chart, falls with in the three sigma bell shaped curve values. This is assuming the process is, itself, stable and within the control limits earlier determined to make the product to customers specifications. At this time the product may meet customer specifications but not always and may drift in and out of tolerance.

A process is said to be capable when 99.5% of its products and all of its variations are within the customer's specifications.

The realization that improvement can still be attained or needed in a process usually occurs at this point when the process is in control but not yet capable. At this point the "screening experiment" is called for to find the contributing factor that leaves the process not yet capable.

SCREENING EXPERIMENT: THE NINE STEPS

The nine steps of the screening experiment are used to improve the process and find the unknown variables needing adjustment to make the process capable and keep all product within customer requirements.

These nine step are:

1. Determine what improvements are needed.
2. Brainstorm for ideas on what affects the variable to be improved.
3. Select the factors to be analyzed and the levels to be used.
4. Randomize the experimental runs.
5. Perform the experimental runs.
6. Separate the effects.
7. Test for significance.
8. Analyze the results.
9. Change the process based on the results of the experiment.

STEP 1. DETERMINING THE PROBLEM

When you realize a critical dimension is not where it should be may be the time to consider using a DOE to find the missing variable and it's required value to manufacture acceptable products. This determination may be based on continual drift of a products required dimensions or specifications. Manufacturing is not able to consistently manufacture repeatable in specification product. At this time the method, materials, tooling, and machine must be evaluated and in a minimum length of time to keep the program on schedule.

This may include, as in this example, an injection molded product, tool cavity dimensions or die size analysis that is needed to manufacture the product. It is very important that all of the tooling variables are deemed correct and during manufacture stabilized. Then, if a change is required only those necessary changed to make process adjustments and improvements.

At this time it is important to evaluate the products manufacturing tooling dimensions that form the product in the manufacturing process. This should be done in all manufacturing process so it is eliminated as a source of the problem and the operator does not have to keep continually adjusting the process to bring the product into the customer's specifications.

Assuming the tooling is correct we can begin with the following example to explore the controlling variable(s) in the manufacturing process for the product.

A glass reinforced injection moldable polyester is to be used to make a pin and the length is not meeting customer requirements. The pin must measure 2.00 inches long (±) 0.003 inches. But, the average pin length is coming out 1.990 inches or 0.010 inches too short. A way must be found to reduce the material shrinkage or pack out the pin in the tool cavity to bring the pin into tolerance.

STEP 2. BRAINSTORMING FOR SOLUTIONS

Different techniques can be used as we have already discussed. These may include an abbreviated quality circle, value analysis/evaluation, cause and effect or fishbone analysis of the process. Also, depending on the seriousness of the problem a special team of engineers and production personnel brought together to solve the problem.

In this situation the "fishbone" diagram was used as a very useful quality analysis method for this group to determine what variables may be affecting the finished length of the pin. From their discussion the problem solving team drew up a list of possible variables that may be the cause of the problem. Remember, no idea is to be left out no matter how absurd. As an example, the group has listed seven variables that should be investigated as causes for the apparent high material and part shrinkage.

But, also prior to this stage the incoming material should have been checked for variability and glass content that will have a definite affect on mold shrinkage of the material in the tool cavity. Now, assuming the tool and material are within specification the variables selected to be investigated, are listed from the "fishbone" diagram.

STEP 3. RANKING THE VARIABLES

From the discussion the variables are ranked as most probable causes but no discussion on "why" is permitted. All possible variables must be listed and then ranked by the team so as not to miss a possible cause of the problem. The screening process alone will show what variables are significant and their importance of pin length. Often the aspects of a few ideas may be combined to narrow the list.

Feasibility, practicality, effectiveness and cost should be considered in ranking the ideas. The groups top seven ideas, ranked were:

1. Mold temperature
2. Cure time
3. Injection pressure
4. Screw back pressure
5. Injection time
6. Flow of material in the tool
7. Secondary hold time

At this point the team decides at what levels the variables will be run during the screening experiment. The variables levels should be selected to show the extreme levels of the equipments operating range. Therefore, the high and low values should be at the edge of the operating window for both the material and manufacturing equipment.

The reason for selecting these variable extremes is to point up major changes. Therefore, as a result, if the change does not show a significant effect on the pin variable and only a small statistical result is obtained, then that variable is almost certainly not significant. Table 1 lists the variables and their low and high values to be used during the screening experiment.

STEP 4. STATISTICAL RANDOMIZING THE RUN

The importance of this step is to select a screening matrix run of (n) times that accommodates $(n - 1)$ variables. In this example an 8 run design

Table 1. Variable Factors.

Factor	High Level (+)	Low Level (–)
A. Tool temperature	370° F	320° F
B. Cure time	60 seconds	30 seconds
C. Injection pressure	2500 psi	1500 psi
D. Screw back pressure	200 psi	50 psi
E. Injection time	12 seconds	6 seconds
F. Flow of material	soft	stiff
G. Secondary hold time	30 seconds	15 seconds

evaluates the 7 variables selected. The matrix designs go up by blocks of four; a 4 run design evaluates 3 variables; a 12 run evaluates 11 variables. For each of the eight trials, the pattern of highs (H's) and lows (L's) in the 8 run matrix in Table 2 dictates whether to use the high or the low level of the variable at the head of each column. As an example, in run Number 1, variables A, B, C, and E will be run at their high levels, while variables D, F, and G will be run at their low levels.

The same would be true if only 2 variables were selected. You use a 4 run design and only have columns A and B with the high and low variables dictated by the pattern shown (*n*) Table 2, columns A and B, for pattern of (H's) and (L's). Also, included is a 12 run variable matrix run in $(n - 1)$ or 11 times shown in Table 3.

Table 2. Seven Run Variable Screening Matrix.

Run	A	B	C	D	E	F	G
1.	+	+	+	–	+	–	–
2.	–	+	+	+	–	+	–
3.	–	–	+	+	+	–	+
4.	+	–	–	+	+	+	–
5.	–	+	–	–	+	+	+
6.	+	–	+	–	–	+	+
7.	+	+	–	+	–	–	+
8.	–	–	–	–	–	–	–

(Adapted from reference [1]).

Table 3. Twelve-run Screening Experiment Design.

Run	A	B	C	D	E	F	G	H	I	J	K
1.	+	+	–	+	+	+	–	–	–	+	–
2.	+	–	+	+	+	–	–	–	+	–	+
3.	–	+	+	+	–	–	–	+	–	+	+
4.	+	+	+	–	–	–	+	–	+	+	–
5.	+	+	–	–	–	+	–	+	+	–	+
6.	+	–	–	–	+	–	+	+	–	+	+
7.	–	–	–	+	–	+	+	–	+	+	+
8.	–	–	+	–	+	+	–	+	+	+	–
9.	–	+	–	+	+	–	+	+	+	–	–
10.	–	+	+	–	+	+	+	–	–	–	+
11.	–	–	–	–	–	–	–	–	–	–	–

(Adapted from reference [1]).

Also, the order of the runs should be randomized to cut down the delay time between High and Low variable adjustments. This means variables related to temperature should start out either low or high with other more easily adjusted by a variables change within the temperature extreme variable selected. This will normally tend to randomize the runs as required during the experiment.

As an example the low temperature trials for tool temperature (2, 3, 5, 8) are run first due to the ease in raising temperature rather than lowering. They are then randomized as (2, 5, 3, 8). The high temperature trials (1, 4, 6, 7) are then run but randomized as (1, 6, 4, 7). The order of the experiment is thus, (2, 5, 3, 8, 1, 6, 4, 7).

STEP 5. RUNNING THE EXPERIMENT

The equipment is set up for the first selected set of conditions and allowed to reach steady state conditions. At this point select one cavity and start to collect samples. The usual sample size is five but more may be taken but never less than five. The cavity selected depends on the runner system built into the tool. If a balance runner system any cavity will usually do. But, if unbalanced usually an extreme cavity is selected due to the flow distance and pressure drop anticipated within the tool at this point in the cavity system. This cavity or cavities will usually be the most difficult to obtain the desired part dimensions due to the mentioned reasons. As a check you can also select a cavity closer to the sprue to see if the variation also occurs within that cavity as the variables are changed.

Once the manufacturing cycle is in equilibrium for the first sample they are collected with the second and successive set of conditions set on the machine and allowed to reach equilibrium and samples collected for each sample cycle.

The molded samples are then measured at the critical dimensions after cooling down for a predetermined time period. The same cooling time for each set of samples is always used to rule out differential post mold shrinkage due to moisture pickup or other temperature or post mold shrinkage variables. The average for the sample dimension is obtained for each run and entered in the "Average" column of Table 4.

The average value (X) is the summation of all sample dimensions (X*n*) in the run divided by the number of samples (*n*) and is the measure of the central tendency.

$$\bar{X} = \frac{X_1 + X_2 + X_3 + --- Xn}{n} \qquad \bar{X} = \frac{\Sigma Xn}{n}$$

The range for each run is then calculated for all run samples in a similar manner and entered in the range column of Table 4. The range value is the difference between the highest and lowest measured dimension for all the samples in the run.

Table 4. Averages and Ranges for an 8-run Screening Experiment.

Run	Average Length, inches	Range of Length, inches
1.	1.992	0.008
2.	2.001	0.005
3.	2.000	0.007
4.	1.990	0.008
5.	1.998	0.009
6.	1.995	0.005
7.	1.998	0.009
8.	1.991	0.006

The time for running the trial will vary depending on the variables selected. It may be as short as one hour to a few days. But, by doing it by changing only one variable at a time it could take even longer and you could miss the variable that has significant effect on the critical pin dimension.

If the job is run on more than one machine normal processing can continue on the other machine. But, if a one manufacturing machine operation production will be lost during the experimental run but can be make up later once the critical variables are determined with all of the products then meeting customer specifications. You will also eliminate bad products or rejects and have a much higher confidence level of being able to accept all the parts produced and reduce inspection time, possible all together.

STEP 6. SEPARATING THE EFFECTS

Each variable during the trial is run in combination with all the other variables to produce test samples for each run. During the analysis, each variable is considered individually whether its' low or high level has had an effect on the products pin length, and by how much.

Evaluation begins by the filling in Table 5 with the values obtained in Table 4, for the "Average Pin Length" of each trial run.

Begin by filling in Table 5, the data table, for run No. 1. The value 1.992 is listed for each variable and where the "H" or high value is indicated from Table 2, mark it (+) plus. Where "L" low, mark it (–) minus. So for run No. 1, your highs are listed (+) in columns A, B, C, and E. Lows marked (–) in columns D, F, and G. This procedure is then continued for each successive run until the completed matrix Table 5, is completed.

Then, for each variable column, the sum of the highs (+) is written in Sum H, and likewise the sum for the lows (–) in Sum L. Their difference is then obtained by subtracting Sum H from Sum L. This value is entered as either (+) or (–) depending on the larger of the high or low Sum values in the (Diff.) row.

The difference value for each variable is then divided by the number of times the variable was changed from high to low, or four times, and entered in the effect row with the same (+) or (–) sign. This will statistically reduced the calculated value to its true value.

Table 5. Matrix for a Seven-variable Screening Experiment Using Pin Length.

Run	A Tool Temp	B Cure Time	C Inject Pressure	D Screw back Pressure	E Inject Time	F Flow Material	G Hold Time
1.	+1.992	+1.992	+1.992	+1.992	+1.992	+1.992	+1.992
2	−2.001	+2.001	+2.001	+2.001	−2.001	+2.001	−2.001
3.	−2.000	−2.000	+2.000	+2.000	+2.000	−2.000	+2.000
4.	+1.990	−1.990	−1.990	+1.990	+1.990	+1.990	−1.990
5.	−1.998	+1.998	−1.998	−1.998	+1.998	+1.998	+1.998
6.	+1.995	−1.995	+1.995	−1.995	−1.995	+1.995	+1.995
7.	+1.998	+1.998	−1.998	+1.998	−1.998	−1.998	+1.998
8.	−1.991	−1.991	−1.991	−1.991	−1.991	−1.991	−1.991
Sum H	+7.975	+7.989	+7.988	+7.989	+7.980	+7.984	+7.992
Sum L	−7.990	−7.976	−7.977	−7.976	−7.985	−7.981	−7.974
Diff.	−0.015	+0.013	+0.011	+0.013	−0.005	+0.003	+0.017
Effect	−0.004	+0.003	+0.003	+0.003	−0.001	+0.001	+0.004

(Adapted from reference [1]).

By using the plus and minus values you can see that by going from a low mold temperature to a high temperature, variable A, decreased the part dimension by 0.004 inches or the mold shrinkage increased on the pin length.

The lower-tool temperature therefore caused a decrease in the part's shrinkage or, more positively, an increase in pin length. Then by reviewing the variable effect line, one can determine how each variable affected the pin length and by what amount. From this data, one can select the significant variable that if changed, may bring part size within the customer's requirements. The next step in the analysis is very important it looks at the normal variation within a system and provides the criteria that determine what variables are significant.

For easier identification, use red markings for plus values and blue for minus or any distinctive color to easily identify the value changes and the tendency to either plus or minus.

Therefore, by reviewing the variable Effect Line, one can easily see how each variable affected the pin length when going from low values to high values and by what amount. From this data one can then select the significant variables that by being changed will improve the process and bring part size within the customers requirements and specification.

The next step is very important in that it looks at the normal variation within a system and provides the criterion that determines what variables are significant.

STEP 7. TESTING FOR SIGNIFICANCE

A variable will be significant if its' Effect as calculated by Steps 1 through Step 6 are larger than the systems normal variation. Therefore, a test is required to test for normal variation.

The criterion that determines what variables are significant is the normal variation within the system. Based on this a test is used to determine the normal variation within the manufacturing process producing the pins and affecting their manufactured length.

Any calculated effect in the matrix will be significant if it is larger than the normal variation computed by the following formulas and procedure. Effects smaller than the normal variation will not be significant.

The MSFE (minimum significant factor effect) is one typically used statistic using the range values from each run in the calculation. The MSFE is developed by multiplying the standard deviation of an effect (range values) by the Students (*t*) value.

STANDARD DEVIATION OF AN EFFECT (Range pin length/run)

$$R = \frac{\text{Sum of the Ranges}}{\text{Number of trials}} + \frac{\Sigma R}{T} = + \frac{0.057}{8} = 0.007$$

The average range (R) is calculated from values in Table 1.

An estimate of experimental error (σ EE) is obtained from:

$$\sigma\ EE = \frac{R}{d_2} = \frac{0.007}{2.326} = 0.003$$

The term d_2 is an estimator used to convert average ranges to standard deviations. It is found in Table 6 or from standard statistical tables. The term k is the number of samples collected in this case it was 5 per trial run. d_2 will

Table 6. Table of d_2 Estimator Values.

Number of Samples N	Estimator (d_2)	Number of Samples N	Estimator (d_2)
2	1.128	16	3.532
3	1.023	17	3.588
4	2.059	18	3.640
5	2.326	19	3.689
6	2.534	20	3.735
7	2.704	21	3.778
8	2.847	22	3.819
9	2.907	23	3.858
10	3.078	24	3.895
11	3.173	25	3.931
12	3.258		
13	3.336		
14	3.407		
15	3.472		

(Adapted from reference [1]).

vary as the number of samples collected. It is important that the same number of samples are collected for each run and they should not vary once selected for the other runs.

The Standard Deviation of an Effect (Sd Effect) is then calculated:

$$\text{Sd Effect} = \frac{2 \times \sigma\ \text{EE}}{\sqrt{N \times \text{k}}} = \frac{2 \times (0.003)}{\sqrt{8 \times 5}} = 0.00095$$

The number of samples measured per run is (k), in this case 5, and N is the number of runs. It is important that the same number of samples are collected for each run, as d_2 will vary as the number of samples collected.

Next one must calculate the confidence level desired. This means to estimate/calculate which effect may be significant and where it lies. This would normally be outside the bell curve, 3 sigma limits.

For a 95.0% confidence level this means only 2.50% of the relative probability remains in each tail of the bell curve. This means that one time in forty an effect will be calculated as statistically significant but will not be significant. Figure 1 shows this in a form more easily understood.

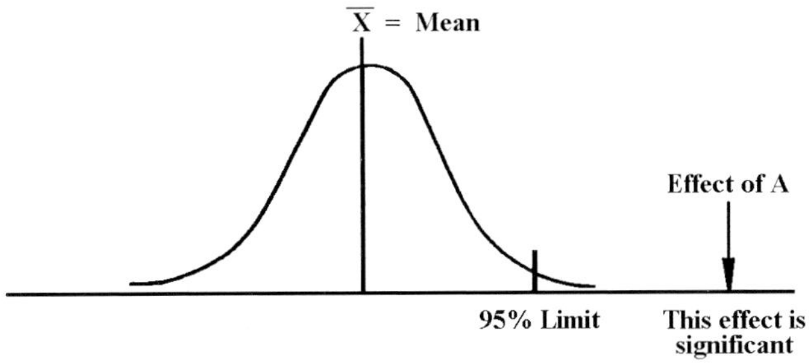

Figure 1. Bell curve with effects noted.

The Student "*t*" Distribution is used to approximate a distribution when sample size is small and where sigma (σ) must be estimated from data that appears normally distributed. Therefore, we consider the sampling distribution of the (t) statistic.

$$t = \frac{\tilde{Y} - \mu}{\sqrt{(N)}n} = \frac{\text{average} - \text{mean}}{\sqrt{\text{std. Deviation of average}}}$$

Figure 2 shows how (t) approaches the normal variate, as (n) number of samples becomes larger. The more samples measured the closer the data will approach the three-sigma distribution curve. Based on the sample size selected, (t) is then determined by calculating the degrees of freedom (df) of the experiment. "T" equals the number of runs and (k) the sample size measured.

Degrees of freedom:

$$df = T(k - 1) = 8(5 - 1) = 32$$

Using Table 7 for a 95.0% confidence level and tracking down the df (degrees of freedom) column to 30, yields an approximate or estimated (t)

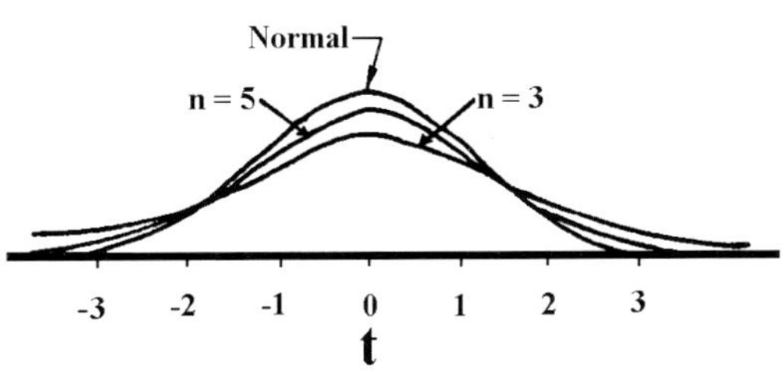

Figure 2. Comparing normal and t distributions.

Table 7. Values of t for Different Degrees of Freedom and Confidence Levels.

Degrees of Freedom d*f*	90%	95%	98%	99%	99.9%
1	6.314	21.706	31.821	63.657	6363.619
2	2.920	4.303	6.965	9.925	31.593
3	2.353	3.182	4.541	5.841	12.941
4	2.132	2.776	3.747	4.604	8.610
5	2.015	2.571	3.365	4.032	6.859
6	1.943	2.447	3.143	4.032	5.959
7	1.895	2.365	2.998	3.499	5.405
8	1.860	2.306	2.896	3.355	5.041
9	1.833	2.262	2.821	3.250	4.781
10	1.812	2.228	2.764	3.169	4.587
11	1.796	2.201	2.718	3.106	4.437
12	1.782	2.179	2.681	3.055	4.318
13	1.771	2.160	2.650	3.012	4.221
14	1.761	2.145	2.624	2.977	4.140
15	1.753	2.131	2.602	2.947	4.073
16	1.746	2.120	2.583	2.921	4.015
17	1.740	2.110	2.567	2.898	3.965
18	1.734	2.101	2.552	2.878	3.922
19	1.729	2.093	2.539	2.861	3.883
20	1.725	2.086	2.528	2.845	3.850
21	1.721	2.080	2.518	2.831	3.819
22	1.717	2.074	2.508	2.819	3.792
23	1.714	2.069	2.500	2.807	3.767
24	1.711	2.064	2.492	2.797	3.745
25	1.708	2.060	2.485	2.787	3.725
26	1.706	2.056	2.479	2.779	3.707
27	1.703	2.052	2.473	2.771	3.690
28	1.701	2.048	2.467	2.763	3.674
29	1.699	2.045	2.462	2.756	3.650
30	1.697	2.042	2.457	2.750	3.646
40	1.684	2.021	2.423	2.704	3.551
60	1.671	2.000	2.390	2.660	3.460
120	1.658	1.980	2.358	2.617	3.373
Infinite	1.645	1.960	2.326	2.576	3.291

value of 2.042. You can extrapolate to get the actual value for a d*f* of 32, that is 2.0378 but since it is so close, I have used the d*f* value for 30.

Therefore, the Minimum Significant Factor Effect is calculated by: Values of (t) are listed in Table 7.

$$\text{MSME} = (t \times 0.95\ df)\ (\text{Sd Effect}) = (2.042)\ (0.00095) = 0.002$$

$$\underline{\text{MSME} = 0.002}$$

Therefore, any calculated effect equal to or less than MSME value, in this example 0.002, will cause minimal or no effect on pin part length. By referring back to Table 5, one can see variables A, B, C, D, and G are greater than the MSME value, and indicate positive effect on controlling pin length.

The larger the effect value, the greater the variable will affect the dimension. But all effects above 0.002 should be considered when analyzing the data and reevaluating the process settings.

STEP 8. ANALYZING THE DATA

When a factor is significant it has a direct effect on the part dimensions. When a factor is significant, it has an effect on the variable of interest that is the part length, for this example.

When the effect is positive (+), the variable increases as the factor is increased from the low level to the high level. When the effect is negative (–), the variable decreases as the factor is increased from the low level to the high level.

Therefore, since variable B, C, D, and G were positive, increasing the variable will have a positive effect on increasing part pin length. With variable A, that was negative, decreasing the variable will increase pin length.

For example in variable C, injection pressure, using the higher packing pressure of 2500 psi versus 1500 psi, increased pin length. With variable A, tool cavity temperature, increasing tool cavity temperature caused greater material or pin length shrinkage thus decreasing pin length. Therefore, decreasing tool cavity temperature will have a lengthening effect on the pin length.

STEP 9. CHANGING THE PROCESS

Due to the screening experiment it was discovered that five of the seven variables had an effect on part pin length. Four optimize length when at their high value and one at its' low value and two with little or no effect on part pin length.

Now would be the time to run one more experiment with the significant values at their upper or lower limits to test their effect on part pin length. If in this test the pin length falls within specification limits consistently the new cycle variable settings would be determined.

This may not often be the case and fine tuning may be required to obtain the final results but the variables and their effects are known and adjustment can now be accomplished with minimum effort and time.

It may in some cases be sufficient to only adjust one of the more meaningful variables to bring part pin size within specifications. But, in a worst case scenario the combined effects may not be the right solution. This means combining them together may move the dimension greater or less than the effect of each added together. This interaction of effects are lost in a screening experiment.

If this occurs, a full factorial experiment 2 to the 7th power, 128 times, is required with each varying in two ways and is beyond the scope of this presentation.

Any properly designed, performed and analyzed designed experiment can yield positive results and the solution to tricky part specification problems. The total process quickly control techniques are then applied to the control of the manufacturing process to maintain the variables within their processing window.

REFERENCES

1. Schleckser, J. "Troubleshooting Technique Shortens Path to Quality." Plastics Engineering July 1987: 35–38.
2. Schleckser, J. Troubleshooting Techniques No. 0267–017–0.5D, Table 4, Rogers Corporation, 1986.

Appendix C

Six Sigma Quality Control SPC Forms and Data

Supplier survey report

TYPE OF SURVEY: DATE OF SURVEY:______________
1.__ PRE-SURVEY
2.__ INITIAL
3.__ FOLLOW-UP

PRODUCT SERVICE:____

COMPANY NAME:____
ADDRESS:____
CITY, STATE & ZIP:____
TELEPHONE NO.: ( )________________ FAX NO.:_____________________

APPROXIMATE NUMBER OF PEOPLE IN WORKFORCE:
MANUFACTURING:________ ENGINEERING:________ QUALITY:________
OTHER:________ TOTAL NUMBER OF EMPLOYEES:________
UNION?:________ NAME:_______________________________
LENGTH OF CONTRACT:______ EXPIRATION DATE:__________________________

PRODUCTS/SERVICES PROVIDED TO: (INCLUDE APPLICABLE PART NUMBERS AND ACTIVE PURCHASE ORDER NUMBER) ATTACH ADDITIONAL PAGES AS REQUIRED.
____
____
____

PERSONNEL CONTACTED:_________________ TITLE OR FUNCTION:___________
____
____
____
____

DISTANCE TO PLANT MILES/TIME:________ PLANT SQUARE FEET:___________

GENERAL CONDITION OF FACILITIES/EQUIPMENT:__________________________
____
____

HOUSEKEEPING OBSERVATIONS:____
____
____

AUDIT RATING:_____% CLASSIFICATION:__ ACCEPTABLE
__ CONDITIONAL
__ MARGINAL
__ UNACCEPTABLE

AUDIT PERFORMED BY:_________________
____
____
____

SUMMARY OF SURVEY

COMPANY:_________________ DATE OF SURVEY:_______ TYPE:________

SUPPLIER SYSTEM ELEMENTS:	RATING	
	POTENTIAL	ACTUAL
1. ORGANIZATION AND MANAGEMENT POLICIES	100	
2. SPECIFICATION REVIEW AND DESIGN ASSURANCE	70	
3. MANUFACTURING PLANNING AND CONTROLS	320	
4. STATISTICAL PROCESS CONTROL	50	
5. MEASUREMENT AND TEST EQUIPMENT	60	
6. FIRST-ARTICLE INSPECTION	30	
7. CONSIGNED MATERIAL	30	
8. HANDLING, PRESERVATION, PACKAGING AND SHIPMENT	40	
9. SUPPLIER PURCHASED MATERIAL CONTROL	90	
10. PERSONNEL TRAINING, CERTIFICATION, EDUCATION AND MOTIVATION	40	
11. RECORDS AND CHANGE CONTROL	55	
12. CONTROL OF NONCONFORMANCE	50	
13. COSTS OF QUALITY	20	
14. CORRECTIVE ACTION AND RECURRENCE PREVENTION	90	
15. QUALITY SYSTEM AUDIT	70	
TOTAL:	815	

SUPPLIER AUDIT RATING (ACTUAL/POTENTIAL X 100) ________ %

SUPPLIER SYSTEM EVALUATION

EVALUATION CRITERIA

1. Organization and Management Policies	Potential	Audit
a. Is there a documented and approved company quality policy?	10	
b. Are functional responsibilities for quality defined?	4	
b.1 Is there a company organization chart showing the relationship of the quality organization to management and other departments?	3	
b.2 Does the quality organization have the independent reporting authority required to be effective?	3	
c. Is the quality function adequately staffed to maintain effective control and assurance?	10	
d. Is the focus of the quality system "prevention" versus to "detection" oriented?	10	
e. Is the machine operator held accountable for rejects?	10	
f. Is the quality system documented in the form of a Quality Manual?	20	
g. Are detailed procedures and work instructions available for use by quality personnel?	20	
h. Is there a system for the review, approval, control, and maintenance of procedures and work instructions?	10	
TOTAL:	100	

Comments: ______________________________

____

____

____

Supplier survey report, continued

2. Specification Review and Design Assurance	Potential	Audit
a. Are contracts and purchase orders reviewed for quality requirements and manufacturability?	20	
b. Are formal design reviews held?	5	
b.1 Are unique part requirements identified?	5	
b.2 Are the critical tolerances and characteristics of product designated?	5	
c. If applicable, are reliability prediction and failure mode and effect analysis performed for new products?	5	
c.1 Is action taken to minimize probability and effect of failure?	5	
d. If applicable, is a safety analysis performed for all new products and/or ASE code and agency approval required?	5	
d.1 Do procedures exist for eliminating the probability of safety related failures?	5	
e. If applicable, are test procedures prepared for qualification tests of pre-production, engineering and production of first articles?	5	
e.1 Are qualification tests witnessed and verified by quality control personnel?	5	
e.2 Are records of qualification tests maintained including date and results of tests?	5	
TOTAL:	70	

Comments: ______________________

____

____

Supplier survey report, continued

3. Manufacturing Planning and Controls	Potential points	Audit points
a. Are routing sheets, operation sheets, and work instructions utilized and checked for compatibility with drawing requirements?	20	
b. Are special workmanship requirements designated on the applicable work instructions?	10	
c. Are process capability studies performed for all new products?	20	
d. Are the instruments used to measure product conformance of adequate precision?	20	
e. Are processes analyzed for trends and possible future corrective action?	10	
f. Does equipment have built-in process control correction for process drift? (Or are alternate methods and equipment used to perform the same function?)	10	
g. Are traceability procedures in effect that identify sources of raw material or component parts?	10	
h. Do environmental controls give consideration to temperature, humidity, vibration and other controllable factors affecting product quality?	10	
i. Does supplier furnish any special processes (e.g., heat treating, plating, welding, etc.) which require certified personnel? j. When required, are certified personnel adequately trained?	10	
k. Are there any special processes that require periodic recertification of equipment?	10	
l. Are production flow charts and control plans developed for all production parts?	5	

Design survey report, continued

3. Manufacturing Planning and Controls Cont.	Potential points	Audit points
1.1 Have adequate inspection stations been established throughout the manufacturing process?	5	
m. Do "route sheets" (e.g. shop travelers or move tickets) accompany parts through the manufacturing process?	10	
n. Do production workers sign off route sheets from operation to operation?	10	
o. Are inspection stations identified on the route sheets?	10	
o.1 If so, are inspection operations stamped, indicating product status at all stages of production?	10	
p. Are written instructions for all manufacturing, assembly, inspection and test operations available at the work stations?	10	
q. Are the instructions clear and easily understood by the operators?	10	
r. Does shop documentation enable traceability to the responsible production department, machine and operator.	10	
s. Is first piece inspection performed?	10	
t. Is roving in-process inspection performed?	10	
u. Is final acceptance inspection performed?	10	
v. Are all inspections performed to written inspection procedures?	10	
w. Do inspection plans include acceptance criteria?	10	
x. Are results of all inspections recorded?	10	
y. Are inspection records used for trend analysis and corrective action?	10	
z. Are current drawings, specifications and purchase orders available to and used by inspection?	10	

Supplier survey report, continued

3. Manufacturing Planning and Controls Cont.	Potential points	Audit points
aa. Is adequate test and inspection equipment available when needed?	10	
ab. Are the test and inspection facilities adequate?	2	
ac. Are visual aids used to define workmanship standards for manufacturing and inspection personnel?	4	
ac.1 Are they available at the work stations?	4	
TOTAL:	320	

Comments: ______________________________

____

____

4. Statistical Process Control		
a. Will statistical process control (SPC) methods be used for ongoing control of the process in "real time"?	10	
b. Have supplier personnel been adequately trained in SPC methods?	10	
c. Are control charts being used for process control?	10	
d. If control charts are used, is corrective action taken when the process shows lack of control?	10	
e. Are valid acceptance sampling plans specified and properly used in inspection and production?	10	
TOTAL:	50	

Comments: ______________________________

____

____

Supplier survey report, continued

5. Measurement and Test Equipment	Potential points	Audit points
a. Does the supplier have a written system for calibration or measuring and test equipment?	10	
b. Do written procedures exist for recall and maintenance of measuring and test equipment?	10	
c. Are employee-owned tools and gages subject to the same controls as those owned by the company?	5	
d. Does the supplier have detailed written procedures for each calibration?	5	
e. Are adequate records of calibration kept on file?	5	
f. Where possible, are labels affixed to measuring and test equipment indicating: date of last calibration, next due date, and by whom calibrated?	5	
g. Is calibration performed under adequate environmental control?	5	
h. If any production tooling is used for inspection or testing, is it included in the calibration system?	5	
i. Are calibrations made against certified higher accuracy standards which have known valid relationships to national standards?	5	
j. Are adequate facilities used for storage of measuring and test equipment?	5	
TOTAL:	60	

Comments: ____________________________________

____

____

Supplier survey report, continued

6. First-Article Inspection	Potential points	Audit points
a. Is there a procedure to assure that initial pre-production samples are submitted to customer for approval?	10	
b. Is there a written procedure for performing first-article verification to drawings and specifications?	10	
c. Are all required production gages and test equipment available at the time of first-article submissions?	5	
d. Are "control plans" completed by the Supplier prior to first-article submission?	5	
TOTAL:	30	

Comments:____________________________________

7. Consigned Material	Potential Points	Audit Points
a. Does the Supplier perform incoming examination upon receipt of any consigned material?	10	
b. Is consigned material uniquely identified and segregated for storage, control and proper use in production?	5	
c. If tests are required, will personnel be qualified to perform such tests?	5	
d. Are suitable records maintained for the control, inventory, and use of consigned material?	10	
TOTAL:	30	

Comments:____________________________________

Supplier survey report, continued

8. Handling, Preservation, Packaging and Shipment	Potential Points	Audit Points
a. Are special handling requirements and procedures available to production?	5	
b. Are parts and materials handled correctly to prevent damage?	5	
c. Are materials and parts correctly identified to prevent intermixing?	10	
d. Are inventory materials and parts protected from damage, corrosion, contamination and age limit requirements?	5	
e. Are procedures available for the control of handling, preservation, packaging and shipping to assure conformance to contractual requirements?	5	
f. Are outgoing shipments checked for: verification of acceptance; damage in handling; conformance to customer requirements and inclusion of documentation?	10	
TOTAL:	40	

Comments: ______________________________
____
____

Supplier survey report, continued

9. Supplier Purchased Material Control	Potential Points	Audit Points
a. Does the Supplier have a formal purchasing function?	10	
b. Are written purchasing operation procedures available?	10	
c. Does the Supplier have formally approved procurement sources?	10	
d. Is source approval based on pre-award survey of the Suppliers and quality history?	10	
e. Do Suppliers purchase orders contain applicable quality provisions such as: chemical and physical analysis, certification of test results, special treatment, source inspection, and other quality data or evidence of acceptability?	10	
f. Does the Supplier's quality system provide for the control of procured items prior to release to inventory?	10	
g. Are all such incoming inspections and tests performed against written inspection plans?	10	
h. Does the Supplier perform source surveillance?	5	
i. Does the quality system provide for early information feedback to Supplier?	5	
j. Is corrective action required of the Supplier on nonconforming supplies?	10	
TOTAL:	90	

Comments: ______________________________

____

____

Supplier survey report, continued

10. **Personnel Training, Certification, Education and Motivation**	**Potential Points**	**Audit Points**
a. Does the Supplier have a formal documented education and training program for personnel responsible for the determination of quality?	10	
b. Is there a certification program for persons performing or inspecting special processes such as: welding, decorating, part assembly and nondestructive testing?	10	
c. Are records of proficiency tests maintained?	5	
d. Are personnel performing the work required to show evidence of periodic certification?	5	
e. Is there an employee participation program for quality or product improvement, such as "Zero Defects" or "Quality Awareness"?	5	
f. Does the Supplier sponsor and promote participation in technical and professional societies, such as SME and ASQO?	5	
TOTAL:	40	

Comments:____

____

____

Supplier survey report, continued

11. Records and Change Control	Potential Points	Audit Points
a. Does the Supplier have a formal release system for drawings and specifications?	10	
b. Is there a formal system to assure that latest applicable documentation is available to purchasing, manufacturing, inspection and testing functions?	5	
c. Is a change order system set up to assure that changes required by purchase order revisions are incorporated?	5	
d. Is adequate control of the distribution and replacement of drawings maintained to assure the removal of obsolete information from production use?	10	
e. Is there a configuration management system which records the configuration status of all delivered items?	5	
f. Are adequate records maintained and analysis performed for inspections and test operations?	10	
g. Does management receive and use quality status reports?	10	
TOTAL:	55	

Comments:____

Supplier survey report, continued

12. Control of Nonconformance	Potential Points	Audit Points
a. Is there a positive system with written procedures for identification, segregation, and disposition of nonconforming material to prevent inadvertent entry into production?	20	
b. Is there a formal Material Review Board?	5	
c. Are material review actions studied for corrective actions?	5	
d. Are records of material disposition used for trend analysis and corrective action?	10	
e. Are formal methods required for any rework or repair actions?	10	
TOTAL:	50	

Comments:______________________________________

13. Costs of Quality	Potential Points	Audit Points
a. Are there procedures for the collection of quality costs?	10	
b. Are detailed quality cost reports analyzed by management?	5	
c. Does evidence exist that quality costs are being effectively collected, analyzed and used for management action?	5	
TOTAL:	20	

Comments:______________________________________

Supplier survey report, continued

14. Corrective Action and Recurrence Prevention	Potential Points	Audit Points
a. Do written procedures define the Supplier's corrective action system?	10	
b. Is evidence of corrective action documented?	10	
c. Is corrective action extended to second-tier Suppliers?	10	
d. Does the corrective action system include analysis of nonconformance trends?	10	
e. Does the corrective action system include analysis of scrap and rework to determine cause?	10	
f. Is timely corrective action promptly taken on root causes of nonconformances?	10	
g. Does the Supplier's management review the effectiveness of the corrective action system?	10	
h. Is customer data put into the corrective system for formal analysis and follow-up?	10	
i. Is the corrective action system used as a basis for product improvements?	10	
TOTAL:	90	

Comments: ______________________________

____

____

Supplier survey report, continued

15. Quality System Audit	Potential Points	Audit Points
a. Does the Supplier have formal internal quality audit procedures?	20	
b. Do the periodic audits address each of the Supplier System Elements outlined in this requirements document?	10	
c. Are audit findings reported to appropriate levels of management?	10	
d. Are records of audits maintained, reviewed, and used as a basis for follow-up audits?	10	
e. Are reports on audit deficiencies maintained and used for corrective action?	10	
f. Are audit trends analyzed and used by management as a basis for corrective action?	10	
TOTAL:	70	

Comments: ____________________________________

____

____

Supplier survey report, continued

Supplier Quality Self - Survey Report

Supplier Name: ______________________ Audit Date: ______________
Address: ______________________
______________________ Auditor(s): ______________
Phone: ______________________ ______________
Fax: ______________________ ______________
Product Type: ______________________ ______________
Contact: ______________________ ______________

		YES	NO	N/A
1.	**Organization**			
1.1	Quality Manager: ______________			
	Reports to: ______________			
	Title: ______________			
1.2	Total number of employees: ______________			
1.3	Total number of Quality Personnel: ______________			
1.4	Is there a Quality Manual? (attach copy)	____	____	____
1.5	Quality System conforms to the following specification: ______________ ______________			
1.6	Years in business: ______ Average Annual Sales: ______			
2.	**Operation**			
2.1	Current Contracts: ______________			
2.2	Major Customers: ______________			
2.3	______ % MIL ______ % Commercial			
2.4	Facilities: List (sq. ft.) ______________ ______________			
2.5	Equipment List (attach)	____	____	____
	Tool List (attach)	____	____	____
3.	**Procurement/P.O. Review**			
3.1	Does Quality review P.O. requirements?	____	____	____
3.2	Is there a system to inform customers of order status?	____	____	____
4.	**Technical Document Control**			
4.1	Do the drawings and other technical documents used for production and inspection purposes reflect the revision on the P.O.	____	____	____
4.2	Are obsolete drawings promptly removed from all points of issue or use?	____	____	____

Supplier quality self - survey report, continued

5. **Purchasing**

5.1 Are documented procedures utilized for the purchase of items which ensure they meet the specified requirements? _______ _______ _______

5.2 Are suppliers qualified via a documented evaluation procedure and qualification criteria? _______ _______ _______

5.3 Is a qualified supplier list maintained? _______ _______ _______

6. **Customer Supplied Items**

6.1 Are there adequate controls for customer supplied items which minimize its loss, damage, or misuse? _______ _______ _______

6.2 Are customer supplied items inspected to ensure their suitability for use? _______ _______ _______

7. **Process Control**

7.1 Are the appropriate procedures available at locations where operations essential to the effective functioning of the quality system are performed? _______ _______ _______

7.2 Are items provided traceability when required? _______ _______ _______

7.3 Are documented work orders/procedures in use? _______ _______ _______

7.4 Are detailed procedures, training, and personnel qualifications available for special processes? _______ _______ _______

7.5 List general workmanship standards used:

____

____

____

____

7.6 Are good housekeeping control in place? _______ _______ _______

8. **Inspection and Testing**

8.1 Are adequate inspection instructions or criteria available to material, in-process, and final inspection personnel? _______ _______ _______

8.2 Are incoming materials adequately inspected and results documented? _______ _______ _______

8.3 Are accepted, rejected, and materials awaiting inspection adequately identified and segregated? _______ _______ _______

8.4 Are in-process and final inspections and tests performed by adequately trained/qualified personnel. _______ _______ _______

8.5 Are required in-process and final inspections and tests adequately identified including their acceptance criteria? _______ _______ _______

8.6 Are results of in-process and final inspections adequately documented? _______ _______ _______

8.7 Are products failing inspections and tests identified and segregated? _______ _______ _______

9. **Inspection, Measuring, and Test Equipment**

9.1 Does the supplier control, calibrate, and maintain gauges and measuring equipment to demonstrate conformance to the specified requirements? _______ _______ _______

9.2 List Standards:

____

____

____

Supplier quality self-survey report, continued

10.	**<u>Control of Nonconforming Product</u>**			
10.1	Does the supplier maintain procedures to ensure that items which do not conform to specified requirements are prevented from inadvertent release to the customer?	_____	_____	_____
10.2	Is responsibility for review, documentation, and authority for disposition of nonconforming items defined?	_____	_____	_____
11.	**<u>Corrective Action</u>**			
11.1	Does the supplier document and maintain procedures for investigating causes of nonconforming items and the corrective action needed to prevent recurrence?	_____	_____	_____
12.	**<u>Quality Records</u>**			
12.1	Does the supplier maintain pertinent documented quality records?	_____	_____	_____
12.2	Are supplier quality records maintained?	_____	_____	_____
12.3	Are quality records stored such as to minimize loss or deterioration and are records readily retrievable?	_____	_____	_____
13.	**<u>Training</u>**			
13.1	Does the supplier maintain procedures for identifying training needs and provide the training to applicable personnel?	_____	_____	_____
14.	**<u>Handling, Storage, Packaging, Preservation, and Delivery</u>**			
14.1	Are written procedures in use to control the quality of items during handling, storage, and delivery?	_____	_____	_____

Six Sigma Team Member Self Assessment Survey

Name: ______________________________

Process: ______________________________ Date: ___/___/___

Rate your knowledge/skill in the following quality or process areas. This will assist the Black Belt in determining what additional training is required for the team. If you have other relevant skills not listed, write them in OTHER.

Subjects	None	A Little	Some	A Lot	Extensive
Failure Mode and Effect Analysis	1	2	3	4	5
Design of Experiments	1	2	3	4	5
Statistical Process Control	1	2	3	4	5
Statistics (e.g., calculations, mean, average, standard deviation, range)	1	2	3	4	5
Basic graphic tools (e.g., use of flowcharts, control charts, histograms, pareto, fish bone diagrams)	1	2	3	4	5
Group Facilitation (e.g., problem solving, conflict resolution, keeping team on track)	1	2	3	4	5
Group Leadership (e.g., decision making, goal settingk motivating team members, time lines)	1	2	3	4	5
Listening skills (e.g., paraphrasing, asking questions, demonstrating sincere interest, empathizing, using nonverbal cues)	1	2	3	4	5
Writing skills (e.g., preparing presentations, briefings, authoring written documents)	1	2	3	4	5
Presentation skills (e.g., delivering presentations;briefings to groups)	1	2	3	4	5
OTHER ______________	1	2	3	4	5
OTHER ______________	1	2	3	4	5
OTHER ______________	1	2	3	4	5

Tally Sheet for Six Sigma Team Member Self Assessment Survey

Team: ______________________________

Process: ______________________________ Date: ___/___/___

Total results of team members self-assessment sheets on this form. Shows areas where extra training is required.

Subjects	None 1	A Little 2	Some 3	A Lot 4	Extensive 5
Failure Mode and Effect Analysis	____	____	____	____	____
Design of Experiments	____	____	____	____	____
Statistical Process Control	____	____	____	____	____
Statistics (e.g., calculations, mean, average, standard deviation, range)	____	____	____	____	____
Basic graphic tools (e.g., use of flowcharts, control charts, histograms, pareto, fish bone diagrams)	____	____	____	____	____
Group Facilitation (e.g., problem solving, conflict resolution, keeping team on track)	____	____	____	____	____
Group Leadership (e.g., decision making, goal settingk motivating team members, time lines)	____	____	____	____	____
Listening skills (e.g., paraphrasing, asking questions, demonstrating sincere interest, empathizing, using nonverbal cues)	____	____	____	____	____
Writing skills (e.g., preparing presentations,briefings, authoring written documents)	____	____	____	____	____
Presentation skills (e.g., delivering presentations;briefings to groups)	____	____	____	____	____
OTHER ______________________	____	____	____	____	____
OTHER ______________________	____	____	____	____	____
OTHER ______________________	____	____	____	____	____

Six Sigma Team Customer & Product/Service List

Team Name: ______________________

Process: ______________________ Date: ___/___/___

Customer	Product/Service Used

Six Sigma Team Customer & Product/Service Affinity List

Team Name: ______________________

Process: ______________________ Date: ___/___/___

Business segment ______________________

Customers ______________________

Business segment ______________________

Customers ______________________

Business segment ______________________

Customers ______________________

Business segment ______________________

Customers ______________________

Business segment ______________________

Customers ______________________

Business segment ______________________

Customers ______________________

Six Sigma Quality Team Customer Interview Form

Team Name:______________________________Date:____________

Customer:________________________________ Phone:________________

E-mail:______________________

Customer's Organization:____

Dept/Group:____

Contacts:___________________________ Position:____________________

___________________________ ____________________

___________________________ ____________________

Interviewer:_________________________ ____________________

____

1. **What products and/or services are provided to this customer? List in order of volume.**

____

____

____

____

____

2. **What are the most important features/characteristics of the products and/or ervices. List for each product.**

____

____

____

____

____

3. **What aspects of our products/services are you most satisfied with? What is most important to your company? List in order of importance.**

____

____

____

____

____

Customer Interview Form, continued.

4. **What product and or service may need improving such as; delivery, design, quality, shipping, etc. List in order of importance.**

____

____

____

____

____

5. **What additional products or services can we provide to improve our position with your company?**

____

____

____

____

____

6. **What products or services do our competitors provide that we do not? How important are they to your company? List in order of importance.**

____

____

____

____

____

Six Sigma Quality Team Product Assessment Form

Team Name:______________________________ Date:__________

Customer:________________________________ Phone:________________

E-mail:______________________

Customer's
Organization:____

Dept/Group:____

Contacts:___________________________ Position:____________________

____________________________ ____________________

____________________________ ____________________

Interviewer:_________________________ ____________________

____________________________ ____________________

PRODUCT or SERVICE CHARACTERISTICS	RATING - IMPORTANCE & SATISFACTION
	Importance Low to High 1 2 3 4 5 Satisfaction Low to High 1 2 3 4 5
	Importance Low to High 1 2 3 4 5 Satisfaction Low to High 1 2 3 4 5
	Importance Low to High 1 2 3 4 5 Satisfaction Low to High 1 2 3 4 5
	Importance Low to High 1 2 3 4 5 Satisfaction Low to High 1 2 3 4 5
	Importance Low to High 1 2 3 4 5 Satisfaction Low to High 1 2 3 4 5
	Importance Low to High 1 2 3 4 5 Satisfaction Low to High 1 2 3 4 5

Six Sigma Quality Team Quality Characteristics Form

Team Name:____________________ Date:__________

Customer:____________________ Phone:__________

E-mail:_______________

Customer's Organization:______________________________

Dept/Group:______________________________

Contacts:____________________ Position:_______________

Interviewer:____________________

Quality Characteristics Worksheet

Quality Characteristics	**Measure(s) of Quality Characteristics**

Six Sigma Quality Team Process Selected for Change Form

Team Name:______________________ Date:__________

Customer:______________________ Phone:__________
E-mail:__________

Product:______________________

Process identified for change:

____
____
____
____

Process improvement goal:

____
____
____
____

Products/services effected:

____
____
____
____

Equipment/Services effected:

____
____
____
____

Possible Other Product lines effected:

____
____
____
____

Customer Impact____

____
____
____

Six Sigma Quality Team Outcome and Output Measures Form

Team Name:________________________________ Date:______________

Customer:____________________________________ Phone:________________

E-mail:____________________

Customer Representative:______________________________________

Product:_______________________________________

Process Improvement Goal to meet customer satisfaction:

____

____

____

____

____

Process Outcome Measures described in terms of name, type of data, method of measurement, historical data exists and the usefulness of the historical data:

____

____

____

____

____

____

Process Output Measures relate to describing the improvement goal and outcome measures in terms of; name, type of data, method of measurement, and if historical data exists.

____

____

____

____

____

____

____

____

Customer Background Information

Process: ______________________________ Date: ___ /___ /___

What products and/or services has this customer acquired or used from your organization in the past?

How often does this customer acquire products/services from you?

How long has this customer been using your products/services?

How much of your budget is related to products/services for this customer?

Does this customer have any pattern in the acquisition or use of your products/services?

Does any complaint data exist to help clarify customer requirements?

Do other customer satisfaction data exist?

Does this customer refer other organizations to you? Who?

✍ Selected Processes

Customer requirements are the basis for selecting processes upon which to focus improvement efforts. Select processes that are thought to impact the product/service characteristics important to the customer. Quality characteristics help to identify what about a process needs improving, along with measuring the impact of improvement efforts. Selecting these processes is part of specifying what is the goal of process improvement. This information can be summarized on the Selected Processes form.

Instructions

1. Briefly describe the process chosen for improvement. Use one form for each process. Note the reasons for selecting this process.

2. Describe the process improvement goal or desired change in the products or services associated with the process improvement effort.

3. Describe the products/services that will be affected by process improvement activities.

4. Identify the desired customer impact of the process improvement effort.

Selected Processes

Process: ______________________________ Date: ___ /___ /___

Process Identified for Change

Process Improvement Goal

Products/Services Effected

Customer Impact

✍ Flow Chart

Flow charts depict the sequence of steps and decisions in a process, and can display the steps of processes at a number of different levels of specificity. Flow charts are useful for identifying boundaries of processes and places where measurements are important. They are also useful as communication devices and training instruments. They also document how the process operated before improvement efforts were initiated.

Instructions

1. Label the process the flow chart represents.

2. Use the Step box to number each process step in its appropriate order. These numbers are useful because steps are often forgotten. You can add a step 1a or 1b later in the form, and the numbering shows that these steps go between step 1 and 2 when you draw the completed flow chart.

3. Map out each step in the process, beginning with inputs from suppliers through to outputs to customers. It is important to map out how the process actually works, not how it's supposed to work. This may require some information from those actually involved in the process.

4. For each step, draw the symbol that applies in the "Symbol" column (see the flow chart symbols for examples).

5. For each step, write a description or explanation in the "Description of Activity" box.

6. Use the information in this form to draw the flow chart on a plain piece of paper.

Flow Chart

Process: ____________________________________ **Date:** ___ /___ /___

Flow Chart Symbols

Start/Stop	Activity	Document
Direction of Flow	Decision Point	Connector (to another diagram or page)

Step	Symbol (Draw symbol)	Description of Activity

✍ Outcome and Output Measures

Process improvement efforts typically begin with information from customers that they are not satisfied or their needs are not being met. It is from here that process improvement efforts are often initiated. Practical experience shows that quality improvement efforts often start with indicators on outcome measures. These outcome measures are then tied back to output measures of a particular product and/or service. From there, output measures are related to actual measures of the process.

This form focuses on outcome and output measures. These two types of measures are important for they relate to different aspects of process improvement. If the goals of the improvement efforts are well-specified in terms of what customers desire, then improvements in the process should lead to improvement on the outcome measures. If output measures of particular products and/or services reflect things important to customers, then stabilization and improvement of the process should lead to improved output measures, and in turn improved outcome measures.

Instructions

1. Describe the process improvement goal. If more than one goal exists, use separate forms for each goal.

2. Describe the outcome measures in terms of name, and the type of data, method of measurement, whether historical data exists on these measures, and the usefulness of that historical data.

3. Describe the output measures that relate to the improvement goal and outcome measures in terms of name, type of data, method of measurement, and whether historical data exists.

Outcome and Output Measures

Process: ______________________________ Date: ___ /___ /___

Process Improvement Goal

Outcome Measures

Output Measures

✍ Data Collection Sheet

This form can be used as a data collection sheet to record the collection of data for any kind of measure. This form can also serve as a model that is modified by the team to best suit the particular data collection effort. Use multiple copies of this form as needed.

Instructions

1. Write the name of the measure and a short description at the top of the form that includes equipment or tools used to measure and any notes on the measurement procedure. If you plan to collect a lot of data, describe the measure and then make copies of the form for data collection.

2. Record the value of the measure in the column labeled "Measurement." Conventions concerning the desired decimal place to record or the rounding of the value should already be clear and specified on the Data Collection Plan form.

3. Record the date of each measure taken, in the "Date" column.

4. Record the exact time each measure was taken, in the "Time" column. The precision of this aspect of the data collection will vary with the nature of the measure and should be decided prior to data collection.

5. Describe the location where each data point was taken in the "Where" column.

6. Record the name of the person who collected each data point in the column labeled "Who."

7. Use this data as input to any of the data analysis forms in this section.

Data Collection Sheet

Process: ______________________________ **Date:** ___ /___ /___

Measure

Measurement	Date	Time	Where	Who

✍ Check Sheet

The check sheet is a basic form that can be used in any data collection effort. Teams often want to construct their own check sheets tailored to their situations. A simple check sheet form is provided as an example.

Instructions

1. Label the measure for which data will be collected.

2. Determine the type of data you are going to be recording: continuous (e.g., weights, measurements) or discrete (e.g., categories of product types, types of customer complaints).

3. If measuring continuous data, indicate the measurement intervals in the "Interval/Category" column.

4. If measuring discrete data, list the categories in the "Interval/Category" column.

5. For each measurement taken, indicate the date and the interval or category in which it falls by checking the appropriate box on the check sheet.

6. The "Total" space allows you to add all the checks in a row and/or in a column. These totals may be helpful for plotting the information as a pareto chart.

7. The data from the check sheet can be summarized in a number of ways, such as with a pareto chart or histogram. Forms for both of these tools follow.

Check Sheet

Process: ______________________________ Date: ___ /___ /___

Measure: ______

Interval/Category	Date										Total
Total											

✍ Pareto Chart of Causes of Quality

A pareto chart can help identify the relative contribution of factors to process variation. Those factors found to have the most affect on process variation should be addressed first.

Instructions

1. Describe the topic or measure that is being plotted (e.g., reasons for customer complaints).

2. Write the names of the categories under the "Interval/Category" heading. Intervals/categories indicate the types of events that will be counted for the variable of interest (e.g., cracks, scratches).

3. Record the number of occurrences (counts) under the "Frequency" heading. This information can be collected using check sheets, and then summarized on this form.

4. Calculate the percentage for each category by dividing the frequency in each category by the sum of all frequencies. Record the answers in the "Percentage" column.

5. Rank the categories by the total value under the "Frequency" column. Record the "Rank" in the right hand column.

✍ Pareto Chart of Causes of Quality (Continued)

6. Use the graph paper to plot your results. Write the names of the categories under the "Interval/Category" heading on the x-axis of the graph paper. List the Interval/Category names in rank order, with the most frequent listed first. Space the heading as is appropriate for the number of different intervals/categories you have. Often it is helpful to draw a slanted line off the side of the graph in the margin and then write the names on the lines.

7. Indicate the measurement scale on the y-axis (vertical axis).

8. Plot the data in order of rank so the data are displayed in decreasing order along the horizontal axis (x-axis).

9. If desired, calculate the cumulative percentage by summing percentages for each interval/category. This also can be plotted on the graph, if desired.

Pareto Chart of Causes of Quality

Process: ______________________________ Date: ___ /___ /___

Topic/Measure

Interval/Category	Frequency	Percentage	Rank
Total			

Pareto Chart of Causes of Quality (Continued)

Process: ______________________________ Date: ___ /___ /___

Topic/Measure: ________

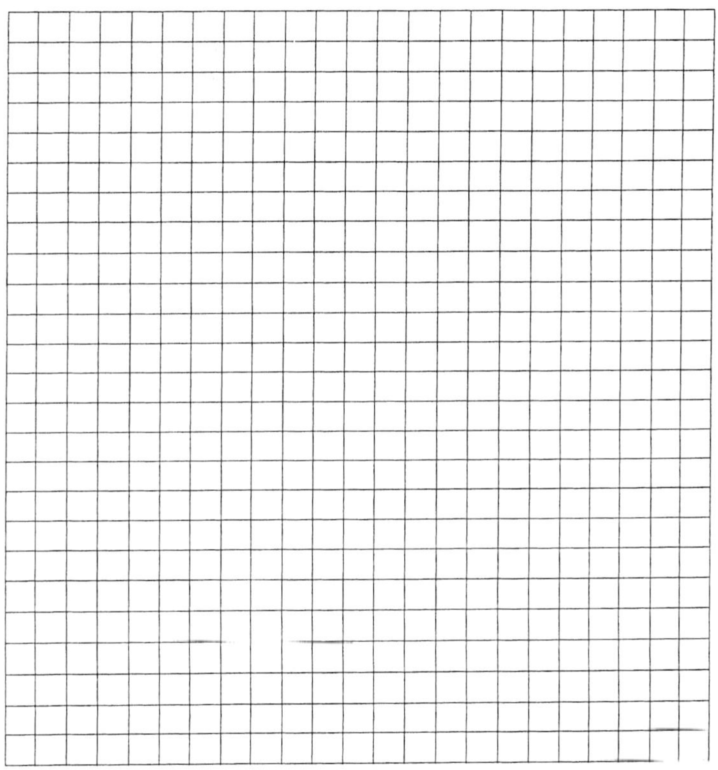

Interval/Category

Measurement Scale

✍ Histogram Worksheet

Histograms can be used to see how much variation exists in a specific process variable.

Instructions

1. Collect and record your data using a check sheet or similar form.
2. Write the topic or name of the measure and a short description in the "Topic/Measure" box.
3. Determine the number of classes into which the data are grouped. The appropriate size of the interval will depend on the data and on the extent to which it is necessary to depict small scale differences between data points.
4. Identify these classes by number under the "Class Number" heading. Record these classes in ascending order, from lowest to highest.
5. Identify the largest and smallest values in the data set for each class. Record this information under the "Class Intervals" heading, the smallest value in the class listed under "Lower," and the largest value in the class listed under "Upper." The mid-value for the class is listed in the "Mid-Value" column.
6. For each data point that falls into a class, record a tally under the "Frequency Tally" column.
7. Total the number of data points in each class. Record this total under the column "Frequency Total."

✍ Histogram Worksheet (Continued)

Instructions

8. Plot the results on the graph paper provided. Indicate what is being graphed on the "Topic/Measure" line.

9. Use the class interval boundaries to define the horizontal axis scale.

10. Use the frequency total values to determine the height of the bar for each class.

11. Draw bars for each class to show the distribution of the data.

Histogram Worksheet

Process: ______________________ Date: ___ /___ /___

Topic/Measure

Class Number	Class Intervals Lower	Upper	Mid-Value	Frequency Tally	Frequency Total

Histogram Worksheet (Continued)

Process: ________________________ **Date:** ___ /___ /___

Topic/Measure: ________________________

Class Intervals

Measurement Scale

✍ Scatter Diagram Worksheet

Scatter diagrams are used to understand the association between two variables. The Scatter Diagram Worksheet records and organizes data for constructing a scatter diagram.

Instructions

1. Define the "X Variable" in the appropriate space on the form. This variable is often thought of as the cause variable, and is typically plotted on the horizontal axis.

2. Define the "Y Variable" in the appropriate space on the form. This variable is often thought of as the effect variable, and is typically plotted on the vertical axis.

3. Number the pairs of x and y variable measurements consecutively. Record each pair of measures for x and y in the appropriate columns. Make sure that the x measures that correspond to the y measures remain paired so that the data are accurate.

4. Plot the x and y data pairs on the graph paper provided. This is done by locating on the horizontal axis the x value, then locating on the vertical axis the y value, and then drawing a point where these two points intersect on the graph.

5. Study the shape that is formed by the series of data points you just plotted. In general, conclusions can be made about the association between two variables (referred to as x and y) based on the shape of the scatter diagram. Scatter diagrams that display associations between two variables tend to look like elliptical spheres to straight lines.

6. Scatter diagrams where the plotted points appear in a circular fashion show little or no correlation between x and y.

✍ Scatter Diagram Worksheet (Continued)

7. Scatter diagrams where the points form a pattern of increasing values for BOTH variables shows a positive correlation: as values of x increase, so do values of y. The tighter the points are clustered in a linear fashion, the stronger the positive correlation, or association between the two variables.

8. Scatter diagrams where one variable increases in value while the second variable decreases in value show a negative correlation between x and y. Again, the tighter the points are clustered in a linear fashion, the stronger the association between the two variables.

9. The actual strength of the association between the two measures can be calculated. See Ishikawa (1982) for more information about the scatter diagrams and correlations.

Scatter Diagram Worksheet

Process: ______________________________ **Date:** ___ /___ /___

X Variable (Horizontal Axis)	**Y Variable** (Vertical Axis)

Order	X Variable	Y Variable	Order	X Variable	Y Variable

Scatter Diagram Worksheet (Continued)

Process: ______________________________ **Date:** ___ /___ /___

x Variable: ____________

y Variable: ____________

✍ Run Chart

Run charts are used to reveal patterns in data over time. They can also be used to document when the process returns to a stable state. Run charts can use any number of different measurement scales, such as frequency counts, percentages, and interval measurements. This form helps you to record and plot your data.

Instructions

1. At the top of the form, describe the unit of measure used to record the data (e.g., inches, degrees). Also describe the measure, the process with which it is associated, and the period of time ("Date") covered by the control chart.

2. At the bottom of the form, record the date, time, and count in the appropriate columns.

3. Compute the center line using either the median or mean. Then plot the center line on the graph.

4. Plot each data point on the graph paper provided. Then connect the dots with a ruler.

5. In interpreting the Run Chart, follow two general rules of thumb. Investigate patterns of nine points above or below the center line, or any upward or downward trends of seven points. For more information, refer to Rodriguez, Konoske, & Landau (1994).

Note: This chart was developed by Rodriguez, Konoske, & Landau (1994).

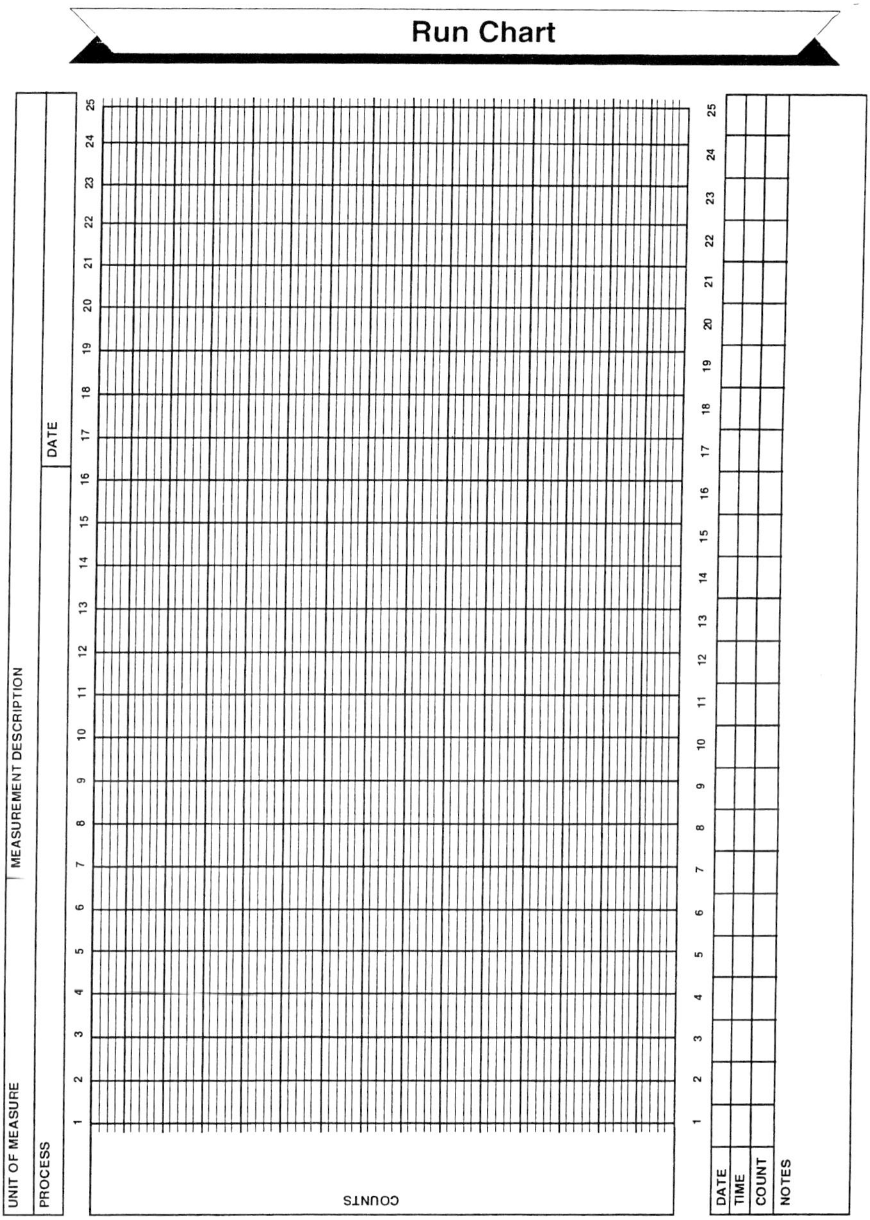
Run Chart
UNIT OF MEASURE
MEASUREMENT DESCRIPTION
PROCESS
DATE
1 2 3 4 5 6 7 8 9 10 11 12 13 14 15 16 17 18 19 20 21 22 23 24 25
COUNTS
DATE
TIME
COUNT
NOTES

✍ Variables Control Chart $\overline{X}$ and R

The control chart serves as a tool to assess and describe process variation. The Variables Control Chart is used with continuous scale data. For more information about this control chart, see Rodriguez, Konoske, & Landau (1994), or Wheeler & Chambers (1992).

Instructions

1. At the top of the form, describe the unit of measure used to record the data (e.g., inches, degrees). Also describe the measure it is associated with and the period of time ("Date") covered by the control chart.
2. At the bottom of the form, list the date and time for each measurement, and the value of the measure. For the first measurement of the first subgroup/sample, record the value of the measure in the row labeled "1" under the column labeled "1." For the first measurement of subsequent subgroups/samples, record these values in the rows labeled 2, 3, 4, 5 under the column labeled "1."
3. Total the values in this first column (labeled "1"), and record this value in the "SUM" box.
4. Indicate the location for each subgroup/sample by calculating the mean. The mean is calculated by dividing the sum of the values by the number in the subgroup/sample. Write this value in the "LOCATION" box.
5. Indicate the variation for the subgroup/sample by calculating the range. The range is calculated by subtracting the smallest (lowest) value in the subgroup/sample from the largest (highest) value in the subgroup/sample. Write this value in the box labeled "VARIATION."
6. Continue recording data in a similar fashion on the bottom of the form. The form supplies space for 25 measurement recordings.

Note: This control chart form was developed by Rodriguez, Konoske, & Landau (1994).

✍ Variables Control Chart ($\overline{X}$ and R) (Continued)

7. Indicate the number of measures in the subgroup/sample on page 2 of the control chart form in the right-hand box (n = __________).
8. Indicate the number of samples on page 2 of the control chart form (k = __________).
9. Fill the appropriate values of the constants on page 2 of the control chart form, selecting the values from the Table of Constants that most closely match your subgroup/sample size (n).
10. Calculate the Upper Control Limit ($UCL_{\bar{x}}$) and Lower Control Limit ($LCL_{\bar{x}}$) for the means using the formulas on page 2.
11. Plot the means and $UCL_{\bar{x}}$ and $LCL_{\bar{x}}$ for the means on page 1 of the form labeled "MEASURE OF LOCATION." The center line can also be plotted ($\overline{\overline{X}}$).
12. Calculate the UCL_R and LCL_R for the ranges using the formulas on page 2.
13. Plot the ranges and UCL_R and LCL_R for the ranges on page 1 of the form labeled "MEASURE OF VARIATION." The center line can also be plotted ($\overline{R}$).
14. Review the rules for defining special cause signals on page 2 of the control chart form.
15. Circle in colored ink any special cause signals.
16. Use the space on page 2 of the control chart form labeled "DATE/TIME" and "DESCRIPTION" to record any notes regarding measurements, calculations, the occurrence of special cause signals, and possible reasons for variations in the process.

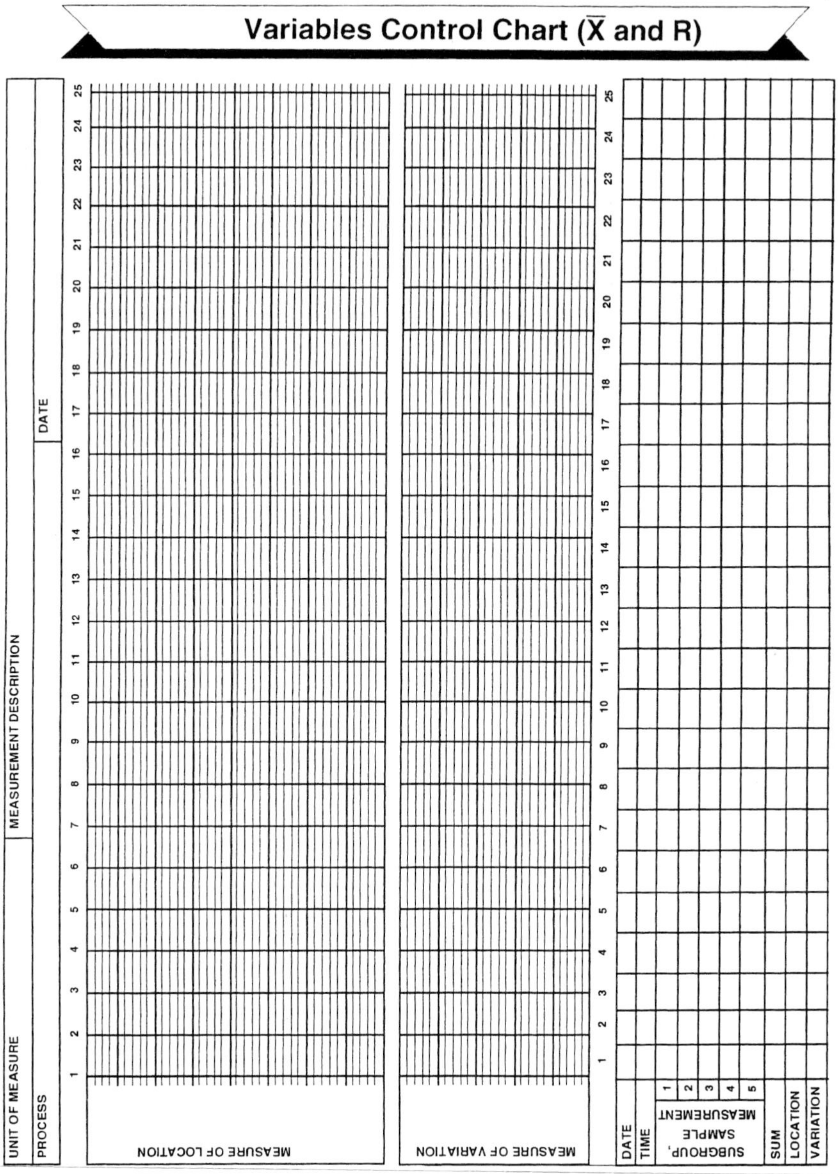
Variables Control Chart ($\bar{X}$ and R)
UNIT OF MEASURE
MEASUREMENT DESCRIPTION
PROCESS
DATE
1 2 3 4 5 6 7 8 9 10 11 12 13 14 15 16 17 18 19 20 21 22 23 24 25
MEASURE OF LOCATION
MEASURE OF VARIATION
DATE
TIME
SUBGROUP, SAMPLE MEASUREMENT
1 2 3 4 5
SUM
LOCATION
VARIATION

Variables Control Chart ($\bar{X}$ and R) (Continued)

Rules for Defining Special Cause Signals

Rule 1

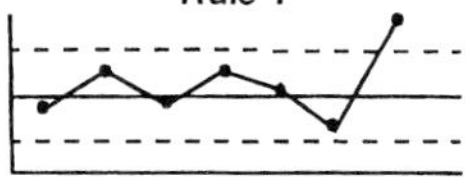

Rule 2

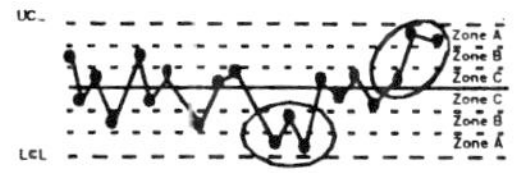

1. Any point outside of the control limits.

2. Two out of three successive points fall on the same side of the center line in zone A or beyond.

Rule 3

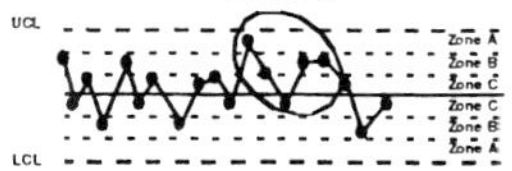

Rule 4

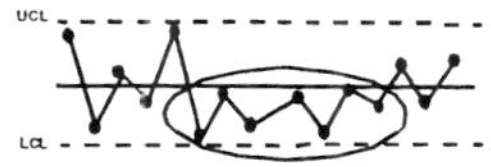

3. Four out of five successive values fall on the same side of the center line in zone B or beyond.

4. Eight successive points fall on the same side of the center line.

Formulas ($\bar{X}$ and R)

n = ______ k = ______

D_4 = ______ A_2 = ______

$\bar{\bar{X}} = \frac{\Sigma \bar{X}}{k}$ = ______

$\bar{R} = \frac{\Sigma R}{k}$ = ______

$A_2\bar{R}$ = ______

$UCL_{\bar{X}} = \bar{\bar{X}} + A_2\bar{R}$ = ______

$LCL_{\bar{X}} = \bar{\bar{X}} - A_2\bar{R}$ = ______

$UCL_R = D_4\bar{R}$ = ______

$LCL_R = D_3\bar{R}$ = ______

Table of Constants

n	A_2	D_3	D_4
2	1.880	—	3.268
3	1.023	—	2.574
4	0.729	—	2.282
5	0.577	—	2.114
6	0.483	—	2.004
7	0.419	0.076	1.924
8	0.373	0.136	1.864
9	0.337	0.184	1.816
10	0.308	0.223	1.777
11	0.285	0.256	1.744
12	0.266	0.283	1.717
13	0.249	0.307	1.693
14	0.235	0.328	1.672

DATE/TIME	DESCRIPTION

✍ Variables Control Chart ($\overline{X}$ and s)

The $\overline{X}$ and s control chart is another common chart used by process improvement teams typically for larger sized subgroups/samples (e.g., $n \geq 15$). For more information about this control chart, see Rodriguez, Konoske, and Landau (1994), or Wheeler and Chambers (1992).

Instructions

1. At the top of the form, describe the unit of measure used to record the data (e.g., inches, degrees). Also describe the measure with which it is associated and the period of time ("Date") covered by the control chart.

2. At the bottom of the form, list the date and time for each measurement, and the value of the measure. For the first measurement of the first subgroup/sample, record the value of the measure in the row labeled "1" under the column labeled "1." For the first measurement of subsequent subgroups/samples, record these values in the rows labeled 2, 3, 4, 5 under the column labeled "1."

3. Total the values in this first column (labeled "1"), and record this value in the "SUM" box.

4. Indicate the location for each subgroup/sample by calculating the mean. The mean is calculated by dividing the sum of the values by the number in the subgroup/sample. Write this value in the "LOCATION" box.

5. Indicate the variation for the subgroup/sample by calculating the standard deviation. The standard deviation is calculated for each subgroup/sample using the following formula:

$$s = \sqrt{\frac{\Sigma(X - \overline{X})^2}{n - 1}} = \sqrt{\frac{\Sigma X^2 - \frac{(\Sigma X)^2}{n}}{n - 1}}$$

6. Continue recording data in a similar fashion on the bottom of the form. The form supplies space for 25 measurement recordings.

Note: This control chart form was developed by Rodriguez, Konoske, & Landau (1994).

✍ Variables Control Chart ($\overline{X}$ and s) (Continued)

7. Indicate the number of measures in the subgroup/sample on page 2 of the control chart form in the right-hand box (n = __________).

8. Indicate the number of samples on page 2 of the control chart form (k = __________).

9. Fill the appropriate values of the constants on page 2 of the control chart form, selecting the values from the Table of Constants that most closely match your subgroup/sample size (n).

10. Calculate the Upper Control Limit ($UCL_{\overline{x}}$) and Lower Control Limit ($LCL_{\overline{x}}$) for the means using the formulas on page 2.

11. Plot the means and $UCL_{\overline{x}}$ and $LCL_{\overline{x}}$ for the means on the graph paper labeled "MEASURE OF LOCATION." The center line can also be plotted ($\overline{\overline{X}}$).

12. Calculate the UCL_S and LCL_S for the standard deviations using the formulas on page 2.

13. Plot the ranges and UCL_S and LCL_S for the standard deviations on page 1 of the form labeled "MEASURE OF VARIATION." The center line can also be plotted ($\overline{s}$).

14. Review the rules for defining special cause signals on page 2 of the control chart form.

15. Circle in colored ink any special cause signals.

16. Use the space on page 2 of the control chart form labeled "DATE/TIME" and "DESCRIPTION" to record any notes regarding measurements, calculations, the occurrence of special cause signals, and possible reasons for variations in the process.

Variables Control Chart ($\overline{X}$ and s)

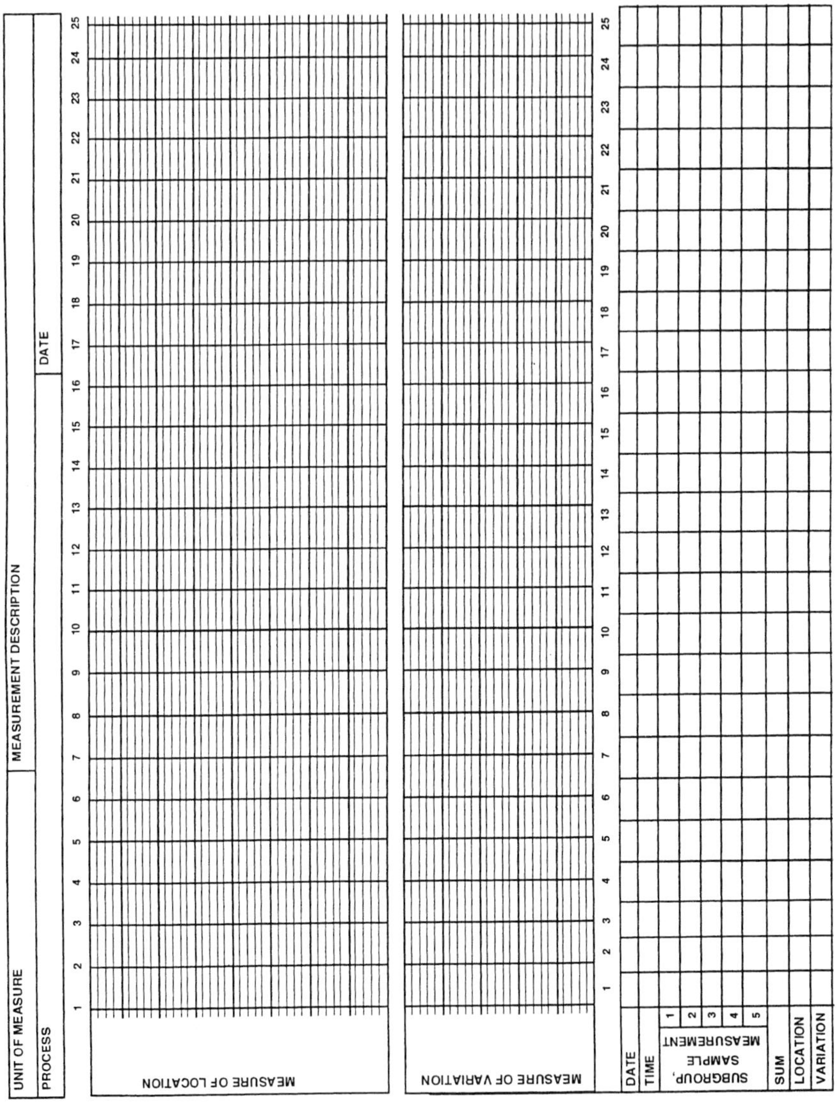

Variables Control Chart ($\bar{X}$ and s) (Continued)

Rules for Defining Special Cause Signals

Rule 1

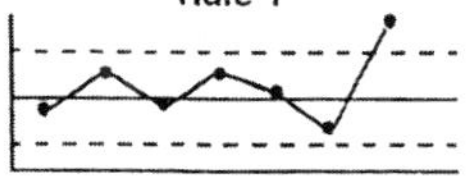

Rule 2

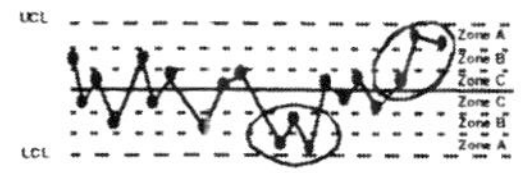

1. Any point outside of the control limits.

2. Two out of three successive points fall on the same side of the center line in zone A or beyond.

Rule 3

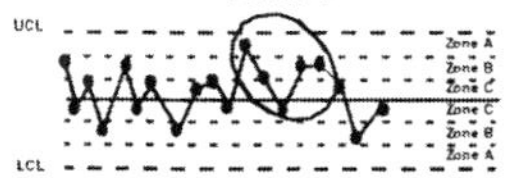

Rule 4

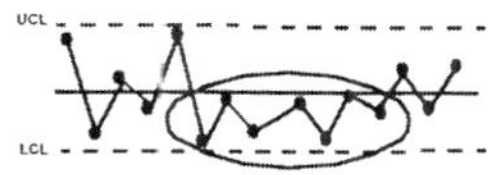

3. Four out of five successive values fall on the same side of the center line in zone B or beyond.

4. Eight successive points fall on the same side of the center line.

Formulas ($\bar{X}$ and s)

$n =$ ______ $k =$ ______

$B_3 =$ ______ $B_4 =$ ______ $A_3 =$ ______

$\bar{\bar{X}} = \frac{\Sigma \bar{X}}{k} =$ ______ $\bar{s} = \frac{\Sigma s}{k} =$ ______

$A_3\bar{s} =$ ______

$UCL_{\bar{x}} = \bar{\bar{X}} + A_3\bar{s} =$ ______

$LCL_{\bar{x}} = \bar{\bar{X}} - A_3\bar{s} =$ ______

$UCL_s = B_4\bar{s} =$ ______

$LCL_s = B_3\bar{s} =$ ______

Table of Constants

n	A_3	B_3	B_4
2	2.659	—	3.267
3	1.954	—	2.568
4	1.628	—	2.266
5	1.427	—	2.089
6	1.287	0.030	1.970
7	1.182	0.118	1.882
8	1.099	0.185	1.815
9	1.032	0.239	1.761
10	0.975	0.284	1.716
11	0.927	0.322	1.678
12	0.886	0.354	1.646
13	0.850	0.382	1.619
14	0.817	0.407	1.593
15	0.789	0.428	1.572

DATE/TIME	DESCRIPTION

✍ Individual Values and Moving Range

This control chart is used when there is only one measurement value; in other words, a sample size of n = 1. It can also be used to measure variation of counts and percentages when each count/percentage is treated as a single observation. For more information about this control chart, see Rodriguez, Konoske, and Landau (1994), or Wheeler and Chambers (1992).

Instructions

1. At the top of the form, describe the unit of measure used to record the data. Also describe the measure, the process it is associated with, and the period of time ("DATE") covered by the control chart.

2. At the bottom of the form record the measurements, with just one measurement listed in each column.

3. Use the formulas on page 2 of the form to calculate the necessary information for the control chart.

4. Plot the center line, the control limits, and the data points.

5. Assess whether the control limits are inflated. If so, use the formulas provided on page 3 to recalculate the control limits, using the median range.

6. Plot the information with the revised control limits on a new copy of page 1 of the form.

7. Review the rules for defining special cause signals on page 2 of the control chart form.

8. Circle in colored ink any special cause signals.

9. Use the spaces on page 2 and page 3 of the control chart form labeled "DATE/TIME" and "DESCRIPTION" to record any notes regarding measurements, calculations, the occurrence of special cause signals, and possible reasons for variations in the process.

Note: This control chart form was developed by Rodriguez, Konoske, & Landau (1994).

Individual Values and Moving Range (X, mR)

☐ Variables data ☐ Attribute data

UNIT OF MEASURE

MEASUREMENT DESCRIPTION

PROCESS

DATE

1 2 3 4 5 6 7 8 9 10 11 12 13 14 15 16 17 18 19 20 21 22 23 24 25

INDIVIDUAL VALUES (X)

MOVING RANGE (mR)

DATE

TIME

X

mR

NOTES

Individual Values and Moving Range (X, mR) (Continued)

Rules for Defining Special Cause Signals

Rule 1

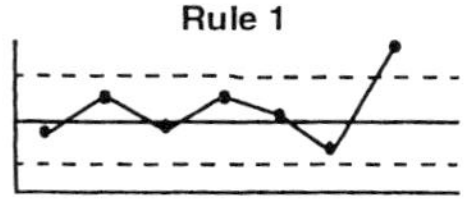

1. Any point outside of the control limits.

Rule 2

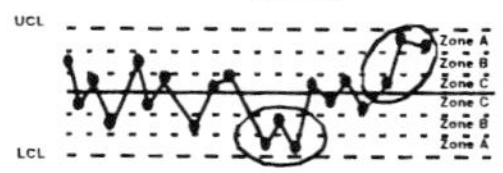

2. Two out of three successive points fall on the same side of the center line in zone A or beyond.

Rule 3

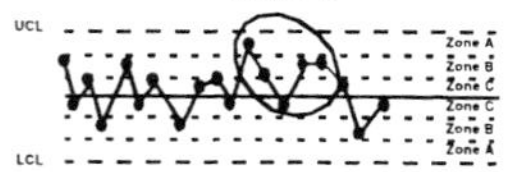

3. Four out of five successive values fall on the same side of the center line in zone B or beyond.

Rule 4

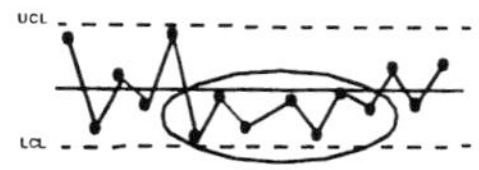

4. Eight successive points fall on the same side of the center line.

Revising Control Limits

If control limits are inflated, and the following two conditions are present, revision of control limits is in order:

(1) Very few signals are present in the original X chart

(2) $2.660\ R > 3.144\ R$

Center line: $\bar{X}$ = ______

$UCL_X = \bar{X} + 3.144\ m\tilde{R}$ = ______ + (3.144) = (______) = ______

$LCL_X = \bar{X} - 3.144\ m\tilde{R}$ = ______ − (3.144) = (______) = ______

Center line: $m\tilde{R}$ = ______

$UCL_R = 3.865\ m\tilde{R} = (3.865)$ (______) = ______

DATE/TIME	DESCRIPTION

Individual Values and Moving Range (X, mR) (Continued)

Rules for Defining Special Cause Signals

Rule 1

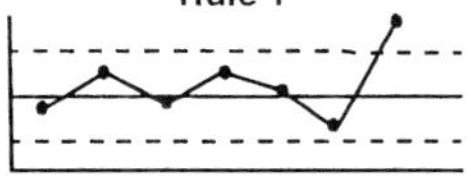

1. Any point outside of the control limits.

Rule 2

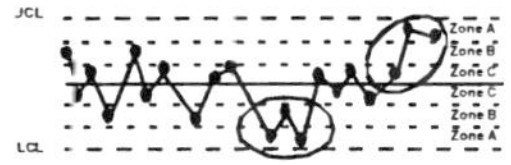

2. Two out of three successive points fall on the same side of the center line in zone A or beyond.

Rule 3

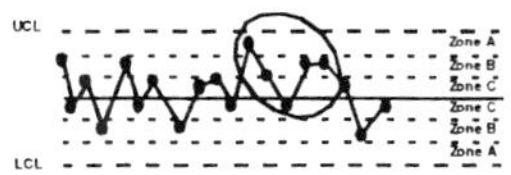

3. Four out of five successive values fall on the same side of the center line in zone B or beyond.

Rule 4

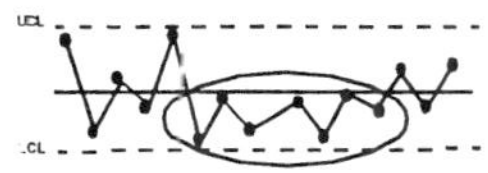

4. Eight successive points fall on the same side of the center line.

Calculations for X chart:

Center line: $\overline{X} = \frac{\Sigma X}{k} =$ ______ = ______

$UCL_X = \overline{X} + 2.660\ m\overline{R} =$ ______ + (2.660) = ______

$LCL_X = \overline{X} - 2.660\ m\overline{R} =$ ______ – (2.660) = ______

Calculations for mR chart:

Center line: $m\overline{R} = \frac{\Sigma mR}{k-1} =$ ______ = ______

$UCL_{mR} = 3.268\ m\overline{R} = (3.268)$ (______) = ______

(Formulas are based on a two-point moving range)

DATE/TIME	DESCRIPTION

✍ Attribute Control Chart

This control chart is used with attribute, or categorical data. This type of data is based on counts or values calculated from counts. Attribute data can be plotted on this form as a np-chart, p-chart, c-chart, or u-chart. See Rodriguez, Konoske, and Landau (1994), or Wheeler and Chambers (1992) for more information and the appropriate formulas for each of these control charts.

Instructions

1. At the top of the form, describe the unit of measure used to record the data, describe the measure, the process it is associated with, and the period of time ("DATE") covered by the control chart. Also check the type of control chart on the right side of the form.
2. At the bottom of the form record the measurements with one measurement listed in each column.
3. Use the appropriate formulas to calculate the necessary information for the control chart. Write the formulas used on page 2 of the form under the heading "Calculations."
4. Plot the center line, the control limits, and the data points.
5. Review the rules for defining special cause signals on page 2 of the control chart form.
6. Circle in colored ink any special cause signals.
7. Use the spaces on page 2 of the control chart form labeled "DATE/TIME" and "DESCRIPTION" to record any notes regarding measurements, calculations, the occurrence of special cause signals, and possible reasons for variations in the process.

Note: This control chart form was developed by Rodriguez, Konoske, & Landau (1994).

Attribute Control Chart

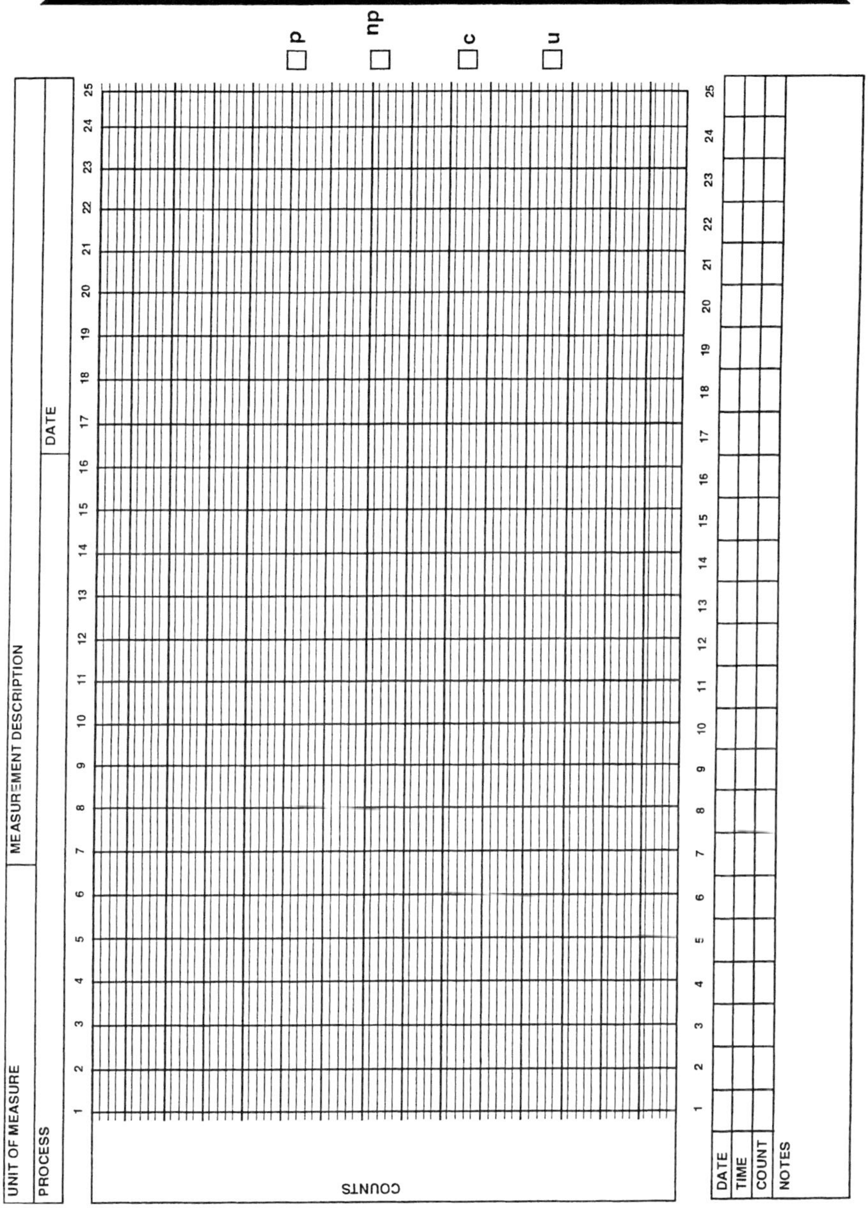

Attribute Control Chart (Continued)

Rules for Defining Special Cause Signals

Rule 1

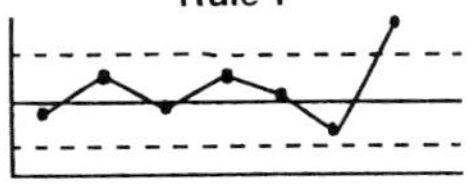

1. Any point outside of the control limits.

Rule 2

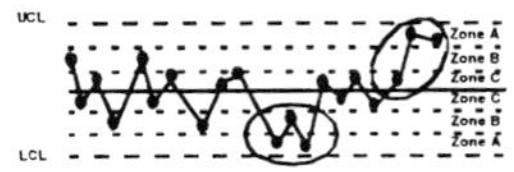

2. Two out of three successive points fall on the same side of the center line in zone A or beyond.

Rule 3

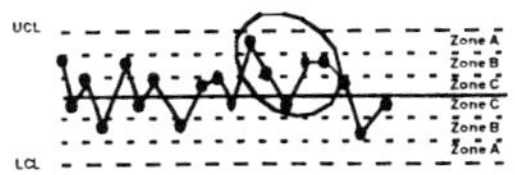

3. Four out of five successive values fall on the same side of the center line in zone B or beyond.

Rule 4

4. Eight successive points fall on the same side of the center line.

Calculations

DATE/TIME	DESCRIPTION

REFERENCE

1. Department of the Navy, "The Process Improvement Notebook", Total Quality Leadership Office, Arlington, VA, TQLO Publication No. 97–01, 134 pages.

Index